C. Jochum M. G. Hicks J. Sunkel (Eds.)

Physical Property Prediction in Organic Chemistry

Proceedings of the Beilstein Workshop
16–20th May, 1988, Schloss Korb, Italy

Sponsored by the BMFT

With 124 Figures and 57 Tables

Springer-Verlag
Berlin Heidelberg New York London Paris Tokyo

Dr. Clemens Jochum
Dr. Martin G. Hicks
Dr. Josef Sunkel

Beilstein Institut
Varrentrappstrasse 40–42, 6000 Frankfurt/M. 90
Federal Republic of Germany

Sponsor:
Bundesministerium für
Forschung und Technologie (BMFT)
Postfach 20 07 06, 5300 Bonn 2
Federal Republic of Germany

ISBN-13:978-3-642-74142-5 e-ISBN-13:978-3-642-74140-1
DOI: 10.1007/978-3-642-74140-1

Library of Congress Cataloging-in-Publication Data
Physical property prediction in organic chemistry : proceedings of the Beilstein Workshop, 16–20th May, 1988, Schloss Korb, Italy / C. Jochum, M. G. Hicks, J. Sunkel, eds. ; sponsored by the BMFT.
p. cm. Bibliography p.
ISBN-13:978-3-642-74142-5 (U.S.)
1. Physical organic chemistry–Congresses. 2. Organic compounds–Congresses. I. Jochum, C. (Clemens), 1949-. II. Hicks, M. G. (Martin G.), 1958-. III. Sunkel, J. (Joseph), 1934-. IV. Germany (West). Bundesministerium für Forschung und Technologie.
QD476.P49 1988 547.1'3–dc 19 88-30801 CIP

Softcover reprint of the hardcover 1st edition 1988

2151/3145-543210

Introduction

For more than 100 years the Beilstein Handbook has been publishing checked and evaluated data on organic compounds. It has become the major reference book for the chemical and physical properties of organic compounds. The prediction of these physical properties was the subject of the Beilstein workshop.

The ability to predict physical properties is for several reasons of great interest to the Beilstein Institute.

It is of primary importance to be able to check the abstracted data for accuracy and to eliminate simple mistakes like typing errors.

Presently all the work whether manuscript writing or evaluation of data is carried out manually. This is very time consuming, with the entry of Beilstein into electronic data gathering and publication, the opportunity for computerized consistency checking has become available.

Contrary to belief, when one examines the Beilstein Handbook or Chemical Abstracts there is a dearth of chemical information. There are a great many compounds but few are well defined resulting in large gaps in the information available to the chemist.

These information gaps could be filled by using algorithmic methods to estimate the properties of interest.

An important question to answer is "What is the chemist's reaction to estimated data?"

Will he accept it for use, within limits defined by the method, or will it be unacceptable and therefore detrimental for the data base.

However if one could partly fill gaps in the data base the increase in the power of the search techniques would be marked.

Unfortunately due to the existence of a vast number of gaps the methods used must be very fast and are therefore unlikely to be of very high accuracy. In this case it is unlikely that the estimated values could be relied upon at face value – the errors would be too big. But as a tool to retrieve molecules with certain properties from the data base, estimated data could be invaluable.

Naturally chemists also want to have access to high quality estimated data. The calculation methods that would have to be used to achieve the required degree of accuracy would probably be too inefficient to be carried out online or at data base creation time. Instead they would need to be carried out offline, on request. Thus the development of program packages to run on work stations and PCs which give reliable, high quality data is of great interest to chemists at large.

With this background the Beilstein Institute organized a workshop on the Estimation of Physical Data for Organic Compounds. This workshop was sponsored by the Bundesministerium für Forschung und Technologie (BMFT) as part of their program supporting information science in Germany. We would like to take this opportunity to thank them wholeheartedly for their generous support.

The aims were to answer the above questions and to determine which methods are of practical use.

To achieve this, an international group of scientists, covering all areas of data estimation were brought together, to give an overview of the field and, with any luck, inspire cooperation and exchange of ideas.

In the three days of intensive lectures and discussions the scientific programme covered the following areas: Thermodynamic Properties, Environmental Properties, Statistical Thermodynamics, Interpolation Methods, Quantum Mechanics, Empirical Methods, Group Contribution Methods, Structure-Activity Relationships and Solubility Determination.

The full manuscripts of the lectures are presented in the following chapters.

Frankfurt, 25th July 1988

Clemens Jochum
Martin G. Hicks
Josef Sunkel

Table of Contents

Index of Authors

Chemical Information – Promotion of Innovation in Science and Technology

Jan Michael Czermak

Bundesministerium für Forschung und Technologie, Heinemannstrasse 2, 5300 Bonn 2, Federal Republic of Germany

An important factor in technological or scientific innovation is the ready availability of relevant information. In chemical research and development data and factual information on chemical species is of prime importance: in chemistry at present we know of about 9 Mil. compounds - this figure is increased annually as a result of the synthesis of another half a million substances. New compounds, new properties, new processes and new methods were reported in the last year in ca. 600 000 articles, with an upward trend predicted for the future. The state-of-the-art is documented in around 40 000 patent applications annually, more than 16% of which are from the Federal Republic of Germany.

Product development in chemical and pharmaceutical research is an iterative process. Between the idea for a new product to its appearance on the market there is a battery of application trials, biological tests and so on. The results of these numerous tests may lead to modification of the product which must then be retested. In pharmaceutical chemistry a new substance has to pass through ca.10 000 test loops before it appears on the market. In statistical terms, the laboratory chemist has to work for about 20 years until his new product is marketable. This overall process can be shortened, and development costs considerably reduced, when factual information is readily accessible - this is equally true for basic research.

Due to the quantity and the growth rate of the chemical literature access to chemical information via conventional methods (i.e. handbooks, card indexes etc.) is time consuming, mostly incomplete and cost intensive. In this area the introduction of modern electronic information technologies has led to significant improvements. A great many databases and information systems for in-house use are being built and operated by the chemical industry. Besides the in-house information and documentation departments in the chemical industry, other institutions have developed whose sole purpose is the collection, evaluation and distribution of specialized chemical information. In the Federal Republic of Germany the activities of the Internationalen Dokumentationsgesellschaft für Chemie mbH (IDC) in Sulzbach, a joint concern of companies of the chemical industry, the Fachinformationszentrum Chemie GmbH (FIZ Chemie) in Berlin, the BEILSTEIN and GMELIN Institutes in Frankfurt are particularly worthy of mention.

As a rule the information systems of private companies are not accessible to the public. Access to stored information at the IDC is restricted to the staffs of the 11 member corporations. As part of its Program for Specialized Information 1985-1988, the Federal Government has initiated a number of measures to support chemical information, securing the information supply to public institutions and universities as well as to small and medium sized chemical firms. The focal points of the policy are as follows:

C. Jochum M.G. Hicks J. Sunkel (Eds.)
Physical Property Prediction
in Organic Chemistry

Participation of the Federal Republic in the development of an international host network **(STN International)**, to supply scientific and technical information to the general public. This international association currently links three host centres in Karlsruhe, Columbus (USA) and Tokyo (Japan) by satellite. The goal of this activity is to support the distribution of up-to-date specialized information, to facilitate access by standardizing retrieval systems and to provide scientists and technicians with the information tools that they need to carry out their day-to-day work. We are involved in the enhancement of MESSENGER SOFTWARE to improve data retrieval from different kinds of factual and numerical databases.

One very important principle of this international association is coordinated labour sharing which leads to cost optimizing in the building of data banks. Only by maintaining this principle have we a long term opportunity to cope with the ever growing flood of information and to utilize it in business, science and technology.

The presence of the Chemical Abstract Service (CAS) databases - the largest collection of information on the chemical literature - in this association has a particular significance for German chemical information services by creating an important "chemical environment" on STN. Such a concentration of chemical information banks on STN allows, for example, chemical information problems to be solved by using searches which overlap several databases and is a further step on the way to an integrated chemical information system

While the American partners concentrate on bibliographic information, the Federal Republic specializes in data and factual information which, by tradition, is intensively treated in Germany. Within this framework the Federal Government's Specialized Information Program 1985-88 has introduced a number of significant emphases in the area of chemical factual database production:

Focal Point "Specialized Scientific Information for Organic and Inorganic Chemistry"

BEILSTEIN ONLINE, GMELIN ONLINE

The largest data collections in organic and inorganic chemistry are BEILSTEIN's Handbuch der Organischen Chemie (comprising over 370 volumes which contain data going back to the beginning of preparative organic chemistry) and GMELIN's Handbuch der Anorganischen Chemie (600 volumes). These printed data-collections are being converted into electronically accessible databases. The objectives are to provide online accessibility of the numerical data and factual information on 9 million compounds and thereby reduce the access time. At the same time the time-lag between excerption and publication, which in the case of the

BEILSTEIN Handbook may amount to upto 25 years, will be dramatically reduced.

The BEILSTEIN database is compiled from two sources. The "Short File" containing unprocessed numerical and factual data extracted from the chemical literature and the "Full File" containing the already published Handbook data. Critical evaluation of the "Short file" data is continuously carried out by the BEILSTEIN scientific staff. The data is compared with previous results, checked for consistency, accuracy and redundancy etc. after which it is transferred to the "Full File". Thus the "Full File" contains evaluated data of the highest quality. The electronic publication of the printed BEILSTEIN Handbook, whose compilation involves exactly this evaluation of data, is also foreseen.

In 1988 the BEILSTEIN project will have reached phase II. The BEILSTEIN ONLINE database will be accessible to the public, on STN International, with the bulk of the handbook heterocyclic compounds. A further implementation, on DIALOG, is also scheduled. It will be interesting to compare and contrast the different implementations and services. In view of the vast amount of data, the BEILSTEIN Information System will not be up-to-date until 1992, unless the publisher receives the funding to finish the integration of the new literature earlier.

The development of the GMELIN ONLINE data system for information on inorganic chemistry has been launched and the bringing of the handbook up-to-date has commenced. As a first step the general index (1924-1979) has been made available online via STN International under the name of GFI (GMELIN Formula Index). Phase I of the activities to build the factual database GMELIN ONLINE started in summer 1987. The implementation period will be probably 8 - 10 years.

Focal Point "Compound- and Factual Data for Chemical Technology"

Database DETHERM

Only comparatively few of the presently known 9 million chemical compounds are produced on an industrial scale. Assuming a world-wide production rate of 50 tons p.a. as a criterion for "industrial relevance", estimates give the number of "relevant" compounds as being between 10 000 and 20 000. There are only few substance and property data available for most of the known compounds. For the "industrially relevant" compounds the detailed knowledge of their thermodynamic data, transport properties, molecular properties, equilibrium data etc. and also safety and protection data is of vital importance.

The availability of such data is not only important for pure substances but also for the technically important mixtures of known and unknown composition.

Within the scope of the Specialized Information Program 1985-88 the data system DETHERM is being extended as an information system for compound-data in chemical engineering. The leading role is taken by DECHEMA, Deutsche Gesellschaft für Chemisches Apparatewesen, Chemische Technik und Biotechnologie e.V. in Frankfurt. Other partners cooperating in the building of the database are from industry, university institutes in Germany and abroad, industrial associations and public research institutes. The activities are financed by industry, industrial associations, the EEC and with funds from the Specialized Information Program, depending on the particular work under development.

Focal Point "Information System on Chemical Reactions"

ChemInform

A database is being created which will be based on the Chemischer Informationsdienst (ChemInform) which at present is jointly published, in printed form, by BAYER AG, Leverkusen and the Fachinformations-zentrum Chemie GmbH, Berlin. On the one hand this will be used to produce the current printed services and on the other, by means of an intelligent retrieval system, allow on-line and in-house access to chemical reaction data.

As in other chemical information projects the basis is a storage of products and educts in the form of a topological matrix - this being the computer readable storage form of chemical structures. This will allow direct access to the corresponding factual data in the BEILSTEIN and GMELIN databases and to the spectroscopic data in the "Informationssystem Spectroskopie". This project is being implemented by FIZ Chemie in Berlin in close cooperation with, and with partial funding by, the chemical industry.

Focal Point "Spectroscopic Information System"

SpecInfo, Combined Spectral Data System

The availability of spectral data and systems for their interpretation is of extreme importance in the area of chemical analysis and structural elucidation. The aim is to design largely automated systems which can deliver structure and substructure suggestions from measured spectra. The "Informationssystem Molekülspektroskopie SpecInfo" is based on this concept. It is being developed by BASF AG, Ludwigshafen, within the scope of the Specialized Information Program. In SpecInfo various spectroscopic methods are used simultaneously for structure elucidation. Besides the general search options in the spectral databases, another feature is the facility for automated interpretation and reconstruction of spectra.

The main difficulty in building an information system on spectroscopy is the acquisition of high quality spectra in an electronically readable form, which by their very nature cannot be extracted from the primary literature. These spectral data are original data which have been measured in industry and university laboratories. The successful building of an extensive and thereby spectra from these institutions. Surveys have found a readiness in industry as well as in universities to participate in the building of a spectral data pool. It is anticipated that eventually, 100 000 spectra from each of the following areas will be available: Nuclear Magnetic Resonance Spectroscopy, Infra Red Spectroscopy, Mass Spectroscopy. Several spectral data projects have already been completed, others are in development.

The SpecInfo system is to be available to the public either in-house or on-line (via STN International).

The focal points described above are predominantly substance oriented: data and facts on properties, reaction characteristics of chemical species and their mixtures are documented. This meets the requirements of applied chemistry where the goal is the production of substances having specified property profiles ("molecular engineering").

The other aim is that of structure analysis and elucidation. Here one tries to identify unknown substances and determine their chemical structure. This can be done by the analytical chemist on the basis of a property profile obtained from spectroscopic and physico-chemical measurements.

The availability of substance data plays a very important role in chemical plant design which depends directly on the properties of the substance to be processed (for example there is a direct relationship between the dimensioning of a heat exchanger and the thermodynamic substance-properties, heat conduction, temperature, pressure and so on).

The questions a preparative chemist asks are "How do I make a substance with the desired properties?" and "What useful purpose can I find for by-products?" The "chemical intuition" of the synthetic chemist will be enhanced by access to a reaction database.

The support of the building of chemical structure oriented databases as described in the above focal points forms part of the strategy of the Federal Government as encompassed by the Specialized Information Program 1985-88. The physical, chemical and spectroscopic properties of substances are determined by their chemical structures. It has been shown that biological effects can also be correlated to structural characteristics.

A systematic evaluation of the data contained in large structure oriented databases in order to reveal structure-activity relationships and structure-property correlations will give new impetus in many fields of chemical and pharmaceutical research.

For a great many substances there are no complete property profiles available because of the lack of experimental data. Since the (conservative) databases are constructed only from published measured data a great many "data gaps" are present. Statistical evaluation of these large amounts of data can reveal interrelationships between the substance data which can be used to develop rules for data prediction. In the future factual databases must be developed which can be used in two ways: either statically, when the stored data are searched and retrieved without further processing, or dynamically, when new data are derived from stored data by additional evaluation and processing. The development of dynamic databases will be a major undertaking in the field of chemical information in the future and forms the theme of this workshop.

The BMFT has employed an independent company to evaluate the effectiveness of the Specialized Information Program, the results of this evaluation will be used in the drafting of a new program. The results of this workshop could make a major contribution to the new program and thus give us a new focal point for our activities in promoting chemical information.

The Importance of Data Estimation for the Beilstein Information System

Clemens Jochum

Beilstein Institut, Varrentrappstrasse 40–42, 6000 Frankfurt 90, Federal Republic of Germany

The Beilstein Information System is a structure oriented collection of physical data and reactions which have been extracted from primary literature. The Beilstein Handbook contains more than 1 million compounds described in over 340 volumes. This information is available in printed form for more than 100 years. The Beilstein-Institute is currently setting up a numerical factual databank based on these 340 volumes of the Beilstein Handbook of Organic Chemistry (literature from 1830 to 1960) and 7.5 million factual records of organic compounds (literature from 1960 to 1980).

Since the Beilstein database consists of a Structure File and a Factual File, the data structure can be divided into two parts: The first part gives a complete topological structure representation of the compounds. The second part defines in more than 400 fields (more than 60 numerical fields) most of the information content of the Handbook and the factual records. This sophisticated and flexible data structure will allow many new search methods currently not available through online hosts. The first part of the computerized Beilstein Database will go online by the end of 1988.

The Beilstein Data Structure.

The articles of the Beilstein Handbook, i.e. the complete factual descriptions of a compound have always been written according to a very well defined structure. Naturally, since the analytical methods and chemical preparations have changed over the last fifty years, the instructions and definitions for the Beilstein manuscript writers have been altered slightly over this period of time. These changes had to be taken into consideration for the definition of a computer-optimized data structure.

After a very thorough analysis of the article structures of the main volume and of all supplementary series, a data structure was defined which allowed the computerized input, storage and retrieval of the Handbook data without loss of information. However, some compromises had to be made: For most organic compounds described in the primary literature, only very little factual information is known. In many cases, only the boiling point, melting point, refractive index and one or two methods of preparation have been described in the literature. These small information compounds (SIC) usually require in their redundancy- and error-free Beilstein presentation only a quarter of a page or less in the Handbook without any loss of information from the primary literature.

A comparatively small percentage of all known compounds (less than 5%) is very important for chemical reactions or

C. Jochum M.G. Hicks J. Sunkel (Eds.)
Physical Property Prediction
in Organic Chemistry

pharmaceutical purposes and has therefore been published widely in the chemical primary literature. Therefore many physical data, preparations and other factual data are known for these large information compounds (LIC). The definition of the data structure had to take the different information contents of SICs and LICs (and all variations in between) into account. The final electronic data structure constitutes a very sophisticated compromise between the information contents of LICs and SICs:

- The database contains the complete Handbook information of the SICs and compounds with an intermediate amount of information.

- For the LICs, only a subset of all Handbook data can be stored in the electronic database. For more information, online users are then referred to the Handbook.

Since the information contents of the Handbook and the database are partially overlapping, Handbook subscribers can access the database at a reduced rate.

Besides the Handbook, the database will have two other sources of information:

1) The literature of the Fifth Supplementary Handbook Series (literature time frame: 1960-1980) has been completely abstracted. This factual information which is contained on approximately 6.5 million file cards (one card per compound per literature citation) is the basis for the Handbook articles of this series. Since this Handbook series will not be completed for a further decade, the online user will be provided with access to the "raw" information. In contrast to the Handbook data, this information contains redundancies (from different articles about the same compound), errors (from the primary publication, which could only be removed by crosschecking with other sources) and missing data (the file cards only contain the 5 most important physical data and the preparation).

2) The factual data of the primary literature from 1980 onwards. The compounds and associated literature data are abstracted in a completely new electronic and paperless way using off-line microcomputers. The abstractor enters the structure graphically and inputs the factual data with a menu-driven program. The design of the data structure took new developments of analytical and synthetic methods into account.

The structure of the database can be divided into two parts: The numeric factual file and the structure file. Since the factual file forms the basis for any kind of data estimation procedures, it is subsequently described in some more detail.

The Factual File has a relational structure and contains three types of fields:

1) **Numerical Fields.** Most numerical parameters can be stored as 2-byte integers, but some physical parameters require a 4-byte floating point format.

2) **Boolean Fields.** These fields store the presence (or absence) of a keyword or a parameter.

3) **Alphanumeric Fields.** Literature citations, comments for a preparation, etc. are stored as character strings.

All boolean fields and most of the numerical fields are searchable separately or in combination.

Chemically, the data structure can be divided into 7 parts:

1) **Identifiers.** These parameters contain the molecular formula and registry numbers for identification and search of the compounds. Three registry numbers are stored for each compound:

   - A structure-independent compound-identifier for the internal organization of the database.

   - A structure-dependent non-unique hash-code, the Lawson-Number. Since this number is structure-related, several structurally closely related compounds can have the same Lawson number. The structure-ordering according to this number is very similar to the Beilstein Handbook Ordering System. The number can be computed on a microcomputer (after having entered the structure graphically) using a rather complex algorithm. Subsequent searching of this number on an online host represents an elegant and inexpensive way of structure and substructure browsing.

2) **Structure-related Data.** These fields include information about the purity of a compound, its possible tautomers or alternative structure representations.

3) **Preparative Data.** This topic includes all preparation-related parameters such as yield, solvents, temperature, pressure, by-products, etc.

4) **Physical Properties.** This division includes most of the numerical fields. It is subdivided into
   - Structure and Energy Parameters (Dipole Moment, Molar Polarization, Coupling Constants, etc.),
   - Physical Properties of the Pure Compound (Colour, Melting Point, Boiling Point, Transport Phenomena, Calorical Data, Optical Properties, Spectral Information, Magnetic Properties, Electrical Properties, Electrochemical Behaviour),
   - Physical Properties of Multicomponent Systems (Solution Behaviour, Liquid/Liquid-, Liquid/Solid-, Liquid/Vapour-Systems, etc.).

5) **Chemical Behaviour.** Reactions of this compound with other chemicals are described under this section. The fields include reaction partners, reagents and reaction conditions.

6) **Physiological Behaviour and Applications.** Use-, Toxicity-, Biological Function- and Ecological Data-Parameters are described under this topic.

7) **Characterization of Derivatives and Salts.**

The Intermediate File.

After input, the data are loaded from the floppy disks into an intermediate file on the mainframe. The data are run through several plausibility checks and when necessary are corrected. In this intermediate file all data from the various input sources are converted to the same file- and data-structure according to our data structure definition.

The Database.

After having been converted to the same file structure the data will be loaded into an Adabas-managed database. More than 10 different commercial database management systems have been evaluated during our systems analysis phase in 1984/85. The two highest scoring systems have been benchmarked with an artificial BEILSTEIN-structured database with 200,000 compounds. Adabas scored best in practically all tests (loading, updating, retrieval etc.).

An Adabas-based update- and retrieval-system is currently being developed in our Institute together with a German software house. This system will be used to make final corrections to the database, to append compounds and retrieve compounds for checking, and for writing Handbook articles.

Why Data Estimation?

Besides many technical and logistical issues there are two general problems when building up such an information system:

1) Avoiding abstracting and input errors. Even double input does not completely avoid input errors and is also too expensive because highly skilled personal is needed for this job. Therefore automatic controls for extracted physical data have to be developed.

2) The information system contains large "holes" of missing data, because these data are unknown from primary literature. This implies for a multi-property range search in the Beilstein Database that compounds containing no data for any one of these properties are not found even if they would fall into the searched range. Compounds also remain undetected on a property search if the property was measured in a parameter range which does not correspond to the search range.

Both problems can be partially solved by employing data estimation methods. For problem 1) input data could be compared with automatically estimated data. If the absolute difference exceeds certain boundaries, a warning can be issued. For problem 2) "filling up" the database with calculated data can lead to much improved multi-parameter search results. Compounds which most likely fall into a certain data range are detected. The estimated parameters allow conclusions on experimental data.

Data About Data

László Domokos

Beilstein Institut, Varrentrappstrasse 40–42, 6000 Frankfurt 90,
Federal Republic of Germany

"Has Beilstein got enough data ?" The answer to this question depends on what the data are used for.

On the one hand the answer is "YES", Beilstein owns the worlds largest collection of factual data of organic chemistry".

As known, the Beilstein Institut has collected for over a hundred years the factual data published in the literature. After a critical evaluation the collected data has been continuously published in the Beilstein Handbook of Organic Chemistry. Up to now more than 250 volumes with more than 200.000 pages have appeared containing about 1.2 Million compounds. The still unpublished but abstracted material contains ca. 6 Million file cards describing more than 3 Million compounds.

This material will be available also via Beilstein Online. The data structure of the online database has the following structure, Figure 1. The basic entry in the database is the compound. The compounds are characterized by their atom-bond structure and by their factual data. The factual data can be grouped into attributes, which can be further divided into fields. Each attribute contains one or more fields. For example, attributes are: "molecular formula", "molecular weight", "preparation", "chemical behavior", "boiling point", "density",

C. Jochum M. G. Hicks J. Sunkel (Eds.)
Physical Property Prediction
in Organic Chemistry

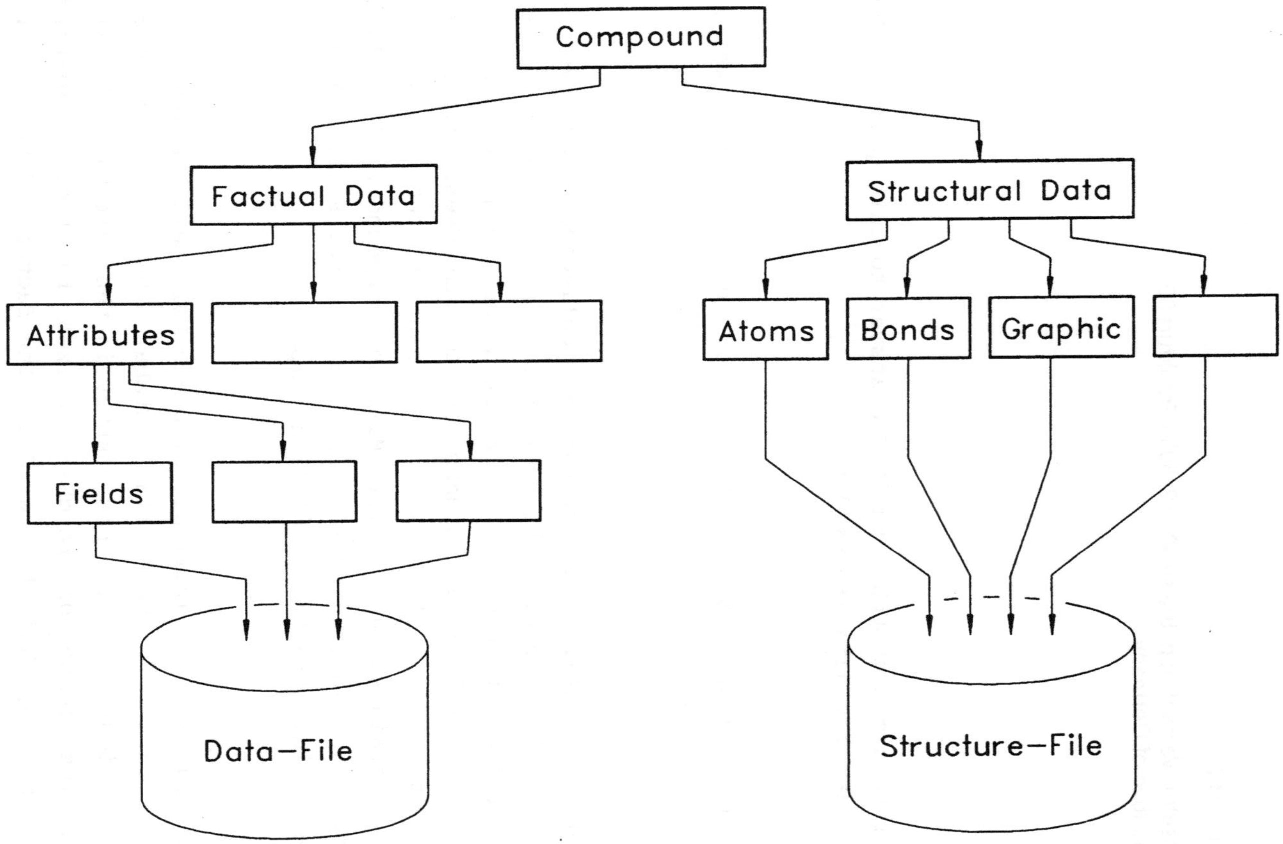

Fig. 1. The data structure of the Beilstein database.

"IR spectrum", etc. The first two attributes in the above example contain only one field, the other ones more than one. For example, the fields of the attribute "preparation" are: "literature reference", "starting material", "reagents", "time", "yield", "solvent", "ambient temperature", "reflux", "temperature range", "catalyst", "irradiation", "pressure range", "other conditions", by-products". The field values can be numeric (real or integer), string or boolean, i.e. 0 or 1, pointing out whether the feature has been published and a more detailed description is available in the referenced literature. The attributes can be classified into different groups, like: physical properties, chemical properties, structure properties, bibliographic data and house keeping data.

The logical data structure contains 382 attributes with 1074 fields. There are a great number of attributes having the same fields. Several of these attributes can be pulled together and handled as a single attribute. After this kind of simplification the physical data structure contains 168 attributes and 401 fields. After having online all the abstracted data, the database will contain ca. 4 Million compounds. The data structure can be visualized as a huge 4.000.000 x 401 matrix, where each row corresponds a compound, and each column to a field, Figure 2. Moreover, most of the attributes and some of the fields can have multiple occurrences characterizing the different methods of preparation, boiling points measured at different pressures, or different chemical behaviors, etc.

On the other hand, generally only very few attributes have been measured and published by the chemists. Therefore the answer to the introducing question can be as well: "NO, Beilstein has only very little data". The huge data matrix is filled only very sparsely and is very unbalanced, Figure 3. There are compounds where almost all attributes are known, even with high multiplicity. However, for the most compounds only 7-10 attributes are known. The same is valid for the attributes. Some of them are known for each compound, but the majority very seldom.

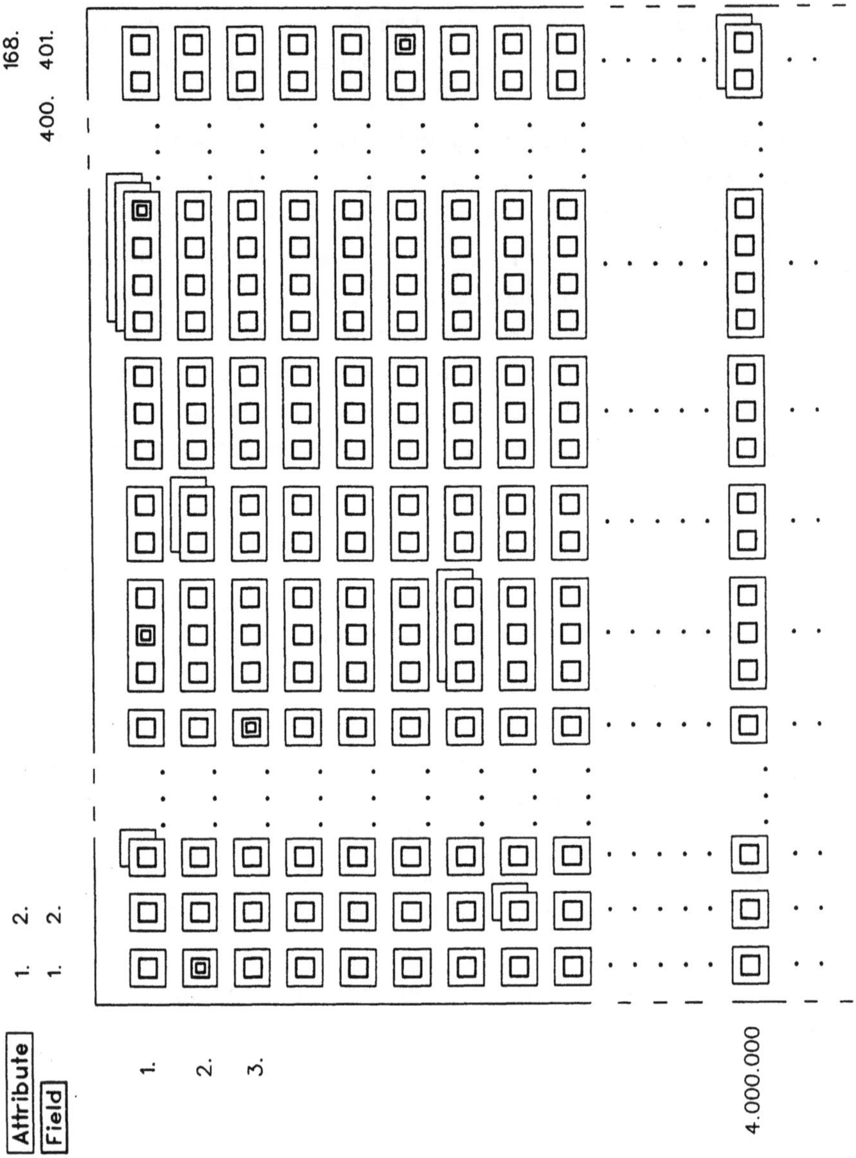

Fig. 2. The data matrix

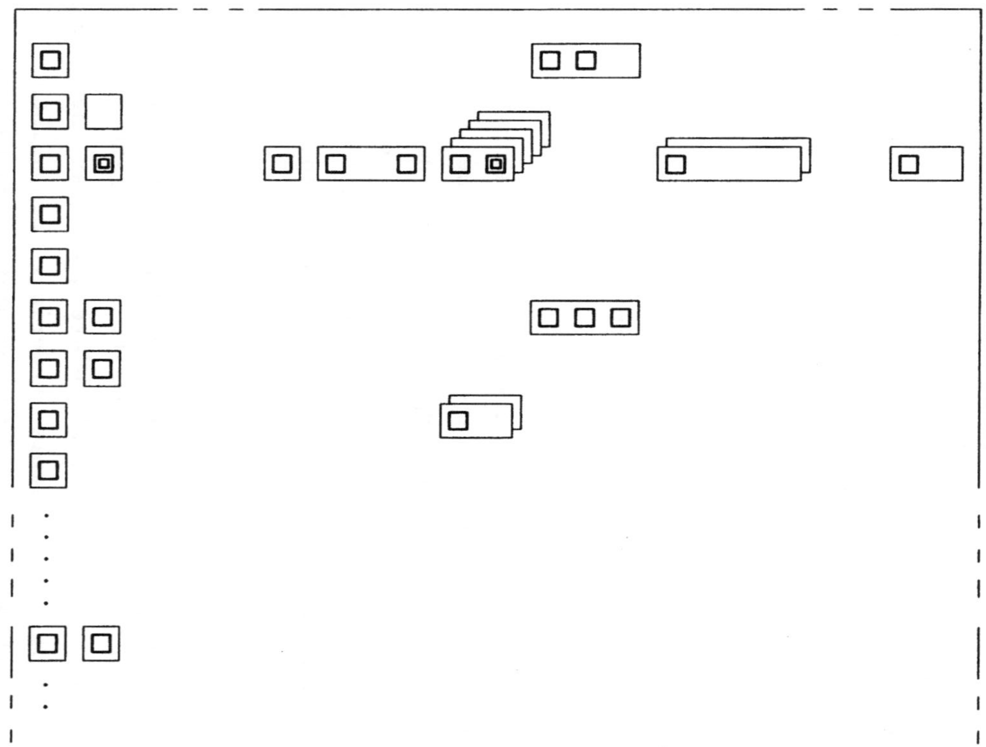

Fig. 3. Data matrix of measured data

Besides the "chemical structure", "chemical name", "molecular formula", "molecular weight", "Beilstein Registry Number" and "Lawson number", which are always present in the Beilstein database, the most frequently known attributes are: "preparation", "melting point", "crystal property description" and "chemical behavior" having a frequency of 93%, 80%, 25% and 16%, respectively. 80% of the attributes are known for less than 10% of the compounds. These numbers refer the heterocyclic compounds of the Handbook material 1840-1960. Acyclic and isocyclic compounds and later publications contain more factual data.

For some applications the low percentage of measured and published data leads to less than optimal use of the online database. Because it is not to be expected that in the near future significantly more experimental data will be available, the possibility of filling up the database with calculated values should be investigated. In the following there are some arguments for using calculated values.

Some physical properties are parameter dependent, like boiling point depends on the pressure, density on temperature, etc. The values published in the literature are given at heterogeneous conditions disabling a simple numerical comparisons of the values. It is rather difficult to formulate a query for online searching using such properties. For example, searching for

"boiling point" (Bp) = 100-105 Celsius,

may result in hits like

Bp = 101 Celsius at p = 760 torr
Bp = 100 Celsius at p = 3 torr
Bp = 102 Celsius at p = 1500 torr,

which are in fact very different results. Moreover, hits can be lost if the boiling point of a compound with

Bp = 100 Celsius at p = 760 torr

was measured only at low pressure and is stored like

Bp = 25 Celsius, p = 5 torr.

In this case recalculating values to a standard parameter value would be of great advantage.

An online database can be used in two different ways :

- a handbook type way, which means to retrieve information about a well defined compound or about a very similar one. In practice, just as in the Handbook, it is done by selecting a compound from the database by an always existing predefined key, like structure, or chemical name, etc. and displaying or printing the required information of the selected compound. Of course, in this case reliable values of the physical and chemical properties are expected.

- the second way, which can not be done with a handbook, is to retrieve a class of compound with one or more prescribed properties. This can be done by formulating queries using several keys simultaneously connecting them by boolean operations like "and", "or", "not". A possible example is to find all compounds having a given substructure, boiling point between 100-110 Celsius at normal pressure, density 1.5-1.8 g/cm3 on room temperature, melting point less than 21.5 Celsius. In many such a cases not the exact numerical values of the physical parameters are important, but to be able to find ALL(!) the compounds having the required properties. The problem is, if such a search query contains for example 2 properties which have an average occurrence in the database of 10%, then only 1% of the relevant compounds are to be expected to be retrieved, because non-existing values do not match the search criteria. In this cases storing calculated values in the database, even with a relative large errors, would be of enormous help, because a complete search could be done. However, knowing that we are searching on a database containing also calculated values with possible significant errors, the query should be formulated less accurately, for example, instead of Bp=100-110 and density=1.5-1.8, Bp=90-120, density=1.3-2.0 to guarantee that all the relevant compounds would be retrieved. However, due to the larger tolerance in the query some unwanted hits would be found as well, which would need to be rejected by other means.

A third advantage of calculated values would be the use in data checking. The path of data into the database is rather long with a lot of possible sources of errors, like experimental error, error in measurement, error in primary publication, error in Beilstein Handbook, error in data input, error in data processing, etc. Significant deviation between calculated and stored experimental values could reveal errors and facilitate error corrections.

Summarizing, there are three promising ways of using calculated values:

1. storing values at standardized conditions.
2. using calculated values in retrieval.
3. revealing and correcting erroneous data.

It is clear, that calculated data must be handled very carefully, and the users must exactly know what kind of data they are using, what kind of data they are searching, and what methods were used for calculations.

The Beilstein data structure has been designed with the possibility of flagging calculated data and describing methods of calculation.

Questions and Issues About the Process of Estimating Properties of Chemicals

Stephen R. Heller

USDA, ARS, BARC-W, Beltsville, MD 20705-2350, USA

Abstract

Questions and issues about property prediction are addressed and discussed. Issues such as reliability, evaluation of predicted values and how to handle different/multiple values predicted from different methodology, and how the data should be presented to the user community are critical to a project of the scale of the Beilstein database of chemicals. Lastly a proposal is presented as to how to initially address large-scale prediction of properties of chemicals.

C. Jochum M. G. Hicks J. Sunkel (Eds.)
Physical Property Prediction
in Organic Chemistry

Introduction

In the "best of possible worlds" (1) a chemist who discovers a new compound would analyze its structure unambiguously and obtain accurate data on at least 50 of its most important chemical and physical properties. We are here in Bolzano because we do not live in the best of possible worlds. The Beilstein Institute exists because, in spite of Candide's optimism, this is not the best of all possible worlds, scientifically or otherwise. The best reason I can give for the lack of published property data about a compound is that most chemical compounds (in fact some 75%) are only reported once in the scientific literature (2). Some 15% are reported twice, which leaves only 10% of the entire known universe of chemicals for which there are more than two literature citations. Thus, it is quite clear that since one probably couldn't even readily obtain an authentic sample of most chemicals in the Beilstein database, the issue of the very costly and time-consuming efforts to experimentally obtain the data desired is really irrelevant. Hence, the very clear reason for this workshop is to find reliable, consistent, and easy to use methods for chemical and physical property prediction.

Property predication is not new. One of the first and probably the best known chemical property predictions was made some 117 years ago. It is quite fitting that the home of Professor Friedrich Konrad Beilstein, St. Petersburg, was also the place which gave rise to property prediction, the subject of this Beilstein Workshop. In 1871, Dmitri Ivanovitch Mendeleev published his prediction (Figure 1) of the existence and properties of Eka-aluminum, Eka-boron, and Eka-silicon (3). Within 15 years, these elements and their properties were discovered and the predictions of Mendeleev were shown to be rather accurate. No doubt those were simpler times and the task the Beilstein Institute wishes to undertake is of much greater complexity than the elements in the periodic table.

Background

The Beilstein Institute factual database now being put into computer-readable form contains some 400 parameters, some of which are shown in Figure 2. As we all know, there are three stages or steps in the creation of a complete database of numerical data. These are:

1. Collect Experimental Data
2. Evaluate Data
3. Fill in Data Gaps

Figure 1: Mendeleev 1871 Property Prediction of Three Elements

Prediction		Determination
Eka* - Aluminum		**Gallium (Discovered in 1875)**
Atomic Weight	68	69.9
Specific Weight	6.0	5.96
Atomic Volume	11.5	11.7
Eka - Boron		**Scandium (Discovered in 1879)**
Atomic Weight	44	43.79
Oxide	Eb_2O_3	Sc_2O_3
Specific Weight (Oxide)	3.5	3.864
Sulphate	$Eb_2(SO_4)_3$	$Sc_2(SO_4)_3$
Eka - Silicon		**Germanium (Discovered in 1886)**
Atomic Weight	72	72.3
Specific Weight	5.5	5.469
Atomic Volume	13	13.2
Oxide	EsO_2	GeO_2
Specific Weight (Oxide)	4.7	4.703
Chloride	$EsCl_4$	$GeCl_4$
Boiling Point - Chloride	< 100 °C	86 °C
Density - Chloride	1.9	1.887
Ethyl Compound	$EsAe_4$	$Ge(C_2H_5O)_4$
Boiling Point (Ethyl Compound)	160 °C	160 °C

* Eka is the Sanskrit prefix for the number one

Figure 2: Examples of Data Elements from the Beilstein Factual Database

Mnemonic	Name
ATC	Atom Count
BF	Biological Function
BP	Boiling Point
BRN	Beilstein Registry Number
CCOL	Crystal Color
CDEN	Crystal Density
CN	Chemical Name
CRP	Critical Pressure
CRT	Critical Temperature
CRV	Critical Volume
DEN	Density
DM	Dipole Moment
ECOL	Ecological Data
ED	Entry Data
ELC	Element Count
ELS	Element Symbol
ENTR	Entropy
FW	Formula (Molecular) Weight
HFOR	Energy of Formation
HFUS	Enthalpy of Fusion
HSUB	Enthalpy of Sublimation
IP	Ionization Potential
IRS	Infrared (IR) Spectrum
LW	Lawson (Classification Scheme) Number
MF	Molecular Formula
MI	Moment of Inertia
MP	Melting Point
MS	Mass Spectrum
NMRS	NMR Spectrum
OA	Optical Anisotropy
ORD	Optical Rotary Dispersion
PHWP	Polarographic Half-Wave Potential
PRE	Preparation
QM	Quadrupole Moment
RAS	Raman Spectrum
REA	Reaction
RN	CAS Registry Number
SFOR	Entropy of Formation
SLB	Solubility
SO	Beilstein Handbook Source Citation
ST	Surface Tension
SY	Synonym
TOX	Toxicity
TP	Triple Point
UP	Update Date
USE	Use
VP	Vapor Pressure

For some 100 years, the Beilstein Handbook activities have been involved with the first two of these stages, that is the collection and extraction from the scientific literature and the evaluation of the data. Now as one looks at the Beilstein Handbook of over 350 volumes the question has arisen, what about the "data gaps". Can something be done to fill in the blanks? To put some perspective on the magnitude of the "data gaps" in one particular area, solubility, it is worthwhile to mention what Horvath (4) said in his book on hydrogenated hydrocarbons:

"Despite the great demand for solubility data by scientists and engineers, experimental values published in the open literature are very limited. Regarding the availability of solubility data for halogenated hydrocarbons in water, Beilstein (4th Supplementary Series, Volume 1/Part 1 (1958), Volume V/Part 1 (1963), and Volume V/Part 2 (1964)) cites 1369 compounds up to six carbon atoms, of which only 61 have information as to their solubility, mostly for a single temperature only."

In some cases methods have been developed for the prediction of a property. In other cases there has been no reported research or activity for predicting a particular property. Prediction should not be confused with prioritization. As Bill Milne at the NIH Cancer Institute will discuss in a later talk, it is often very useful to know the relative values of a given property. While this is definitely property estimation, it is not the sort of estimated values which would fill the "data gaps" in the Beilstein database. Also, in this matter of relative values of a given property, it could be harmful if non-experts see relative numbers and mistakenly take these values to be real and absolute numbers.

While predictive methods or procedures can be developed without data they can only be tested if one has data. Such data must also have several attributes. First, they must be accurate and precise to test a given hypothesis on how to predict a numeric data value. Second, there needs to be a reasonable number of data points so that there is some "weight of evidence" to the prediction. While two points will mathematically give a straight line, in chemistry (as well as other disciplines) more data are required before the scientific community feels the predominance of evidence is correct. Third, the data should cover a broad range of chemical classes to have maximum any predictive value. In preparing this presentation I searched the literature from 1967 to early 1988, using the Online Chemical Abstracts database, for property predictions. I also talked with some colleagues about this subject. If this literature search and these discussions were all one knew about organic chemistry it would seem that the field was concerned only with hydrocarbons and simple mono-functional groups compounds with fewer than 10-15 non-hydrogen atoms. Most predictive methods are useful as teaching examples, rather than being able to fill in the "data gaps" in the Beilstein Handbook or elsewhere. My research is concerned with the creation of a database of chemical, physical, and other properties of pesticides for use in models which will predict possible contamination of our nation's groundwater. With few exceptions (methyl bromide and 1,2-dibromoethane (EDB) being the only two I have found so far) most pesticides contain over 10 non-hydrogen atoms and 2-3 elements in addition to carbon and hydrogen. Thus, for this research to proceed new or expanded predictive capabilities are needed. Before this can occur, at least in the pesticide chemistry field, one needs better data for the existing compounds. One should not try to extrapolate from a vacuum into the real world.

How does one choose which method for property prediction to use? Are the known methods valid for the entire range of organic chemistry? Of course not. Have the authors explicitly stated what the limitations of the methods are? What is the reliability of a given method? What are the error ranges for the predicted data? Scientists are notorious for reporting calculated numbers to beyond the range of significant figures. What is necessary to be sure predicted data are properly presented? How was the method developed? If the method is based on calculation of another piece of data for input, what is the reliability and error range associated with the input data?

One of the first and most important properties being collected for the Agricultural Research Service (ARS) database is the aqueous solubility of a pesticide. Highly soluble materials are rapidly distributed in the soil and can be easily transported to the water table. There are a number of methods for estimating solubility. Five basic methods are described by Lyman and his colleagues (5) Chapter 2, Table 2-1. However most give an estimated value at only one temperature (25 ^{o}C), and "few have actually been presented (and tested) as predictive tools" (5, page 2-1). Furthermore, issues such as "relative merits, applicability, and accuracy" of these methods had not been reviewed prior to the work of Lyman and his colleagues (5, page 2-2). For example, the PC-GEMS program (6) first calculates an octanol/water partition coefficient (LogP) from a two dimensional structure input. It then calculates a melting point from the two dimensional structure and the calculated LogP value. From there the water solubility is calculated. As seen in Figure 3, the predicted water solubility is given to two significant figures, without any justification that the prediction is that accurate. Another estimation program, CHEMEST (7), which is based on work published in the book by Lyman and his colleagues (5), provides for calculations of 11 different properties using 36 different methods. This system, CHEMEST, fares much better with respect to providing information about the limits and accuracy of the methods and the calculation or estimation errors. Figure 4 shows the same calculation of solubility using the CHEMEST program.

Figure 3: Sample Property Estimation From PCGEMS Program

```
        21:40:01                    Sunday                 05/01/88

                           PHYSICO-CHEMICAL PROPERTIES
                           ---------------------------

Smiles Notation            = CCCO
Chemical Name              = Propanol
Molecular Formula          = C3H8O                     Calc. from Smiles
Molecular Weight           = 60.10                     Calc. from Smiles
Physical State *           = Liquid                    User Entered
LogKow                     = 2.9 -01                   User Entered
Water Solubility           = 8.59E+04 mg/L             Equation 13N
Melting Point              = -8.5E+01 (C)              Grain and Lyman
Vapor Pressure             =    48.33 mm at 25.00(C) Antoine
Boiling Point              =    82.33 (C)              Meissner
Henry's Law Constant       = 4.89E-05 atm m3/mol       Method 1
Bio Concentration Factor = 9.78E-01                    Kow (Method 1)
Adsorption Coefficient     =     1.00                  Kow, Eqn. 4-10

* Estimated MP or BP does not change entered physical state.

                                         Press any key to continue
```

(The output shown is exactly as it appeared on the computer monitor.)

Figure 4 - Sample Property Estimation From CHEMEST Program

```
**********************************************************
*                                                        *
*   CHEMEST ........... CHEMICAL PROPERTY ESTIMATION     *
*                                                        *
*   FILE: ITALY.TST  DATE:  1-May-88   TIME: 21:44:28    *
*                                                        *
**********************************************************

      CHEMICAL NAME/IDENTIFICATION ... Propanol
============================

      WATER SOLUBILITY ESTIMATION:
---------------------------
SOLUBILITY :  8.59E+04 MG/L

      ESTIMATION ERROR:
----------------
METHOD ERROR      : X    1.6
PROPAGATED ERROR : X    1.0
TOTAL ERROR       : X    1.6

      METHOD IDENTIFICATION:
---------------------
METHOD USED   : 1
EQUATION USED : 13 in Reference 15


      KEY INPUT:
 -------
ACID GROUP IN CMPD.?    : NO
OCTANOL-WATER PRT. CF. :      0.290     L
PHYSICAL STATE AT 25 C : L
```

(The output shown is exactly as it appeared on the computer monitor.)

Both programs use the same procedures, but CHEMEST provides the user with information on the error associated with the method.

Recently I was told (8) of a chemical company which decided to re-run some partition coefficient data for several acid anilide pesticides. The chemicals were from their own company, as well as from other manufacturers. They designed and ran the experiments very carefully, with the proper quality assurance and quality control. When they compared their results to the predicted values using the CLOGP program (9), there were sufficient differences to warrant some concern. As a result, they discussed the matter with the world's authority in the field, Professor Corwin Hansch. After seeing how the experimental data were collected and what values the CLOGP predictive program generated, the CLOGP program was revised by Hansch and his coworkers to take into account high quality experimental data.

How many and what type of compounds were used to test the validity of the method? How accurate were the data used to develop and/or test the predictive technique? How does one prove that the predictive method used is accurate? In a talk at this workshop Peter Jurs from Penn State University (USA) will describe some of his research activities in this field, including a recently reported study (10) on predicting olefin boiling points from molecular structure. Jurs wisely chose these compounds because there were a reasonably large number of compounds (123) available and the data were of high quality. The method he has developed for the class of compounds studied appears to have solved the problems encountered with earlier prediction techniques (7,11) which were not able to handle many isomeric compounds. In contrast, the data in the Beilstein Handbook, while evaluated, are collected in a random fashion in terms of a class or series of compounds. Also, with rare exceptions the data come from many sources, published over many years, and using many different experimental conditions. Thus, it may not be possible to obtain a large number of similar compounds with enough identical properties from the Beilstein database to assure that a predictive method is properly tested and evaluated. Unquestionably this is a handicap which must be overcome. Without sufficient and good experimental data, predictive methods must be viewed very carefully.

Assuming satisfactory answers to these questions, let us now proceed to the question of data quality or reliability. How should the predicted results be evaluated? Since all the experimental data which go into the Beilstein Handbook are evaluated, it is reasonable to assume that methods must be developed to evaluate the predicted data. What does one do in the case where two (or more) methods are believed to be scientifically valid, yet yield different answers? We are beginning to develop a series of expert systems for data evaluation. The process will be based on our SELEX expert system (12) which provides objective and consistent evaluations of published data on the selenium content in foods. Our first data property expert systems will be in the areas of solubility and vapor pressure evaluations.

Once agreement is reached that a number for inclusion in the Beilstein database, how will be it noted or tagged in the database? Will it be clear to the user that the information or numeric value is not an experimental value, but rather in a predicted or calculated value? Certainly a clearly marked reference citation should suffice, but how can one be sure that an entry transferred from the Beilstein database to a report (or a value from the ARS Pesticide Property database used as input for some model or other purpose) is properly referenced and properly used? Should experimental data be in a separate section of the Beilstein database to help assure the user notices the difference? Should there be a notation in the record saying, for example, "No experimental data, please see predicted value given below"?

Should interpolated data be noted as such, as compared with extrapolated data? How will the evaluation criteria take such differences into account? What will happen when an experimental value for a particular parameter is found? Will the experimental value automatically replace the predicted value?

What if the values are far apart? In some cases errors can be a few percent or less (for interpolation), but can orders of magnitude (for extrapolation). What might this imply about the method used for the prediction? Will there be a notation in the record that the newer, experimental value is a replacement value? Should the original predicted value be kept in the database?

When an experimental value is found to be considerably different from the predicted value, what should be done about the predicted values for other, similar compounds in the database (or the other compounds in the database which have data values predicted by this method)? If the reliability of a method comes into question later, how easy will it be to change all the records in the database found in a number of online systems which use this method?

One should also ask if the method is automated. If not, can it be automated? For a method to have any possible practical application for the Beilstein Institute, and be used with such a large database as the Beilstein Structure Registry Connection Table database, computerization is essential. Is the two-dimensional structure sufficient for input into the prediction method?

Responsibility

Who is responsible for the predicted data? When an error is found, should it or must it be quickly corrected? Certainly a computer program can regenerate a large set of predicted values in a very short time. Can this be done as a practical matter and will this be done even if it is costly? If it is done, what guarantee is there that the online vendors of the Beilstein database will quickly replace the older or incorrect data with the corrected or new data? Dealing with scientific data implies a greater responsibility than is normally taken for with bibliographic abstracts.

How will the scientific community accept a database with many entries of predicted values? How will Government agencies, in the US, Europe, and elsewhere accept such data? What will be the effect of such data have on patents and patent rights? Will predictions be considered, under any conditions, as "prior art"?

Proposal

Now that the less positive aspects of property prediction have been raised, I would like to propose some possibilities for future work. The research falls into two distinct areas. The first is creation of collections of high quality databases for a series of class of multi-functional group compounds. This is essentially what we are doing with the ARS Pesticide Database, since accurate values for parameters such as solubility and vapor pressure do not exist for most pesticides. Results from solubility experiments run under conditions of different temperature, pH, and ionic strength will give us the necessary data input for the wide range of agricultural conditions which exist. We then hope to use these results as the foundation for developing accurate predictive methods. In other areas, such as bio-medicine and pharmaceutical chemistry a parameter like the octanol-water partition coefficient (LogP) may be considered to be of high priority and importance. LogP also has been proposed to be of potential value in the prediction of aqueous solubility (4,9).

I would hope that the Beilstein Institute, with the support of the German Government and others, would fund several such projects as prototypes to see how useful some high quality data for a number of parameters will be in creating as broad a predictive strategy as possible.

Conclusion

This presentation discusses several difficult issues associated with the wide-scale use of predicted property data for a large and chemically diverse database. While the overall state-of-the-art is in its infancy, and is quite limited in its current application, this workshop has taken the first and bold step in looking into the question which Clemens Jochum asked in his October 1986 letter of invitation to all the workshop attendees - "Is it possible to fill the data gaps in the millions of Small Information Compounds in the Beilstein database?". As we hear the many excellent research activities described in lectures by experts in their fields over the next three days, I hope that we will all remember to ask some of the questions and address some of the issues mentioned in this presentation, so that the goal of the Beilstein Institute can be reached.

Acknowledgements

I would like to thank my colleagues at ARS, D. Bigwood, S. Rawlins, D. Wauchope, and C. Helling for their valuable suggestions. I would also like to thank D. Lide and L. Gevantman (NBS) and G. W. A. Milne (NIH) for their insightful comments and thoughts on numeric data, data evaluation, and data quality. Lastly, I would like to thank my son Matt for his contribution of suggesting the Mendeleev prediction of new elements while we were studying for one of his high-school chemistry tests.

REFERENCES

1. Dr. Pangloss in Chapter 1 of Candide, Voltaire (1759).

2. Y. Wolman, "Chemical Information - A Practical Guide to Utilization", 2nd ed., J. Wiley & Sons, New York (1988).

3. D. Mendeleev, Ann., Suppl. VIII, 133-229 (1871).

4. A. L. Horvath, "Halogenated Hydrocarbons", M. Dekker, New York (1982).

5. W. J. Lyman, W. F. Reehl, and D. Rosenblatt, "Handbook of Chemical Property Estimation Methods" McGraw-Hill, New York (1977). This book is, at present, out of print.

6. PC-GEMS (Personal Computer version of the Graphical Exposure Modeling Program) is available from Ms. Cathy Turner, US Environmental Protection Agency, TS-798, Washington, DC 20460 USA. The software will be provided free of charge so long as one sends a sufficient number of formatted 360K or 1.2 MB 51/4 inch floppy disks. The 1986 manual (Publication # SGC-TR-13-88-003) is also available at no charge.

7. The IBM PC version of CHEMEST is available for $585 from TDS (Technical Database Services) Inc., 10 Columbus Circle, New York, NY 10019 (Telex: 6714962).

8. D. Gustafson, Monsanto Chemical Co., St. Louis, MO 63198, private communication.

9. MedChem Project, Chemistry Department, Pomona College, Claremont, CA 91711.

10. P. J. Hansen and P. C. Jurs, Anal. Chem., 59, 2322-2327 (1987).

11. R. C. Reid, J. M. Prausnitz, and T. K. Sherwood, "Properties of Liquids and Gases" 3rd ed., McGraw-Hill, New York (1977).

12. D. W. Bigwood, S. R. Heller, W. R. Wolf, A. Schubert, and J. M. Holden, Anal. Chim. Acta, 200, 411-419 (1987).

REFERENCES

[illegible] Paradoxes [illegible] Cascades [illegible]

[illegible] Wiley & Sons, New York (1988).

[illegible]

Numeric Features of the Beilstein Database on STN

Andreas Barth

FIZ Karlsruhe, 7514 Eggenstein-Leopoldshafen, Federal Republic of Germany

Abstract:

Beilstein's handbook of organic chemistry is currently implemented as a database on STN, starting with the first part containing the heterocyclic substances.

It is the intention of this paper to give an overview of the capabilities of this database on STN with special focus on the numeric features. At first, an introduction to the database is presented starting with an outline of the database design. Following, the various possibilities to search and retrieve the documents from the Beilstein database are discussed. In addition to the normal search for the content of a field it is also possible to search for the name of the field, thus offering an alternative way to access the file.

Several new software features are required to support especially the numeric fields of this database. Most numeric properties are associated with an experimental uncertainty, which must be taken into account by the database loading and retrieval software. This new feature is called numeric range search capability. The problem is briefly described together with a possible solution. In general, a numeric interval can be entered as a range with a lower limit and an upper limit. However, many scientists are more familiar with the notation of a measured value and its associated uncertainty. A feature which allows the specification of a range in terms of a value plus/minus a tolerance offers additional user support. Another important function for numeric databases is the conversion of physical units. Especially in the older literature many units are found which are no longer used mainly due to a standardization process. To overcome the difficulties which are inherent to the use of different units, a function is required which allows the user to work with his own set of physical units and lets the software worry about the transformation to the units used in the specific field of this database. At last, some future developments and requirements are outlined.

C. Jochum M. G. Hicks J. Sunkel (Eds.)
Physical Property Prediction
in Organic Chemistry

Disposition:

1. Introduction

Beilstein's handbook contains the largest collection of critically reviewed data of organic compounds covering information like: constitution and configuration, occurrence, isolation from natural products, production, modes of formation and purification, structure and energy data of the compound, physical properties, physical properties of multi-component systems, chemical behaviour, characterization and analysis, salts and derivatives.

Data from four different information sources will be input to the Beilstein database:

1. The printed handbook-series from the basic series up to the fourth supplementary series (H to E IV) covering the literature from 1830 to 1960 contains critically reviewed data for more than 1.5 mio. organic substances.
2. The material for the fifth supplementary series (E V) covers the period from 1960 to 1980. A careful perusal with respect to precision and consistency will be performed exactly as for the handbooks of the previous editions.
3. The handwritten excerpts for the E-V series forms another valuable source for the database yet these data have not been crosschecked.
4. Machine readable computer excerpts being available since 1980 build the last information source. Except for some simple error and redundancy checking these data are directly loaded into the database.

Obviously, there are two different classes of data distinguished by their reliability. These two classes of data form the full file (1 + 2)and the short file (3 + 4). In the forthcoming Beilstein database on STN the handbook data (sources 1 and 2) will be indicated by a special flag.

2. Beilstein Database

2.1 Survey of Database Design

Handbook and Datastructure:

The content of the database can be roughly divided into factual data including structures and data references. For many properties a reference to the original literature is given instead of quoting the property value itself. The datastructure is hierarchically organized comprising three levels: substance, property and measurement. A document in the file consists of all information which is available for a specific organic substance (=structure). The individual properties (=entities) generally consist of a set of parameters (=attributes). A set of parameter values results from a measurement and is associated with one or more literature references. For a certain property there may be multiple sets of parameters due to different experimental results.

In order to register the substances in the Beilstein Institute a special registry service has been set up using the Beilstein Registry Number. This registry number is a purely numeric quantity and has nothing to do with the CAS Registry Number. The Beilstein Registry Number (BRN) is used as the primary key to the Beilstein database both for the structure as for the factual part.

Each entity has a unique name and abbreviation, e.g. 'Enthalpy of Formation' (HFOR). The attributes have a name which is not necessarily unique, e.g. 'Temperature'. The corresponding acronym consists of two parts: abbreviation of the entity, a period and the abbreviation of the attribute. In the above example, the parameter 'Temperature' of the entity 'Heat of Formation' is abbreviated as HFOR.T. Since an experimental result consists of a set of values (property + parameters) it is necessary to connect the property and its parameters with a proximity operator. In this case the P-operator is used. The application of this operator is illustrated in the following example:

=> SEARCH 20 < HFOR < 100 (P) 25 <= HFOR.T <= 30.

Here, the 'Enthalpy of Formation' is requested at a restricted temperature range. The physical properties are presented in standardized units which are identical for SEARCH and DISPLAY.

In the case of reference information, a bound phrase (Controlled Term) indicates the type of data which is available in the literature. In general, the bound phrase is identical with a property name.

Description of Entities:

The spectrum of information can be divided into four groups: Identification of Substance (IDE), General Data (GEN), Physical Data (PHY), and Chemical Data (CHE). The IDE-group consists of all information identifying the chemical substance including the Beilstein Registry Number (BRN).

Each substance is assigned a unique BRN and is identified by its chemical structure (STR). in addition, the following set of dictionary fields may be present:
- Molecular Formula (MF)
- Chemical Name (CN) and Synonyms (SY)
- Chemical Name Segments (CNS) (search only)
- Element Symbol (ELS) and Periodic Group (PG) (search only)
- Single Element Counts (element symbol) (search only)
- Total Element Count (ELS) and total Atom Count (ATC) (search only)
- Formula Weight (FW)
- Charge (CHA) and Number of Components (NC) (search only)
- Lawson Number (LN).

To locate the corresponding hardcopy article in the Beilstein handbook, a source number (SO) is given. The data from the short file, i.e. which is not yet published in the Beilstein handbook, has a reference to the original literature.

In the case of physical data there is a broad and rich spectrum of properties available in the database. We will restrict ourselves here to a few illustrative examples. The entity 'Dipole Moment' (DM) has the following structure:

Field Name	Field Code	Unit	
Dipole Moment	DM (1)	D	(Debye)
Temperature	DM.T	Cel	(°C)
Method	DM.MET	-	
Solvent	DM.SOL	-	

The most important parameter is the temperature since mosts properties are temperature-dependant. A display of the Dipole Moment contains also the literature references.

Chemical entities like Preparation (PRE) and Reaction (REA) have the same structure but more parameters. Preparation is presented here as an example:

Field Name	Field Code	Unit
Preparation	PRE (1)	
Educt	PRE.EDT	-
Reagent	PRE.RGT	-
Time (2)	-	-
Yield	PRE.YLD	%
Solvent	PRE.SOL	-
Ambient Temp. (2)	-	Cel
Reflux (2)	-	-
Temperature (2)	-	Cel
Catalyst (2)	-	-
Irradiation (2)	-	-
Pressure (2)	-	Torr
Detail (2)	-	-
By-product	PRE.BPRO	-

Notes:

1. This code is also used for the diplay/print of the property.
2. Field is not searchable.

The abbreviation PRE can only be used for the display or print of the entity. The chemical parameters educt, reagent, solvent, catalyst, by-product and yield are searchable. Yield, temperature and pressure are numeric fields and have an associated physical unit. Irradiation and reflux are just flags which indicate whether the reaction was irradiated or reflux has been used.

Special Indexing Rules:

Some fields require special indexing rules. The total Element Count (ELC) is generated from the Molecular Formula (MF) and is equal to the sum of different elements. It is created in order to limit range searches of element counts to a maximum or minimum number of chemical elements. In addition, an Atomic Count (ATC) is build representing the total number of atoms in the molecular formula. For each chemical element a special field is generated containing the number of occurences of this element in the sum formula. A typical search query could be to look for a range of organic substances with two oxygen atoms and at most one other heteroatom:

```
=> SEARCH  5 <= C <= 15  AND  H >= 12  AND  O = 2  AND  ELC <= 4.
```

Each element symbol appearing in MF is indexed in an additional field Element Symbol (ELS). This serves mainly to search for the presence of elements in chemical substances. The fields ATC, ELC, and ELS are not present in the CAS Registry database.

Chemical names are present for the title compounds, the educts, and the products. They are contained in several different fields:
- Chemical Name (CN)
- Chemical Derivative (CDER)
- Isolation from Natural Product (INP)
- Preparation:
    - Educt (PRE.EDT)
    - Reagent (PRE.RGT)
- Reaction
    - Reaction Partner (REA.RP)
    - Reagent (REA.RGT)

The chemical names are indexed in the above fields as complete names. In the Basic Index (BI) a parsing at special characters, e.g. hyphen and parenthesis, is performed.

2.2 Search and Retrieval Capabilities

Textretrieval:

Text data consists either of bound phrases or of free text. Accordingly, text may be indexed as phrases or single words. In both cases the field content is sorted in a lexicographical order. Using the Messenger command language the following set of operators may be used to connect the content of text fields:
- logical operators: AND, OR, NOT
- proximity operators: Link (L), Paragraph (P), Sentence (S), Word (W), Adjacency (A)
- meta operators: character masking or truncation symbols ('?', '#', '!')

The content of all fields containing free text is indexed ("searchable") both in individual fields and the basic index (BI). Thus, users do not have to worry about all the specific field names and their acronyms. A nice feature of the Beilstein database is the capability to search for the presence of fields, e.g. to seek for all substances for which the toxicity is reported:

```
=> SEARCH  toxicity/FA
```

The entity names are indexed in 'Field Availability' (FA).

Numeric Retrieval:

In the Messenger system there is a numeric data type which is used to represent both integers and real numbers. The content of these fields is sorted in numerical ascending order. Experimental values are often given as the lower and upper values of the measurement, e.g. 100 <= BP <= 102, indicating the boiling point has been reported to be 101 ± 1 degree Celsius (see section 3.2). In some cases only the range of the measurement is available, like 100 <= IRS <= 900 meaning the infrared spectrum for this substance has been measured in the range between 100 and 900 nm. For both cases it is strictly recommended to perform only range searches and not searches of a single point. As for text searches a set of operators is available:

- logical operators: AND, OR, NOT
- proximity operators: Link (L), Paragraph=Parameter (P)
- compare operators: '=', '<', '=<', '>', '>='.

There are some problems associated with numeric searches in a database. At first, there is the problem of complete recall or interval intersection. This problem has been solved (see section 3.2) and a numeric range search capability will be available for the Beilstein database. The parameter dependency of physical properties is another point which has been taken care of in the design of the database. It is solved by combining the values of the property and its parameters via the P-operator. Finally, there is the relation between a physical property and the corresponding unit and the possibility of units conversion which has to be solved. STN will provide such a feature within the next future (see section 3.4).

Structure Retrieval:

In the Beilstein database the structures of the chemical substances are uniquely represented by a connection table, i.e. a matrix containing the atoms and their bonding values. The input connection table delivered by the Beilstein Institute has been converted to the CAS Registry III format in order to implement the structure file in the Messenger retrieval system. As a result, the stereochemical information is not available in the Messenger version of the database. However, all the structure retrieval features of the Messenger system can also be used to search the Beilstein database, e.g.

- full, sample and range search
- exact, substructure and family search
- structure query modelling (using Beilstein Registry Numbers)
- screen search
- combination of answer sets from structure and dictionary search.

2.3 Some Illustrative Examples

In this section we will discuss two search and display examples from the Beilstein database. More examples are given in the appendix.

Example: Ascorbic Acid (/CNS)

A search for ascorbic acid can be performed using the field Chemical Name Segments (/CNS) which contains the individual segments of the complete chemical name.

```
=> search ascorbic?/CNS (W) acid?/CNS
                     4 ASCORBIC?/CNS
                 22479 ACID?/CNS
L3                   4 ASCORBIC?/CNS (W) ACID?/CNS
```

```
=> display L3 3
```

```
L3   ANSWER 3 OF 4

BRN  89123  Beilstein
MF   C6 H8 O6
CN   L-ascorbic acid
     L-Ascorbinsaeure
SY   (S)-5-<(S)-1,2-Dihydroxy-aethyl>-3,4-dihydroxy-5H-furan-2-on
FW   176.13
SO   4-18-00-03038; 4-18-00-03038
LN   19316
```

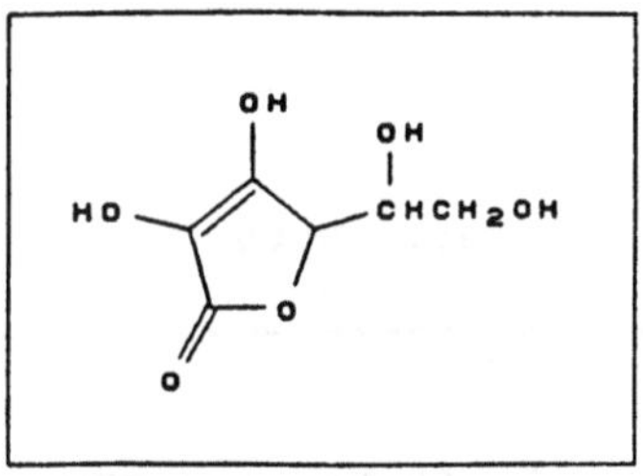

L-ascorbic acid is contained in the answer set and it has the preliminary Beilstein Registry Number 89123. The display of the document shows the substance identification information including the chemical structure.

<u>Example:</u> Boiling Point(/BP)

In general, a search for a numeric entity like the boiling point should be specified as a range search. In this example, however, we will search for a specific boiling point value (100 °C) at a given pressure (760 torr).

```
=> s 100/bp (p) 760/bp.p
            295 100/BP
           1849 760/BP.P
L1            8 100/BP (P) 760/BP.P
```

An answer set of 8 substances is retrieved. The first compound is a derivative of furyl:

```
=> d
```

```
L1   ANSWER 1 OF 8

BRN  37741  Beilstein
MF   C13 H14 O3
CN   <4-Ethoxy-phenyl>-<2>furyl-methanol
     <4-Aethoxy-phenyl>-<2>furyl-methanol
FW   218.25
SO   4-17-00-02146
LN   17475; 298
```

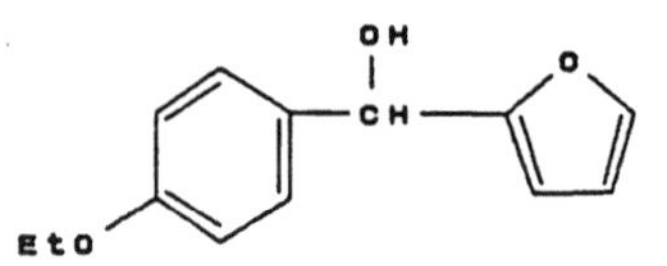


```
---> Handbook Data <---
Boiling Point:
Value (BP)                Press.(BP.P)              Ref.  Note
(Cel)                     (Torr)
98.00 - 100.00------------+760-----------------------+1----+-----

Reference(s):
1. Mndshojan et al., Doklady Akad. Armjansk. S.S.R. 29 Nr. 1 <1959> 41
   CA: 1960 7673
```

In this case, the display shows the substance identification information and the boiling point. This is an example for the dynamic display function which has been developed for the Beilstein database. When a display command is entered in this file without specifying a display format, this function will generate a display consisting of the substance identification information and the data related to the query. This means simply that the user always sees what he has asked for.

3. Special Features Supporting Numeric Retrieval

3.1 Property Thesaurus and Beilstein Datastructure

Thesaurus Capability:

A thesaurus function will be implemented in the Messenger system in the near future. The capability is discussed here because it is planned to add a thesaurus file to the Beilstein database.

A Thesaurus is any kind of dictionary. It may be used in various different ways, as a
- predefined descriptor list
- dictionary of synonyms
- classification or structuring of descriptors.

In the first case, the thesaurus is used to have a terminological control of the data at file building time. Here, the last case is more important, since it allows for a hierarchical structuring of the scientific concepts. Before we apply this to the Beilstein datastructure, it is necessary to give a short introduction into the thesaurus function.

A thesaurus function must provide the user with the following capabilities:
- search via preferred property names or less-preferred synonyms
- display property definitions, parameters and field codes
- determine the preferred unit and other valid units for a property
- browse through a property hierarchy, to find broader and narrower properties and property classes.

The data in a thesaurus database consists of a main term, a relationship code plus a corresponding level and a related term, e.g.

```
zero point energy ------ SY (synonym) ------ energy of zero point.
(main term)             (relationship code)  (related term)
```

Typical relationship codes may be hierarchical (Narrower, Broader), equivalent (Use, Used For), or associative (Related Terms). In addition, notes and other information associated with the main term can be included. This may be illustrated by the following extract from the printed ENERGY thesaurus (NT1 means Narrower Term of level 1, BT1 corresponds to Broader Term of level 1):

```
Thermodynamic Properties
   BT1 Physical Properties
   NT1 Chemical Potential
   NT1 Critical Pressure
   NT1 Enthalpy
     NT2 Absorption Heat
     NT2 Adsorption Heat
     NT2 Mixing Heat
     NT2 Reaction Heat
       NT3 Combustion Heat
       NT3 Dissociation Heat
       NT3 Formation Heat
```

In the Messenger system the EXPAND command will allow the user to display the relationships of a specific term, and the SEARCH command will generate the expanded queries including the related terms which have been requested by the user.

Application to the Beilstein Datastructure:

How can the thesaurus function be used in connection with the Beilstein database? The answer is that the datastructure has a hierarchical structure and consists of a set of controlled descriptors. Application of the thesaurus function allows for searches of property names wihout knowing the content of specific fields. According to the user requirements given above, it can also be used to get information about the property names, synonyms, physical (standard) units and the corresponding field codes. It can also be used to inform the user about the content of the database in general.

The Beilstein datastructure is organized in the following way:

```
- (All) Substance Data
  - Identification of Substance
  - General Data
  - Chemical Data
  - Physical Data
    - Thermodynamic Data
      :
      - Enthalpy of Formation
      - Enthalpy of Hydrogenation
      - Entropy of Sublimation
      :
    - Molecular Properties
    :
```

Here, each lower level is indented by two spaces with regard to the higher level. Using the relationship codes BT (Broader Term), NT (Narrower Term), SY (Synonym), USE, and UF (Used For), a simple thesaurus can easily be constructed. Let us assume that we are asking for the term 'Physical Data' plus all narrower terms. Performing an EXPAND command could lead to the following result (extract):

```
=> EXPAND  Physical Data + NT/PH       (PH = Property Hierarchy)

E1        427  -->  PHYSICAL DATA/PH
E2         81     NT1  THERMODYNAMIC DATA/PH
E3         17       NT2  ENTHALPY OF FORMATION/PH
                         (FC = HFOR, Unit: J/g*cm**3, Parm: Temp.)
E4          0          SY  FORMATION ENTHALPY/PH
E5          0          UF  HEAT OF FORMATION/PH
E6         62       NT2   ENTROPY OF FORMATION/PH
                       (FC = SFOR, Unit: J/g*cm**3, Parm: Temp.)
********  END  *********
```

If the user requests all substances which have thermodynamic data available he can simple search for E2. The EXPAND list contains also the valid field code, the standard default unit of the field and the set of parameters. If there are only data references in the database instead of numeric values this may be indicated in the note below the term.

3.2 Numeric Range Searching

The retrieval of numeric data deviates considerably from normal text retrieval. In the latter case the user requests an exact match of the character string he has specified. Taking into account the features of right and left truncation and character masking the actual text search process retrieves all documents containing the specified string with varying characters embedded in broader contexts. In the case of numeric fields with single values for each entry, the procedures for indexing and retrieval is similar to the case of text fields. A publication year which is associated to a published document constitutes an example for this case. All numeric operators can be applied and yield the correct results.

In the case of experimental data an uncertainty of the measured property must be taken into account. The uncertainty which is incorporated in the process of measurement may originate from different sources. Examples for this incidence of uncertainties are: principal uncertainty of the measurement, inaccuracy of the method, inaccuracy of the instrument, calibration error, impurity of the substance, reading error, or external sources of error. Due to the presence of various sources of error it has been established to add an estimated uncertainty to the experimental result. In the literature two different notations are found:

(1) value ± uncertainty [unit]
(2) lower limit to upper limit [unit].

Since both notations are completely equivalent we can restrict ourselves in the following analysis to the second type.

Due to the inherent uncertainty in the nature of numeric data a numeric search query must be formulated as a range search. A user who specifies a numeric range actually expects an interval intersection to take place. The standard retrieval software interprets the endpoints of the numeric range of the input query as the starting and ending points for the search process, i.e. a record becomes an answer only if one of the indexed values (endpoints) lies within the range of the input interval. This is not really an interval intersection. If the stored interval contains the searched interval as a subset, the corresponding document is not included in the answer set. This may be illustrated by the following example:

```
=> SEARCH  100 - 120/IRS   (Infrared Spectrum)

               80   90   100  110  120  130  140  150
           ...|....|....|....|....|....|....|....|...
                        <--query-->
 90 -  94           <->                                 no hit
 98 - 108               <---->                          hit
104 - 110                  <-->                         hit
110                          x                          hit
120 - 124                          <->                  hit
 90 - 130           <------------------->               hit, miss
```

As can be seen from the picture, the last interval (90-130) is missed. Obviously, this is not what the user expects to happen, at least it is rather puzzling. To overcome this problem, the Messenger system will be enhanced by a numeric range search capability which will be invisible for the user. When a user enters a numeric range the Messenger software will retrieve automatically all the indexed numeric ranges which overlap with the query range.

3.3 Tolerance Specification(1)

The result of a measurement can be specified either in the form value plus/minus uncertainty or as a numeric range with lower and upper limit. When we deal with property values both forms are mathematically equivalent and the property can be indexed in either form. A user who wants to search this property can always specify his query using the range formulation. However, it is more convenient to frame a query as value plus/minus tolerance, e.g. (2)

```
=> SEARCH   BP = 100 ± 4.
```

This query means that the user is interested in those substances which have a boiling point of 100 (unit: degree Celsius), but he also accepts substances boiling a little bit above or below (±4°C). It should be noticed that this formulation makes no restrictions about the acceptable measurement error. If the database contains a boiling point with a lower limit of 50 and an upper limit of 96 the corresponding substance should be retrieved as a hit.

The uncertainty which is associated with the property value may be expressed in one of the following ways:

- absolute value, e.g.
  412.765 ± 0.005 (<==> 412.760 to 412.770)
- percentage value, e.g.
  100 ± 5%   (<==> 95 to 105)
- last digit uncertainty,e.g.
  84.198   (<==> 84.197 to 84.199).

In all these expressions it is implicitly assumed that the value and the uncertainty are represented in the same units. If mixed units are requested, they have to be specified for both 'values', e.g. 4 hrs. ± 5 min. Finally, it should be recognized that the tolerance specification must result in the same hits as the range specification, i.e. it must retrieve all substances whose indexed property range has a non-zero overlap with the range that results from the range specification.

Notes:

1. In the moment, the Messenger software does not allow to use tolerance specifications. It is planned, however, to implement this capability in the future.
2. It is important to distinguish between the concept of 'uncertainty' and 'tolerance'. An uncertainty is inherent to the result of a measurement while a tolerance specifies an acceptable level of deviation from the requested search result.

3.4 Units Conversion

Except a few dimensionless constants all physical properties consist of a numeric value and an associated unit. The physical entity is invariant to a change of the unit system. Although, a lot of standardization has been done, several different unit systems are currently in use depending upon the area of application. For numeric databases it is very important to use a standard unit system which is popular in the user community and is identical for all numeric databases. This implies that a query with numeric values for a set of properties and its parameters has the same meaning in all files. Let us consider two databases which contain chemical substances and their corresponding boiling points. If the entries in the boiling point field are given in two different units, like degree Celsius (°C) and Kelvin (K), the following numeric range search

```
=> SEARCH  100 - 150/BP
```

yields substances with a boiling point in degree Celsius in the specified range in one file and in Kelvin in the other. The consequence is a disjoint set of answers. In general, numeric databases are quite small and the number of substances that are covered is very limited. Thus, it is often necessary to search in more than one file and the problem of default units may accur quite often.

An important question is the choice of an appropriate unit system. Obviously, the SI system is the most favoured one. Its seven basis units are the meter, kilogram, second, ampere, kelvin, candela and mole. Other units like volt or newton are derived from these. In a numeric database like the Beilstein file a meter or kilogram is not a very popular unit since we are dealing with molecular properties and not with macroscopic properties. For the Beilstein database the following set of units has been established as the default units and will be used for other numeric databses on STN as far as possible:

Basic units:

length (1)		mm, cm, ...	
mass	gram	g	
time	second	s	
electric current	Ampere	A	* not used *
temperature (2)	degree Celsius	Cel	
amount of substance	mole	mol	
luminous intensity	candela	cd	* not used *

Notes:

1. Several different units are used depending upon the application. Volumes are given in cm**3.
2. Currently the temperatures are given in °C (degree Celsisus) since this is the unit favored by organic chemists. This will be changed, however, to K (Kelvin) when the units conversion feature is available. Properties depending upon the temperature have to be changed, too.

Derived units:

angle	degree	deg
concentration	percentage	%
	gram/liter	g/l
density		g/cm**3
electric potential	Volt	V
energy	Joule/mole	J/mol
entropy		J/mol*K
heat capacity		J/mol*K
moment	Debye	D
pressure (1)	Torr	Torr
wavelength	nanometer	nm

As far as possible, display and search units are identical. For the display of the properties the units are taken as delivered by the database supplier.

In addition to the standardization of units the capability to convert physical units will also be provided soon after the release of the first part of the database. The feature will allow the user

- to specify units with numeric search terms
- to display the default units for a property
- to set the system unit for a numeric property according to his convenience
- to select a common standard for the system units.

To illustrate the units conversion feature a few examples are presented in the following.

Example: Search with default units

When the user does not specify any units, e.g.

```
=> SEARCH   100<= BP <=150
```

the implied unit will be the STN customer unit. The default unit for all temperatures is Kelvin (K).

Example: Search with user-specified units

To override the STN customer units one can specify a valid unit in the search command:

```
=> SEARCH   100 - 150 Cel/BP.
```

In this case, degree Celsius is used instead of Kelvin.

Notes:

1. Currently the pressures are measured in Torr. This will be changed, however, to Pa (Pascal) when the units conversion feature is available.

Example: Change unit via SET-command

When a user wants to override the units for all subsequent searches, he may change the units via the SET-command:

```
=> SET UNIT BP=F

SET COMMAND COMPLETED

=> SEARCH 100-150/BP
```

The units conversion feature also allows to specify a system of units, e.g. MKS, CGS or SI.

3.5 Special Purpose Software

The Messenger software allows search and retrieval of data which is stored in a database, it does not support the generation of new data. A Messenger interface is currently under development which allows the integration of external software packages, e.g. so-called data generation procedures. This is a computer program which calculates data on the request of the user. These data values are not searchable, simply because they do not exist until they are calculated.

Adding special software packages to Messenger will enable the user to estimate or calculate data which is not in the database. The new tool is important for the creation of missing data, i.e. the empty field problem, but it can also be used for a further processing of data. Examples for data generation procedures are:

- interpolation or extrapolation of data, curve fitting
- statistical calculations
- simple quantum chemical calculations
- evaluation of structure property correlations
- computation of bond values
- thermodynamic properties or functions (esp. of mixtures and solutions)
- spectrum estimation (via link to C13NMR).

To illustrate these ideas, an example for a spectrum estimation of an organic substance is presented using the C13NMR database. Currently, there is no link between Beilstein and C13NMR and therefore it is necessary to rebuild the structure in the latter file.

Example: C13-NMR Spectrum Estimation for a Chemcial Substance

The C13-NMR spectrum can be estimated from the chemical structure which could be build using the Messenger structure command or by uploading a structure from a remote workstation with STN Express.

```
L1 STRUCTURE CREATED

=> RUN l1
     :
L2 ESTIMATION CREATED

=> d l2

Estimated Spectrum:
---------------------------------------------------------
C1    S     210.0   ppm  +-  5.0  ppm    (interpolated)
C2    T      39.5   ppm  +-  1.1  ppm    (3 lines)
:
C9    S     147.5   ppm  +-  0.7  ppm    (4 lines)
```

The spectrum is estimated by using only the structural features of the chemical substance. For a further analysis of the accuracy of the estimation procedure, the substances of the C13NMR database are required.

It can be concluded that the link between Beilstein and C13NMR is an important tool for the analytical chemist. Concerning the problem of missing data, it can be stated that the new interface feature of Messenger supporting data generation procedures offers an alternative approach to updating the database with evaluated data. No additional storage would be needed for the nonempirical data and no false data could enter the database. Only the user who runs the data generation procedure will obtain the resulting information.

4. Future Developments and Requirements

Finally, I would like to mention a few requirements which are important for numeric databases. These features, however, are not part of any ongoing Messenger enhancement project for numeric files.

- Link to other numerical and chemical databases
  A large part of the information contained in Beilstein, handbook as well as database, are just data references to the original literature. An example is NMR spectrum: The user is informed that a NMR spectrum has been measured with a specific nucleus (H, C, etc.) and he also gets the corresponding reference. In this case a crossover to the C13NMR database would be highly convenient for the user. A link between the Registry and the Beilstein file essential for an effective searching in chemical databases.

- Overlap function for numeric range search
  Currently, it is not planned to introduce an overlap function in the Messenger software and it is probably less important for the Beilstein database than it is for databases containing materials. A standard question in this area is the following: 'Find all materials covering at least 90% of a given temperature range.' Using an overlap function this could be formulated as => SEARCH OVLP(600-800/NUM) >= 90.

- Enhancement of table display
  Enhancing the display and print of tables to make the layout more compact and flexible. In the moment the tables are rather broad and they contain all columns even if some of them do not contain any data. A user-defined grouping of physical properties into a single table is also rather convenient for the user.

- Graphical display
  In many cases the graphical display of a quantity is more illustrative than a large amount of numbers, e.g. a substance can easily be identified from the characteristic features of an infrared spectrum. The results from the special purpose software packages may also require graphical features.

- Uploading and downloading of data
  The ability to upload and download data in standardized formats is another important tool for users of numeric databases.

It is obvious that numeric databases require special tools supporting the various user needs. The features which are currently under development on STN will provide an important step in this direction. However, it is important to note that numeric and factual databases like the Beilstein file require a certain effort of work for maintenance and improvement including the development of new capabilities.

5. Appendix: Examples

The following search and retrieval examples are taken from the Beilstein database. They are listed here without further explanation.

Example: Ascorbic Acid (/CNS)

```
=> s ascorbic?/cns (w) acid?/cns
              4 ASCORBIC?/CNS
          22479 ACID?/CNS
L3            4 ASCORBIC?/CNS (W) ACID?/CNS

=> d 3 fa


L3   ANSWER 3 OF 4
```

Code	Field Name	Occur.
MF	Molecular Formula	1
CN	Chemical Name	1
SY	Synonym	1
FW	Formula Weight	1
SO	Beilstein Citation	1
LN	Lawson Number	1
PRE	Preparation	35
MP	Melting Point	7
REA	Chemical Reaction	114
RSTR	Related Structure	3
INP	Isolation from Natural Product	6
CT	Skeletal Characteristics	2
DM	Dipole Moment	3
CT	Crystal Phase	1
CSYS	Crystal System	1
CSG	Crystal Space Group	1
CLP	Crystal Lattice Parameter	2
CDEN	Crystal Density	3
CT	Calorific Data	1
GFOR	Gibbs-Energy of Formation	1
CT	Optics	3
ORP	Optical Rotatory Power	20
IRS	Infrared Spectrum	5
RAM	Raman Maximum	2
EAS	Electronic Absorption Spectrum	16
EAM	Electronic Absorption Maximum	2
MSUS	Magnetic Susceptability	1
CT	Electrical Data	1
CT	Electrochemical Behaviour	2
DE	Dissociation Exponent	17
RDXP	Redox Potential	6
SLB	Solubility	2
CTM	Liquid/Solid Systems	3
CTM	Adsorption	4

```
=> d 3

L3   ANSWER 3 OF 4

BRN   89123  Beilstein
MF    C6 H8 O6
CN    L-ascorbic acid
      L-Ascorbinsaeure
SY    (S)-5-<(S)-1,2-Dihydroxy-aethyl>-3,4-dihydroxy-5H-furan-2-on
FW    176.13
SO    4-18-00-03038; 4-18-00-03038
LN    19316
```

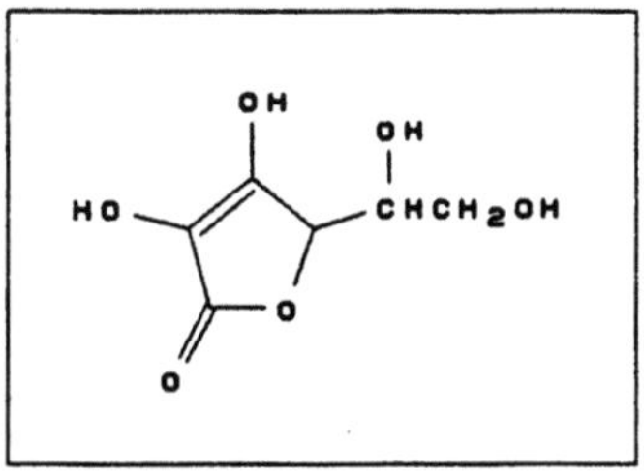

Example: Inflammatory Inhibition (/BF)

```
=> s entzuendungshemmend?/bf
L4           6 ENTZUENDUNGSHEMMEND?/BF
=> d 2 ide bf pre
```

```
L4   ANSWER 2 OF 6

BRN  19118  Beilstein
MF   C13 H9 N3 O2
FW   239.23
SO   5-26
LN   30004; 5219; 1762
```

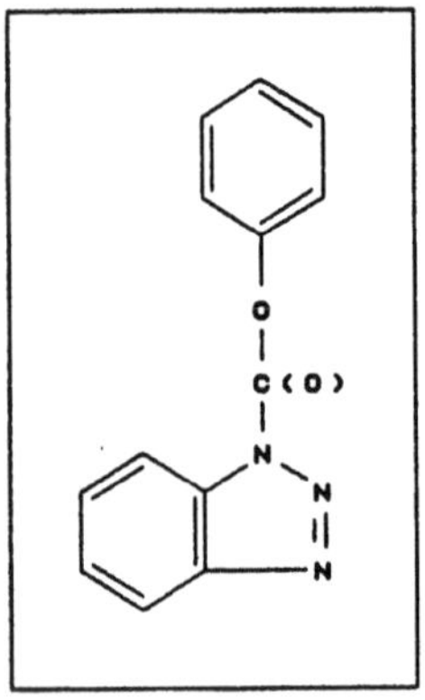


```
Biological Function:
BF   Entzuendungshemmende Wirkung
     Reference(s):
     1. Kreutzberger, van der Goot, J.Heterocycl.Chem., 12, 1975, 665,
        CODEN: JHTCAD

Preparation:
PRE
     Reference(s):
     1. Butula et al., Synthesis, 1977, 704, CODEN: SYNTBF
     Note(s):
     2.  BRN=17784 , Phenol, Triethylamin
PRE
     Reference(s):
     1. Kreutzberger, van der Goot, J.Heterocycl.Chem., 12, 1975, 665,
        CODEN: JHTCAD
     Note(s):
     2. Chlorameisensaeure-phenylester, Benzotriazol
```

<u>Example:</u> Boiling Point Range (/BP)

```
=> s 100/bp (p) 760/bp.p
            295 100/BP
           1849 760/BP.P
L1            8 100/BP (P) 760/BP.P

=> d


L1   ANSWER 1 OF 8

BRN  37741   Beilstein
MF   C13 H14 O3
CN   <4-Ethoxy-phenyl>-<2>furyl-methanol
     <4-Aethoxy-phenyl>-<2>furyl-methanol
FW   218.25
SO   4-17-00-02146
LN   17475; 298
```

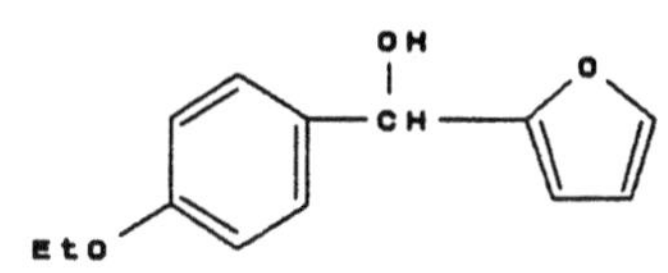


```
---> Handbook Data <---
Boiling Point:
Value (BP)                  Press.(BP.P)                Ref.  Note
(Cel)                       (Torr)
98.00 - 100.00-----------+760----------------------+1----+-----

Reference(s):
1. Mndshojan et al., Doklady Akad. Armjansk. S.S.R. 29 Nr. 1 <1959> 41
   CA: 1960 7673
```

<u>Example:</u> Preparation of Substance (/PRE)

```
=> s xanthen-9-one/pre.edt and rot?/cpd and 150<mp<160
            38 XANTHEN-9-ONE/PRE.EDT
          1194 ROT?/CPD
          4689 150<MP<160
L6           1 XANTHEN-9-ONE/PRE.EDT AND ROT?/CPD AND 150<MP<160

=> d fa


L6   ANSWER 1 OF 1

     Code     Field Name                             Occur.
     --------+--------------------------------------+------
     MF       Molecular Formula                         1
     CN       Chemical Name                             1
     FW       Formula Weight                            1
     SO       Beilstein Citation                        1
     LN       Lawson Number                             1
     PRE      Preparation                               8
     MP       Melting Point                             5
     REA      Chemical Reaction                        17
     DM       Dipole Moment                             2
     CPD      Crystal Property Description              2
     CSYS     Crystal System                            1
     CSG      Crystal Space Group                       1
     CLP      Crystal Lattice Parameter                 1
     EAM      Electronic Absorption Maximum            12
     MSUS     Magnetic Susceptability                   1
     CTM      Liquid/Solid Systems                      3
```

=> d

L6 ANSWER 1 OF 1

BRN 2114 Beilstein
MF C13 H8 O S
CN Xanthene-9-thione
Xanthen-9-thion
FW 212.27
SO 5-26; 4-17-00-05301; 2-17-00-00382; 0-17-00-00357
LN 18039

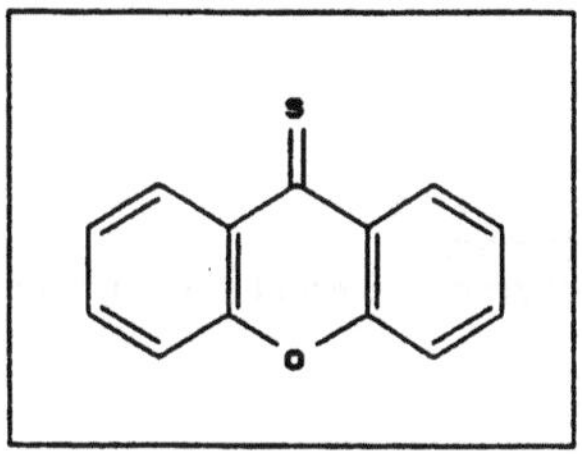

---> Handbook Data <---
Preparation:
PRE
Educt: Xanthen-9-one, Thionylchloroide
Reag: Thioacetic acid, Benzen
Reference(s):
1. Mann, Turnbull, J.Chem.Soc. 1951 757, 760, CODEN: JCSOA9

---> Handbook Data <---
Crystal Property Description:
CPD rot
Reference(s):
1. Mann, Turnbull, J.Chem.Soc. 1951 757, 760, CODEN: JCSOA9

Melting Point:

Value (MP) (Cel)	Solv.(MP.SOL)	Ref.	Note
---> Handbook Data <---			
152.00 - 154.00	benzene	1	
---> Handbook Data <---			
156.00		2	
---> Handbook Data <---			
155.00 - 156.00		3	
156.00 - 158.00		4	

Reference(s):
1. Mann, Turnbull, J.Chem.Soc. 1951 757, 760, CODEN: JCSOA9
2. Graebe, Roeder, Chem.Ber. 32 <1899>, 1689, CODEN: CHBEAM
3. R. Meyer, Szanecki, Chem.Ber. 33 <1900>, 2580, CODEN: CHBEAM
4. Scheeren et al., Synthesis, 1973, 149,150, CODEN: SYNTBF

<u>Example:</u> Refractive Index Range (/RI)

```
=> s 1.275<ri<1.290 (p) 20/ri.t and gelb?/cpd
              3 1.275<RI<1.290
           3899 20/RI.T
              2 1.275<RI<1.290 (P) 20/RI.T
           5624 GELB?/CPD
L2            1 1.275<RI<1.290 (P) 20/RI.T AND GELB?/CPC

=> d

L2  ANSWER 1 OF 1

BRN  79648  Beilstein
MF   C9 H2 Br3 N O4
CN   Heptafluoro-2-undecafluoropentyl-tetrahydro-furan
     Heptafluor-2-undecafluorpentyl-tetrahydro-furan
SY   3,8,X,x-tetrabrom-6-nitro-cumarin; 3,8,X-tribrom-6-nitro-cumarin
FW   427.83
SO   2-17-00-00361; 4-17-00-00117; 2-17-00-00361
LN   16828
```

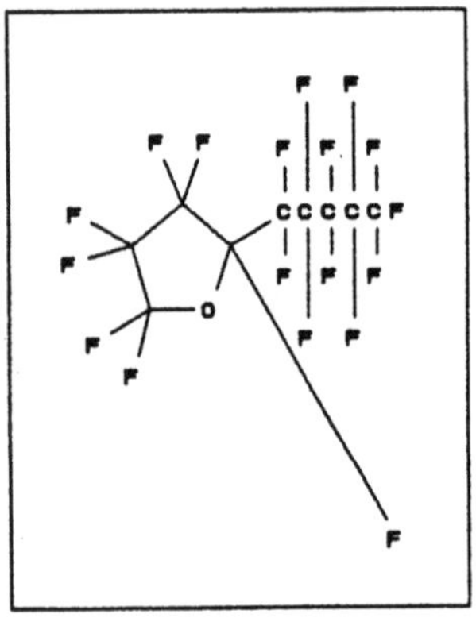

---> Handbook Data <---
Crystal Property Description:
CPD gelbliche Nadeln
Reference(s):
1. Dhar, J.Chem.Soc. 117 <1920>, 1000, CODEN: JCSOA9

---> Handbook Data <---
Refractive Index:

Value (RI) (--)	Wavel.(RI.W) (nm)	Temp.(RI.T) (Cel)	Ref.	Note
1.28340	589.00	20.0	1	

Reference(s):
1. Kolenko et al., Trudy Inst. Chim. Akd. Uralsk. S.S.R. Nr. 15 <1968> 59, 62, 64
CA: 70 <1969> 8742d2

<u>Example:</u> Substructure Search

```
=> dis l1 que sda
L1                STR
```

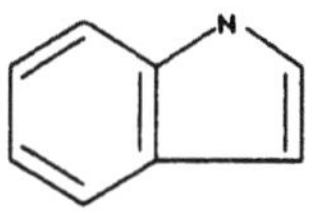


```
NODE ATTRIBUTES: NONE
GRAPH ATTRIBUTES:
RING(S) ARE ISOLATED OR EMBEDDED
NUMBER OF NODES   IS 9
               ****CONNECTIONS****
NOD SYM          NOD/BON  NOD/BON  NOD/BON
  1  C             2 RN     6 RN
  2  C             1 RN     3 RN
  3  C             2 RN     4 RN
  4  C             3 RN     5 RN     7 RS
  5  C             4 RN     6 RN     9 RS
  6  C             5 RN     1 RN
  7  N             4 RS     8 RS
  8  C             7 RS     9 RD
  9  C             8 RD     5 RS
=> search
ENTER LOGIC EXPRESSION OR QUERY NAME (END):l1
ENTER TYPE OF SEARCH: (SSS), FAMILY, OR EXACT:sss
ENTER SCOPE OF SEARCH: (SAMPLE), FULL, OR RANGE:full
FULL SEARCH INITIATED 13:16:28
FULL SCREEN SEARCH COMPLETED -   2474 SUBSTANCES TO ITERATE
 53.7% PROCESSED   1328 ITERATIONS                             235 ANSWERS
100.0% PROCESSED   2474 ITERATIONS                             412 ANSWERS
SEARCH TIME: 00.00.26
L2              412 SEA SSS FUL L1
=> d fa 259


L2  ANSWER 259 OF 412

     Code    Field Name                              Occur.
     --------+---------------------------------------+------
     MF      Molecular Formula                         1
     CN      Chemical Name                             1
     FW      Formula Weight                            1
     SO      Beilstein Citation                        1
     LN      Lawson Number                             1
     PRE     Preparation                               1
     MP      Melting Point                             1
```

```
=> d all 259

L3  ANSWER 259 OF 412

BRN  9486  Beilstein
MF   C15 H8 Cl2 N4
CN   2,4-Dichlor-6-(9-carbazolyl)-1,3,5-triazin
FW   315.16
SO   5-26
LN   30298; 24442
```

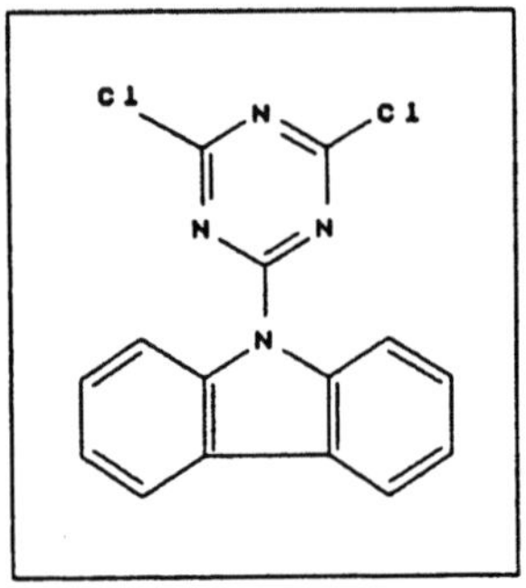


```
Preparation:
PRE
     Reference(s):
     1. Pat. No.: 2954377, US, Degussa, 1960
        CA: 4547i, 55, 1961
     Note(s):
     2. Carbazol, Cyanurchlorid

Melting Point:
```

Value (MP) (Cel)	Solv.(MP.SOL)	Ref.	Note
237.00		1	2

```
Reference(s):
1. Pat. No.: 2954377, US, Degussa, 1960
   CA: 4547i, 55, 1961

Note(s):
2. goldgelb
```

The Thermodynamics Research Center Databases on Original Measurements and Evaluated Data

Kenneth N. Marsh and Randolph C. Wilhoit

Thermodynamics Research Center, Texas A&M University,
College Station, TX 77840-3111, USA

Abstract

The Thermodynamics Research Center (TRC) maintains several large collections of thermodynamic and physical property data on organic compounds. This data is being transferred to computer accessible databases. The TRC Source File contains measured data from the scientific literature including citations to sources of data, compounds names, CA registry numbers, sample descriptions, numerical data, uncertainty estimates and quality codes for most thermophysical properties. The TRC Selected Data File contains values of selected properties (either evaluated or estimated) published in the TRC Thermodynamic Tables. This database will be used for the automatic generation of camera-ready hard copies. Both of these databases will be described.

Introduction

Carefully evaluated data on the thermophysical properties of a diverse variety of organic compounds are required for the development of predictive methods. Further, it is essential that predictive methods be verified by comparison with evaluated data on compounds not used in the development.

The Thermodynamics Research Center operates five continuing projects for the collection and compilation of thermodynamic properties of organic and non-metallic inorganic compounds. A description of these projects follows.

TRC Thermodynamic Tables - Hydrocarbons. This set is a continuation of the American Petroleum Institute Research Project 44 started under the direction of F. D. Rossini in 1942. These tables give the selected best values of various properties, based on measurements reported in the scientific literature when available, or on theoretical calculations or correlations when experimental data is not available. Over the years these tables have been issued as two supplements per year, each supplement containing approximately 75 loose-leaf data sheets. The current set consists of about 3000 sheets on some 35 properties for approximately 4000 hydrocarbons and sulfur compounds of importance to fuel technology. At the top level the tables are organized by types of properties identified by alphabetic codes. Thus, for example, a-tables contain density and refractive index at 20

C. Jochum M. G. Hicks J. Sunkel (Eds.)
Physical Property Prediction
in Organic Chemistry

and 25 °C, melting point, normal boiling, and dT/dp at the normal boiling point. The k-tables contain values of parameters in the Antoine equation which are used to calculate boiling points at a series of standard pressures, for the ka-tables from 0.05 to 1500 Pa, for k-tables from 20 to 200 kPa, and kb-tables from the boiling point at 0.2 MPa to the critical temperature. The h-tables contain values of second virial coefficients, while the p-tables contain thermochemical data at 25 °C. Within each group of properties the tables are organized by classes of compounds which are identified by a finding number which identifies the elements and a table number which identifies a particular homologous series or group of isomers. A reference sheet listing references to sources of data accompanies each table of numerical values.

TRC Thermodynamic Tables - Non-Hydrocarbons. This set contains selected values of properties of other types or organic compounds and nonmetallic inorganic compounds in a manner similar to the Hydrocarbon tables. The Tables began in 1957 under the sponsorship of the Manufacturing Chemists Association. The current set contains approximately 2440 sheets on 2700 compounds.

In order to keep the sets current and useable, subscribers must file the supplements of new and revised tables in the appropriate place as they are received. Because of the open-ended nature of these sets and the complexity of the organization this requirement had caused much trouble and confusion in past years. The 1986 reissue of the complete set of valid sheets included for the first time page numbers and a comprehensive formula index. All new tables are numbered correspondingly. This simplifies the filing operation considerably but does not eliminate it. The index allows one to determine if property data exists for a particular compound.

Properties of Chemical Compounds. The Thermodynamics Research Center is a Data Center of the Office of Standard Reference Data of the National Bureau of Standards and its projects on the Properties of Chemical Compounds contains reviews and selections of data on several groups of compounds completed during the past 24 years. The results have been published in the Journal of Physical and Chemical Reference Data. Major groups of compounds reviewed so far include aliphatic alcohols, halogenated ethanes, organic oxygen compounds containing 1 to 4 carbon atoms and, most recently, organic amines. The contribution of the organic section of the Bulletin of Chemical Thermodynamics, an annual bibliography on thermodynamic data, is part of the NBS project.

International Data Series. IDS Series A is a collection of tables of thermodynamic properties of binary non-electrolyte mixtures. The publication is similar in some respects to a research journal. Data are contributed by authors and evaluated by an international panel of editors. The Editor-in-Chief is Dr. Henry Kehiaian. Each sheet presents values of a particular property, in SI units, on a particular system in a pre-defined format. Both original experimental data and extracts of previously published data are included along with information on measurement technique, purity of materials, and measurement accuracy. The series began in 1973 and four supplements are issued per year. The Series at present contains over 3000 sheets.

DIPPR Project 882 - Evaluated Data for Mixtures. The Thermodynamics Research Center has recently started a project sponsored by the Design Institute for Physical Property Data of the American Institute of Chemical Engineers. This project involves the evaluation of transport and selected physical property data for mixtures and will produce up to 250 tables per year. In previous years the Thermodynamics Research Center has also prepared a number of special compilations for organizations such as the American Petroleum Institute, The Gas Processors Association, and the Department of Energy.

This paper describes our two databases for pure components, the Source File and the Selected Data File, future developments in terms of intellegent interfaces, and our plans for the extension of the Source File to mixtures.

Steps in the Preparation of Numerical Data Compilations

The first step in the compilation of experimental data is the location of sources of such data in the primary literature and the extraction of the pertinent information. It is then processed through a series of steps to produce the final results. The first step in the process is normalization.

Normalization refers to the conversion of the data to some common basis. It usually involves conversion to a consistent set of units and reference states and may include recalculation with a consistent set of auxiliary data such as atomic weights and fundamental constants. It may also include corrections for variations in instrument calibrations. Normalization is required for intercomparison of related data from different investigators.

The assignment of uncertainties is a critical step which is difficult to describe rigorously. A statistical analysis plays a major role but sometimes a subjective judgment, based on a knowledge of the state-of-the-art at the time of publication, the limitations of the experimental technique, the information provided by the author, and the reputation of the investigators, can form the basis of the final judgment.

A smoothing process converts sets of experimental data to some regular function of the appropriate independent variables. This may be done graphically by reading values from a curve drawn through or near the plotted data points, or analytically by fitting data to a smoothing function according to some criterion. The analytic procedure usually consists of a least squares fit to the parameters in a smoothing function. It is common practice to derive statistical weighting factors for the fit from the assigned uncertainties. The function may be entirely empirical or based on theoretical considerations. The independent variables may be state variables such as temperature, pressure, and composition, or may include parameters describing molecular structure or other properties. The purpose of smoothing is to reduce the effect of random experimental errors, to reduce the size of the original data set or to represent the data by a mathematical function, to generate tables at regular increments of the independent variables, and to help in the intercomparison of data from different sources.

Data values are sometimes adjusted to satisfy some known relationships among various properties. This kind of internal consistency is especially important for thermodynamic data because of the many rigorous relationships implied by thermodynamic principles. For example, vapor pressure is related, through the Clapeyron equation, to the enthalpy of vaporization and the volumetric properties of co-existing condensed and gas phases. The enthalpy of vaporization is in turn related to the entropy and heat capacity of these phases. Users of thermodynamic compilations often derive properties other than those directly tabulated. Even if each tabulated value is within the assigned uncertainty tolerance, derived values may be unreliable if the tabulated set is not consistent.

In principle, such relationships can be incorporated into the smoothing process by treating them as mathematical constraints on the fitting criterion. They can also be accommodated by simultaneously fitting sets of related properties to the same set of parameters. Thus an equation of state might be fit simultaneously to P-V-T data, enthalpy of vaporization, and changes in heat capacity with pressure. Thermochemical properties for sets of chemical reactions involving common reactants and/or products are related by a set of linear algebraic equations. An entire set might be simultaneously fit by incorporating such constraints.

In order to be able to make data evaluations economical and efficient, it is essential to be able to access all the data readily. Such access is best achieved through a comprehensive computer database containing an extensive set of original data on many compounds and many thermophysical properties.

Automation of Data Compilations

Most of the data compilation steps identified above, from the literature search to the composition and publication of tables, are amenable to computer automation. The exceptions are the identification and interpretation of pertinent data from the literature, the assignment of uncertainties, and the final selections of the best values. The evolution of practices at the Thermodynamics Research Center over the years reflects this situation. For a number of years all the steps were carried out manually. Literature searching was done by direct scanning of journals, reports, and Chemical Abstracts. Our database was in the form of handwritten 3x5-inch index cards, each containing the value of one property of a particular compound along with the reference, sample description, and comments. Around one-half million of these were collected during the first 45 years of the project. Tables were composed by manual entry on various typesetting devises.

A major problem has been keeping appropriate records on all these activities. To generate a new table a compiler will accumlate all of the available information related to that table and then carry out the evaluations, smoothing, and selections. To revise a previously issued table, it is necessary to gather both the data available for the previous version and the newer data accumulated since that time. The whole process of evaluation and selection must be repeated for all the values in the table, even though some of them may not change. It is important that the person making such revisions understand the procedures and reasoning of the previous compiler even when 10 to 30 years has elapsed. With the passing of time the record-keeping chores become rapidly more burdensome because both the number of extant tables and the size of the pool of data in the literature increases.

In recent years we have moved toward automating as much of this process as possible. The purpose is to allow the compilers to concentrate on the critical tasks of interpretation, evaluation, and selection of data. Although we make extensive use of literature searching services such as those offered by Chemical Abstracts Service we cannot rely on them exclusively. The index terms available in such searches do not efficiently identify experimental property data. In principal direct text searches of articles, such as ACS JOURNALS ON-LINE could be useful. Unfortunately tables of numerical data, which are our main goal, are not yet included in these on-line databases. These automated searches do furnish an initial screening, but much manual searching is still required.

TRC Computer Databases

In order to help automate the intermediate compilation steps, the Thermodynamics Research Center has implemented two large computer-readable databases during the past three years. The TRC Source File contains numerical values of physical and thermodynamic properties extracted from the chemical literature. The TRC Selected Data File contains the selected values that have been published in the printed versions of the TRC Thermodynamic Tables - Hydrocarbons and Non-Hydrocarbons. Both are organized by a commercial database management system which can retrieve data in either a heirarchal or relational mode. The results can be displayed directly on the console screen or sent to external files for printing or further processing.

The schema of the TRC Source File is illustrated in Figure 1. Each box in the figure represents a type of record. The records are indexed by certain sets of key values. The solid lines show the heirarchal relationships among the key values. The set of key values at any level includes all those at the next higher level. Records of type 1 are indexed by the Chemical Abstracts Registry Numbers. They also contain the molecular formula and a systematic name. Additional names, as many as needed, are kept in records 4 and 5. Provision is made for Wiswesser line notations, but not many are included at this time. References are identified in records of type 3. They are keyed by the year of publication and a sequence number within the year. Each of these records contain the reference citation,

Schema TRC Source File

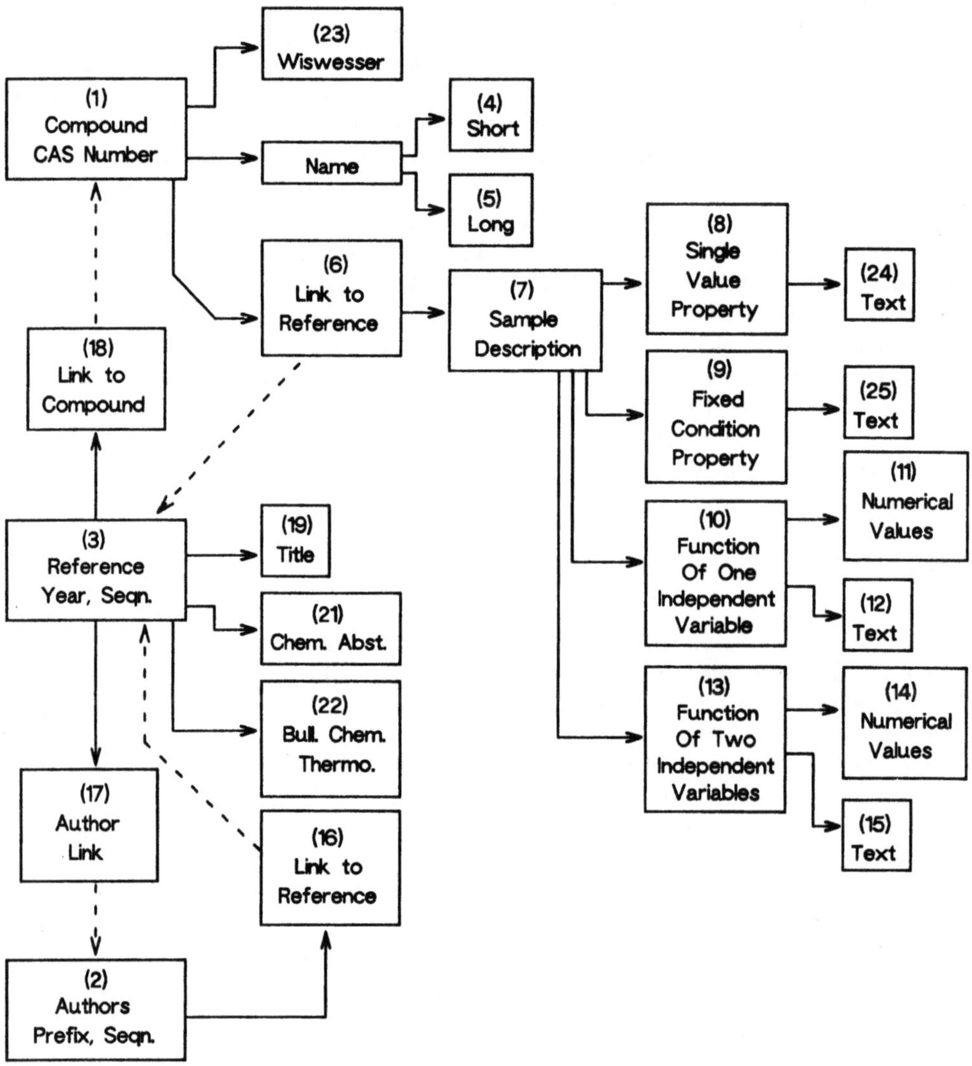

Figure 1.

a comment field, and key values for up to three authors. Additional authors are identified in records of type 17. Titles of articles, and references to Chemical Abstracts and the Bulletin of Chemical Thermodynamics are stored in records 19, 21, and 22. Names of authors, keyed by the first six letters of their last name and the initials are stored in records of type 2. Key values of references written by a given author are stored in records of type 16. Registry numbers of compounds reported in a particular reference are kept in records of type 18.

The remaining records are concerned with the numerical data and their descriptions. Record 6 identifies the references which contain data on a particular compound. Following down this heirarchy, record 7 describes the sample (preparation, purification, and purity) for a particular compound in the reference identified in record 6. Records 8, 9, 10, 11, 13, and 14 contain numerical values of measured properties of pure compounds and their attributes. The various properties are identifed by a list of 35 basic symbols. Record 8 stores properties, such as a triple point, a normal boiling point, or a critical property, represented by a single number. Properties that are functions of some variable, such as temperature or pressure, are stored in records 9, 10, and 11. These correspond to measurements along a two-phase coexistence line, or a single phase at constant pressure or temperature. If only a few values are given record 9 is used, if more extensive tables are given records 10 and 11 are used. Properties that are functions of two independent variables are stored in records 13 and 14. Text comments are stored in records 12, 15, 24, and 25.

The data in the Source File have been normalized. They are converted to SI units as they are entered. Each one is also accompanied by a preliminary estimate of uncertainty. The records identify the phase or phases present, constraints if any, conditions of measurement, and other information useful for evaluation. At present the database contains 20,000 registry numbers, 28,000 references, and around 90,000 numerical values. The total number of records of all types is 370,000.

The schema for the TRC Selected Data File is illustrated in Figure 2. It is designed primarily to regenerate the tables as printed in the TRC Thermodynamic Tables - Hydrocarbons and Non-Hydrocarbons. Record type 1 contains the Chemical Abstracts registry number, formula, and the single name used in the printed tables. Data for the various tables are stored in records of type 6. They are keyed by the table number that appears in the title of each printed table. Data for the specific reference sheets are stored in records of type 9. The references are identified by the same key used in the TRC Source File. Records of type 2 identify the tables in which data for a particular compound may be found. The other records contain the information needed to regenerate complete printed versions of the tables. At the present time we have accumulated about half the data in the extant sets. The software which drives the printer to produce new tables also merges the data and references from the most recent tables directly into this database.

Schema TRC Selected Data File

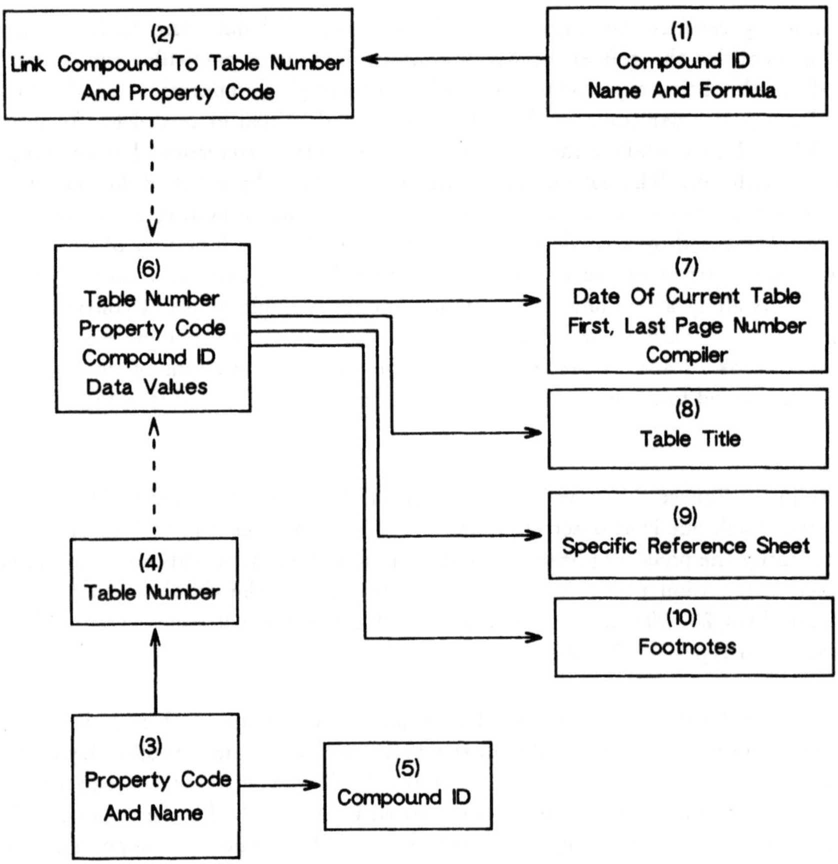

Figure 2.

Goals for Further Automation

Our immediate goal is to link the various steps more closely together. This is illustrated in Figure 3. Experimental data, from both the chemical literature and existing TRC hard copy files, are entered into the TRC Source File. We are currently entering literature data as it is published and we are attempting to arrange for exchange of data with databases being maintained at other locations. Whenever an appropriate set of related data is entered, the evaluation and selection process can begin. The box labeled "smoothing" represents a group of linear and non-linear least squares fitting programs. The data to be examined are extracted from the Source File and displayed as text or graphs, or analyzed by programs which utilizes the accuracy estimates to determine weighting in a least squares procedure to produce smoothing equations and appropriate scatter plots to indicate the inconsistencies in the data. The selections, along with the references to sources, are then placed in the Selected Data File. Data sheets for the two TRC Thermodynamic Tables projects can be generated automatically at this point. It will probably be economically feasible to produce complete sets of frequently revised tables in this way rather than supplements. This will then eliminate the problem of requiring subscribers to file the data sheets. It will also give us the practical options to revise the format and organization of the tables to fit special purposes.

Of greater importance is the possibility of distribution of machine readable files from both the Source File and the Selected Data File. We have conducted an experiment along this line for the past two years. Vapor pressure data extracted from our files have been sent to Numerica, an on-line service offered by Technical Database Services, Inc. It is in the form of two files. One from the TRC Source File contains the direct experimental data, estimated uncertainties, and references. The current collection contains about 20,000 compound-reference combinations. The other contains parameters for a smoothing equation for vapor pressure from the TRC Selected Data File and from other compilations. From this, the Numerica system generates tables of vapor pressure at designated temperatures (or boiling points at designated pressures) in various units. A rating of accuracy is included. Parameters for nearly 6000 compounds are included. We are also experimenting with a floppy disk version of this data set.

The identification, interpretation, evaluation, and transcription of numerical data from the literature is a slow step which cannot be easily automated. However, duplication of this work among the databases being maintained around the world can be reduced by mutual exchanges of data in machine readable form. TRC has been involved in the development of two means for the ready transfer of data. One technique of data transfer, called COSTAT, can be used to transfer data between many data centers and is independent of the database. The second method, which is specific to the TRC database, is a menu driven program which prompts for information on reference, compound, purity, and data and puts the information in a form suitable for direct entry into the database.

Data Flow For TRC Compilations

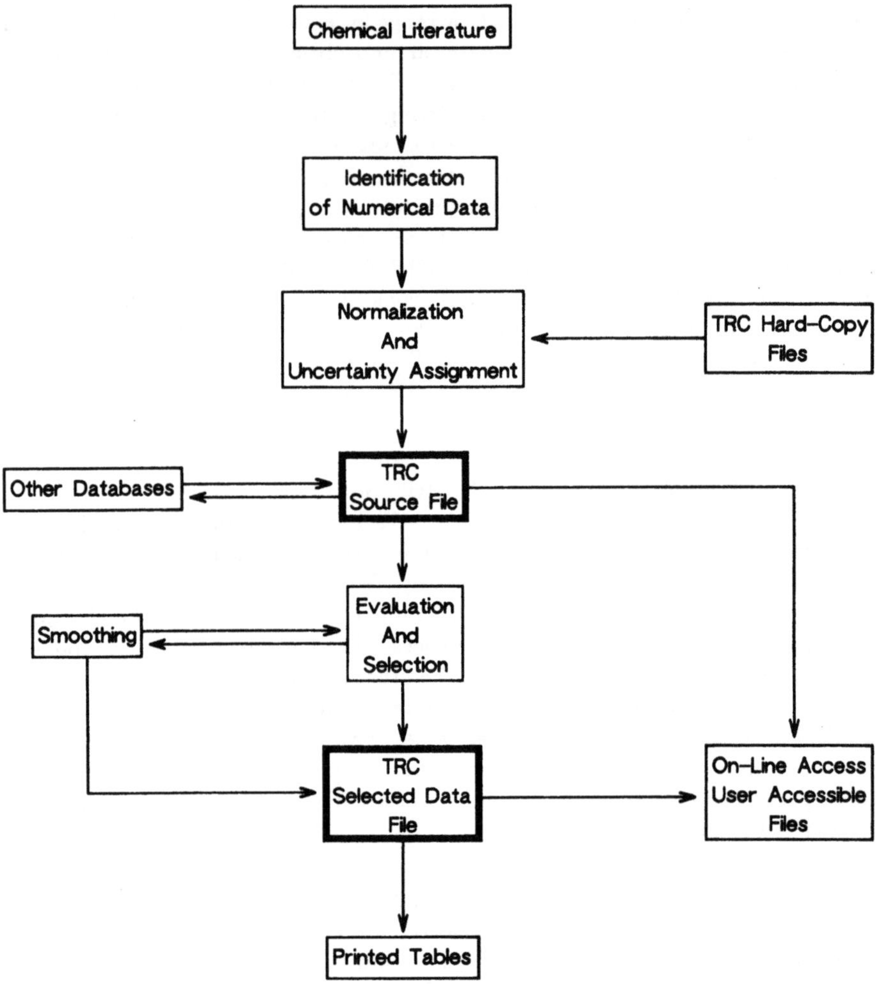

Figure 3.

Interaction With TRC Databases Through Use Of SIR DBMS

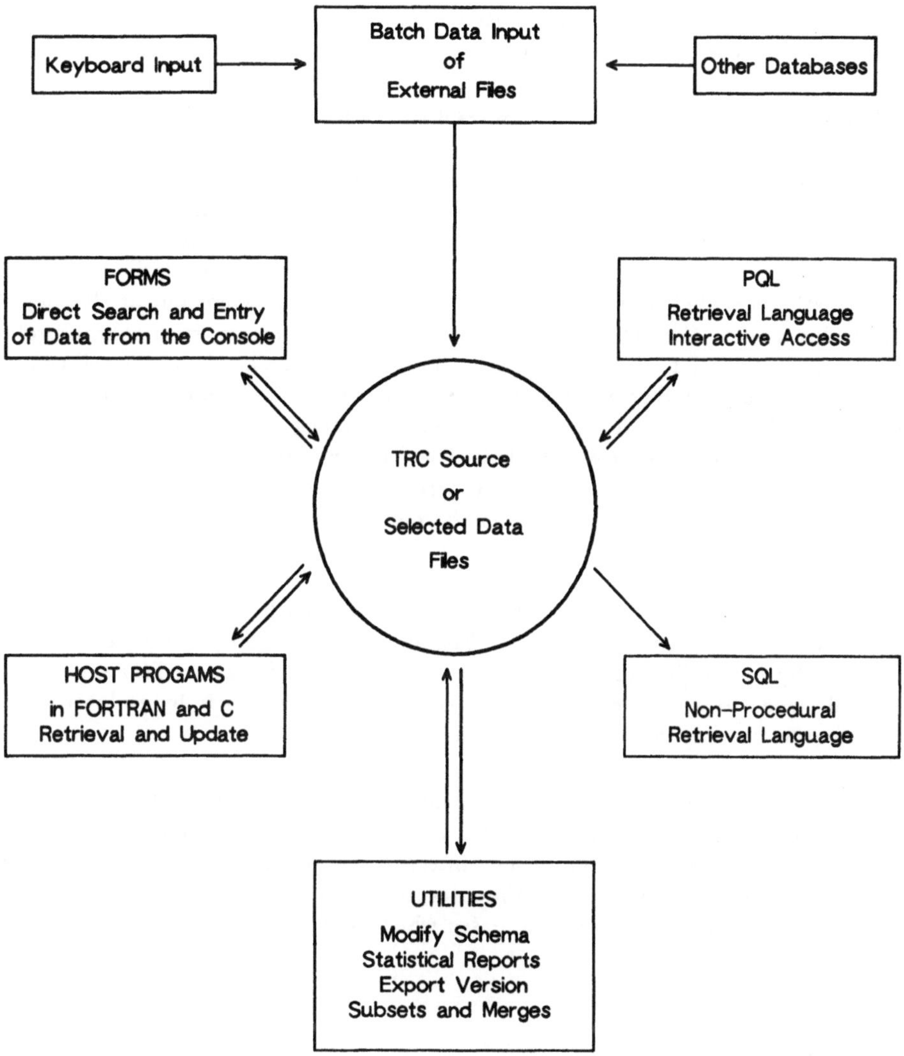

Figure 4.

Future Developments

The present Source File will be extended to include mixture data. The structure of the database allows the inclusion of additional records to include mixture information. The major difficulty is in allowing for all the possible combinations of phases, properties and components. Different records will be required for two-, three-, and multicomponent mixtures. Data accumulated from the International Data Series and the DIPPR project will be added to the database and will be used to produce the various tables and computer-readable files required for these projects.

TRC personnel are also involved in a contract with NBS-OSRD for the development of an expert system for the evaluation of thermodynamic data on pure compounds. Aside from collecting data from the literature, one of the most time consuming tasks a data evaluator encounters is determining the precision of the data (which often differs from that claimed by the author) and the selection of the appropriate smoothing equation. This selection process usually relies upon past experience and invokes many calculational procedures which must be repeated after small adjustments to weighting parameters or after rejecting a particular model. Certain aspects of the this selection process, such as interpretation of reports, identification of the significant data, and assignment of uncertainties can only be done by humans. Other aspects such as intercomparison of values, rejection of data, statistical analysis, testing for internal consistency among thermodynamic relationships, and graphical displays can be automated. These latter aspects can be codified into a set of computer procedures which, in current terminology, can be called an expert system. To be effective such an expert system in thermodynamics must be interfaced to a large database of archival data on thermodynamic properties. The database should contain all the information pertinent to the targeted set of compounds, systems, and properties, and should be continually updated as new information is published. The system we propose will extract information from the Source File and do the majority of the data evaluation automatically, only calling on the evaluator to make crucial decisions. This expert system will be designed to learn the reasons for particular decisions made by evaluators during the evaluation process and hence will evolve with time.

The TRC is also involved with the development of an intelligent interpreter that will access the Selected Data File. Such an interpreter will query the database and determine if data is present. If data is present, appropriate interpolation procedures will calculate data for specific conditions. If no data exists, the interpreter will use other methods for predicting the property or will access other databases which could have alternative computational procedures.

References

1. *TRC Thermodynamic Tables – Hydrocarbons*, Thermodynamics Research Center: The

Texas A&M University System, College Station, Texas.

2. *TRC Thermodynamic Tables - Non-Hydrocarbons*, Thermodynamics Research Center: The Texas A&M University System, College Station, Texas.

3. International Data Series – Selected Data on Mixtures Series A. Thermodynamics Research Center: The Texas A&M University System, College Station, Texas.

Computer-Aided Selection of Chemicals for Biological Testing: Estimation of Biological Activity

G. W. A. Milne and L. Hodes

Developmental Therapeutics Program, Division of Cancer Treatment, National Cancer Institute, Bethesda, MD, USA

The United States National Cancer Institute manages a program whose goal is the discovery and development of drugs to be used in the treatment of cancer. This Program, which began in 1955, has examined approximately 500,000 materials for anticancer activity and it continues to screen much smaller numbers of compounds per year. The large database that has been built as a result of this work was used between 1980 and 1985 as the basis of an experiment in which the potential biological activity of a chemical structure was estimated before the compound was acquired and tested. The accuracy of such estimates and the impact upon the overall program is described.

INTRODUCTION

For over thirty years, the U.S. National Cancer Institute has been systematically screening chemical compounds for possible anticancer activity (1). The number of compounds tested annually has generally been between 10,000 and 13,000. This is considerably less than the number of new structures that are identified each year (2) and it follows that a selection process of some sort must be an integral part of the overall screening effort. A number of non-scientific decisions, such as availability of the compound, are used in the process. In the final analysis however, selection comes right down to choosing which of two structures is more likely to have anticancer activity. The more accurately this question can be answered, the fewer compounds will have to be tested to uncover a potential drug - a fact that has significant financial implications. The mechanics of an estimative approach to this question are the subject of this paper.

METHODS

The screening system used by the NCI between 1975 and 1985 had two stages (3). In the first stage, the test was against a mouse leukemia, designated P388. In this test, the compound is administered at different dose levels to mice that have previously been inoculated with leukemia cells. The assay measures the life spans of the drug-treated mice and compares them to those of untreated control mice. The result is expressed as the (treated) life span divided by the (control) life span, or %T/C. For a drug that has no effect, the %T/C should be 100. Any anti-leukemia activity will result in a higher %T/C and the current NCI definition of a "positive" compound is one whose %T/C is reproducibly over 128%. Because all compounds were tested in this P388 system, the NCI collected a large

C. Jochum M. G. Hicks J. Sunkel (Eds.)
Physical Property Prediction
in Organic Chemistry

database of such biological data (4) and it was decided that this could be used as the basis of the statistical estimation scheme that is described here.

The other necessary ingredient for a structure-activity study is the chemical structure. All chemicals in the NCI system are registered as they take their place in NCI's Drug Information System (DIS, 5) with an NCI registration number, known as the "NSC Number" (6). The chemistry file of the DIS contains all compounds that have ever been in the NCI program and it is searchable in real time on the basis of structure or substructure. This search system relies upon connection tables which can be used as a source of machine-readable atom-centered fragments that constitute the molecules in the system. Thus the atom-centered fragment, a notation which describes each non-hydrogen atom along with its non-hydrogen neighbors and the bonds to those neighbors, can be derived for every atom in a molecule. It was elected to attempt some sort of a

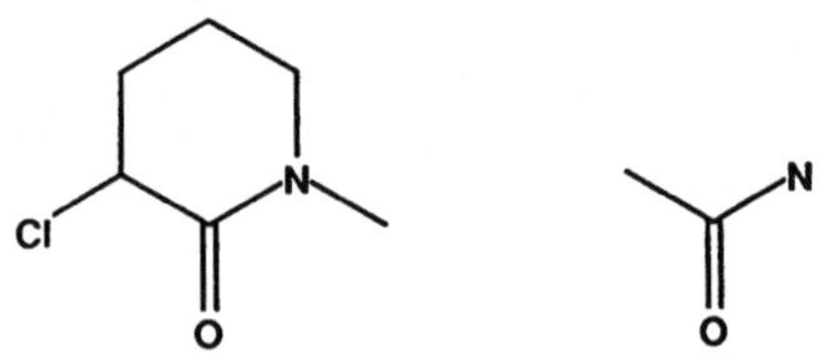

Structure Atom-Centered Fragment

correlation between these fragments and the molecular biological activity.

Once a definition of activity in the P388 system is established, it is a fairly simple matter to define any compound in the database as "active" or "inactive". Then each structure can be redefined in terms of the atom-centered fragments it contains and each such fragment can be tagged as having been derived from an active or an inactive molecule. It is clear that virtually any fragment could be present in some active compounds as well as in some inactive compounds. What this method reveals however, is the frequency of occurence of a fragment in the entire database and the proportion of those occurrences that are associated with active or inactive compounds.

If a new compound is now broken into atom-centered fragments, each fragment can be looked up in such a table and its frequency of occurrence as part of an active compound, as opposed to an inactive compound, can be determined. A "weight" can be derived from these frequencies. The weights of the various fragments which make up a molecule can be used to predict the likelihood that the molecule will show some biological activity. Similarly, the frequency of occurrence of a fragment in the full database can be looked up and the various occurrence frequencies of fragments which comprise a molecule can be used to make an estimate of the molecule's structural novelty with respect to the database. These two criteria can be used separately or in combination to make a determination as to the desirability of acquiring and testing the compound.

RESULTS

This technique has been tested in numerous ways on numerous occasions (7) and no attempt will be made here to describe all of these. Rather, a typical set of studies designed to answer some specific questions will be provided (8). A group of about 35,000 compounds for which clear and definitive screening data are on record was selected and used as a training set. The 35,000 compounds could be separated into 120 highly active, 2,000 moderately active and 33,000 inactive compounds (9). Then test sets were formed by taking every fifth compound from each of these sets and with each cut, the remaining 80% was used as a training set, i.e. to generate the statistics that support the model. Every test set member was then treated as a new structure and entered into the model. Finally, within each of the test sets, the structures were ranked by score and the results for one test set are shown in the Table below. For the 24 "highly active" compounds, 18 (75%) were ranked in the highest decile and 4 more showed up in the second and third deciles. In 395 of the moderately active compounds (10), the spread was broader, with 130 (32%) in the highest decile and the others spread through the nine remaining deciles.

Decile	Highly Active Compounds	Moderately Active Compounds
10	18	130
9	2	64
8	2	41
7	0	30
6	0	29
5	2	28
4	0	27
3	0	19
2	0	15
1	0	12
Totals	24	395

The clear indication from this test was that the model does succeed in discriminating between active and inactive compounds. The discrimination is not extreme, to be sure, but overall, 34% of the active compounds showed up in the top 10% of the ranked compounds. The relative enrichment of active compounds in the higher deciles is exciting, because it suggests that while screening the nine highest deciles would cost 10% less than testing all ten of them, the yield of active compounds would at the same time be reduced only by about 4%.

In a second test, the performance of the model was compared directly to that of an experienced medicinal chemist. Nine hundred and ninety eight newly acquired compounds were examined by the chemist, who designated 312 as possibly active. Then the same 988 compounds were passed through the model which ranked them according to probability of activity and also according to structural novelty. The 312 compounds identified by the chemist were then contrasted with the 312 compounds ranked highest by the model and also with 312 compounds drawn from the 988 at random. The results are summarised in the Table below.

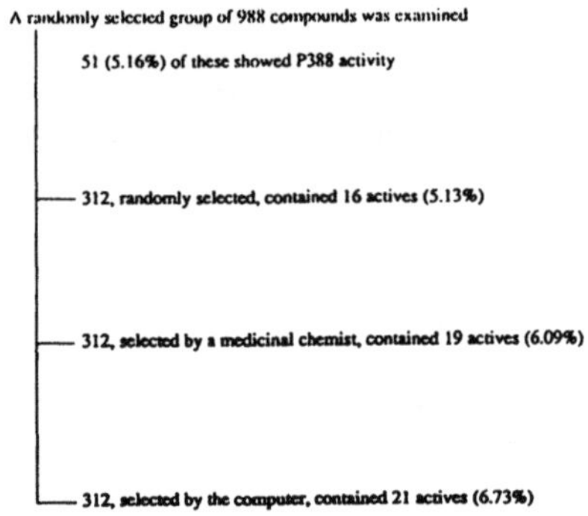

The 312 randomly selected compounds contained 16 active compounds (5%). The 312 selected by the chemist contained 19 actives but the model did even better, identifying 21 actives out of its 312. Thus the chemist's performance was 18% better than a random selection and the computer model outperformed the random selector by 31%.

The chemist's 19 compounds had considerable, but not complete overlap with the 21 compounds promoted by the model, as shown below. The same 12 active

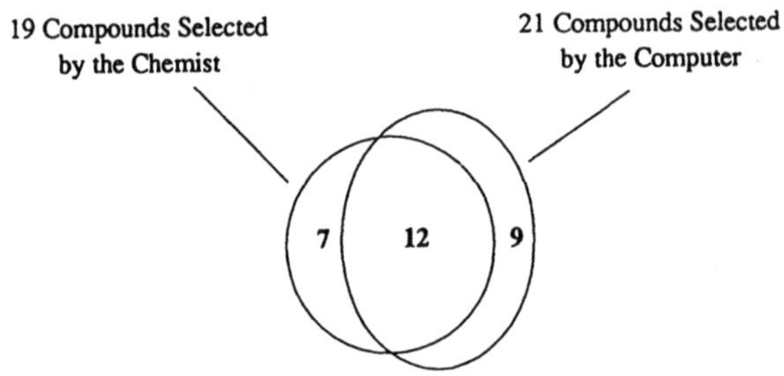

compounds were predicted by both to be active. The chemist also designated a further 7 compounds which were not in the computer's highest rank and the computer likewise designated 9 compounds which the chemist did not identify.

SHORTCOMINGS

This estimative approach represents in one way, a major advance over random selection of compounds to be screened. From a different perspective however, it may be seen to be very primitive. Specifically, it rests heavily upon three assumptions, all of which, arguably, are suspect. First, the atom-centered fragments are handled with the assumption that they are independent, although interaction between different fragments is known to take place. In a simple example, neither the aminoethyl or the chloroethyl moieties, by themselves show any particular antitumor activity. If however, they are combined in the 2-chloro-1-aminoethyl unit a well-recognized cytotoxic fragment results:

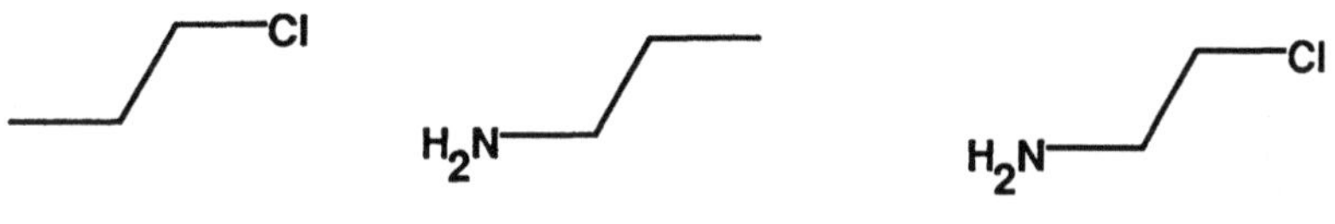

Chloroethyl Aminoethyl 2-Chloro-1-Aminoethyl

An effort has been made to examine models based upon fragments that are larger than the atom-centered fragments - the so-called ganglia augmented atoms. Such larger fragments do indeed provide a higher degree of

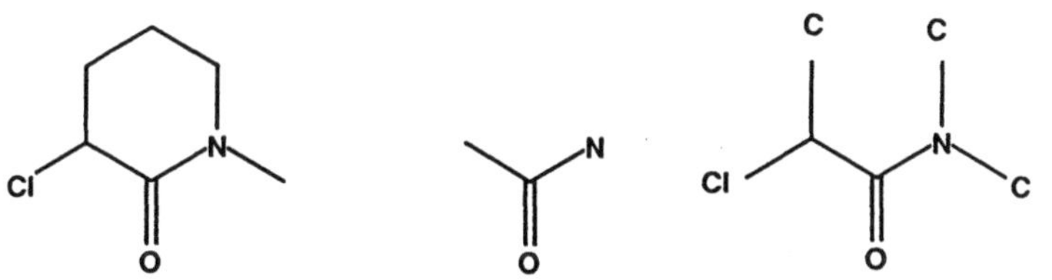

Structure Atom-Centered Fragment Ganglia Augmented Atom

discrimination (11) and a model which employed such larger fragments was established in 1980.

A second shortcoming of this particular method stems from the fact that the model is derived unambiguously from data of the P388 leukemia testing. The assumption that it might nonetheless predict for other tumor types, never strong, is now known to be generally invalid. Most anticancer agents are in fact active against P388, but many of the compounds which show activity against P388 have no other useful anticancer activity. This is not really a deficiency of the model so much as a problem with the entire Program. Drugs that are effective against mouse leukemias are discovered regularly by this Program; some of these are active against human leukemias, but in general, this activity does not extend to other cancers, such as solid tumors. Models can be built to predict for other cancers; we have established, for example, systems that predict for activity against melanoma and L1210 leukemia and they seem to function about as effectively as the P388 model. Unfortunately, concerning most of the important tumors, there is too little reliable data to support a good statistical model of this sort.

The third major problem for this model is its inability to handle atom-centered fragments that it has not previously encountered. This results simply from a gap in the knowledge base and an ideal solution to this problem does not exist. As a practical matter, a compound containing a previously unencountered atom-centered fragment is generally given a structural novelty score that is so high that it overrides the lowest score obtainable for predicted activity and thus guarantees selection and testing of the compound.

IMPLICATIONS

The level of screening at NCI, 10,000 compounds per year, was high enough to leverage the gains that might be realised from even a minor improvement in the selection process. Thus, as shown below, random selection and testing of 10,000 compounds will, as a matter of experience, result in approximately 450 positives. An increase of 20% in the efficiency of the selector would produce 540 positives. Secondary testing however, is very

Compounds Entering Primary Screen	Yield Of Positives	Compounds Entering Secondary Screen
10,000	4.5%	450
10,000	5.4%	540
8,333	5.4%	450

expensive and the NCI is budgeted to handle only 450 positives from the P388 screen. To identify 450 positives at the higher level of selector efficiency would require testing only 8,333 compounds per year in the P388 screen. Each test in this screen costs about $250 and thus the annual savings from a 20% increase in efficiency would be $416,750.

The immediate cost of setting up and running the model is trivial - a few thousand dollars at most - and the most serious operational cost is the risk of false negatives, i.e. that the model will trigger rejection of compounds that actually possess anticancer activity. Mitigating this danger is the fact that, as can be observed from the data presented here, the probability of activity diminishes continuously as one descends through the ranked list of candidate structures. Those at the lowest deciles are likely to be rejected, but they are the compounds that are least likely to have any activity. Even if one of these compounds possessed some activity, it is likely to be quantitatively inferior to those compounds that place in the higher ranks.

A different question concerns the competition between predicted activity and structural novelty. An earlier school of thought had held that given the scale of this screening program, it should be possible to examine all known "types" of chemical compounds. This idea was probably naive; the NCI database currently contains perhaps 5% or all known compounds and new, previously unencountered structures are continually being found. Nevertheless, a reasonable prima facie case can be made for a novel structure, as opposed to a minor variation on a known structure. Novel structures, however, are at a disadvantage vis-a-vis the model, and so the Program cannot rely purely upon the predicted probability of activity. Rather, an attempt is made to combine this score with a score based upon degree of structural novelty before making a selection decision.

The operation of the model as a front end to the DIS selection and acquisition modules is described below. It is implemented in such a way that the acceptable probability of activity can be adjusted by NCI staff. Thus the goal of 450 P388 actives per year can be approached by either raising this cutoff and increasing the number of incoming candidates, or lowering the cutoff and decreasing the number of candidates.

USE OF THE MODEL WITHIN A WORKING SYSTEM

In the normal course of operations, NCI monitors the literature and also meets upon a regular basis with many chemists around the world in an effort to keep up with the appearance of new compounds which may be of interest to the screening program (12). Samples tested by the NCI must be provided at no charge for testing (13) and our experience shows a 40% response to requests for free samples for this purpose. Accordingly, to receive 10,000 samples, we must request about 25,000 and to find 25,000 compounds of sufficient interest, we generally examine twice that many. When a compound is felt to be of interest to the Program, its structure is entered into a "Pre-Registry" section of the DIS. There, its connection table is generated as are the atom-centered fragments. These data are used first to determine whether or not the structure has ever been in the system previously. If the compound is new to the DIS it is then examined by the Hodes model for both structural novelty and probability that it may be active. The general sequence of operations is shown in the inset. Every batch of compounds is then sorted by their scores, with the highest scores first. Next, the computer consults the rest of the system to learn of the current cutoff point, which is a point determined by both the Hodes ranking and the number of compounds in the batch that will pass. All structures above the cutoff point are recommended for acquisition and the remainder are "de-selected". A report of this process is then printed for review by a staff chemist; an example of such a Report is shown below. The chemist may disagree with any of the proposed actions. If the chemist takes no action, concurrence is assumed and the

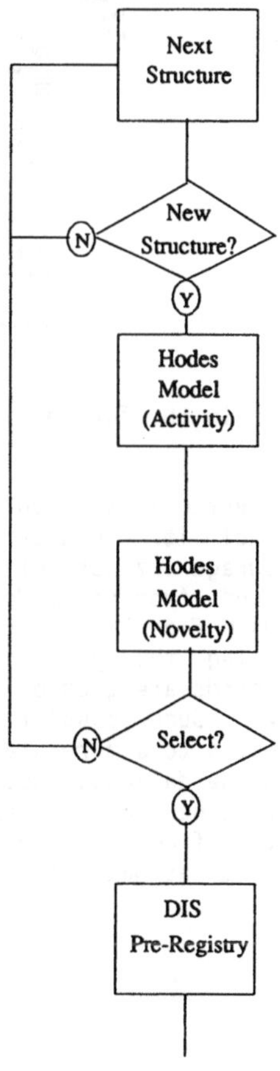

listed "selections" will become active acquisitions. The computer prints request letters and does the bookkeeping necessary to monitor the requests. All the "de-selections" are dropped from consideration, but a record of each of them is kept so that, if the structure is considered

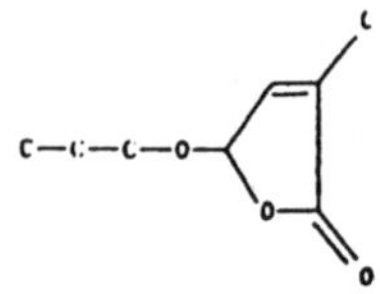

```
9219681

TID: 9219681-A
HODA: I
HFLG: OFF
HODN: 92
SONH: NO
OSWT:
HODM: 1
HODT: A
HDTS: 0
HOD1: I
HD1S: 12.29
HOD2: I
HD2S: 5.53
HOD3: H
HD3S: 4.46
HODL: J
HDLS: 24.21
HDAS: 32.3
NMAT:
MATC:     85278-Z\SIMILAR\SELECTED
         230337-Y\SIMILAR\SELECTED
         230340-C\SIMILAR\SELECTED
         230341-F\SIMILAR\SELECTED
         249797-A\SIMILAR\SELECTED
         271634-N\SIMILAR\SELECTED
         284621-F\SIMILAR\SELECTED
         286312-R\SIMILAR\SELECTED
         289946-R\SIMILAR\SELECTED
         329968-T\SIMILAR\SELECTED
         329969-U\SIMILAR\SELECTED
         350598-S\SIMILAR\SELECTED
         402957-G\SIMILAR\SELECTED
         600195-T\SIMILAR\SELECTED
         600199-X\SIMILAR\SELECTED
         600201-Z\SIMILAR\SELECTED
         601078-F\SIMILAR\SELECTED

SSPL: U.S.DEPT.AGRICULTURE
SLBN: 1951-A
CRIC: Y01I-3;Y01M-5;Y02I-4
LOGP: 0.94
LPFL: CALCULATION COMPLETE
DTSL:
DTEN: 03-MAY-84
OTCK: P.LIWANAG
DTCK: 4
CONF: OPEN
ACAT: DFDV
EMRS: NO
COMI: 905
MOLF: C18 H12 O3
STXT:
DSOR: 14-MAY-84
NSOR: 280
SREC: SELECT
MTXT:
UHAZ:
HFST:
QRSN:
```

again in the future, its having been de-selected can be made known at that time.

The system works smoothly, and a minimum of manual effort is necessary to control it. Concurrence with computer-generated decisions tends to encourage laziness of course, but an absolute score for each compound, derived from the model is carried as part of its permanent record and can be used from time to time to monitor the quality of the new compounds entering the system. A positive aspect of the model is that, whether its standards are good or bad, they are applied consistently from week to week. Such consistency is impossible with human intervention and contributes to a more stable baseline. A disadvantage, as noted above, is that the database underlying the model is, of necessity, large. This means that as new test systems are introduced (14) the model cannot nimbly switch from one to another and this unwieldiness is currently causing problems as major changes in testing strategy are going on.

REFERENCES

1. The starting point for this effort is generally regarded as the Congressional authorization, in 1955, of $5 million for the establishment by NCI of a Drug Development Program. See DeVita, V. T.; Oliverio, V. T.; Muggia, F. M.; Wiernik, P. W.; Ziegler, J.; Goldin, A.; Rubin, D.; Henney, J.; Schepartz, S. "The Drug Development and Clinical Trials Programs of the Division of Cancer Treatment, National Cancer Institute". Cancer Clin. Trials, 1979, 2, 195-216.

2. On 23 June, 1986, the CAS Registry contained 7,908,232 entries (STN International online "News"). This number is currently increasing by about 750,000 compounds per year.

3. Beginning in 1986, NCI began to develop an in-vitro screening method which makes no use of the older P388 screen, or of the second stage, which involved more refractory tumors, also in mice.

4. Murine lymphocytic leukemia, strain P388, is the preliminary screen against which all compounds have been tested since 1976. The number of compounds for which there are P388 data in the DIS is now in excess of 200,000. Full details of all screening procedures are published by NCI in "Instruction 14" (1985). Copies of this booklet may be obtained from the Information Technology Branch, DTP, DCT, NCI, Bethesda, MD, 20205.

5. Milne, G. W. A., and Miller, J. A. The NCI Drug Information System. I. System Overview. J. Chem. Inf. & Comp. Sci., 26, 154-159, (1986); Milne, G. W. A., Feldman, Alfred, Miller, J. A., Daly, G. P. and Hammel, M. J.: The NCI Drug Information System. II. The DIS Pre-Registry. J. Chem. Inf. & Comp. Sci., 26, 159-168, (1986); Milne, G. W. A., Feldman, Alfred, Miller, J. A. and Daly, G. P. The NCI Drug Information System. III. The DIS Chemistry Module. J. Chem. Inf. & Comp. Sci., 26, 168-179, (1986); Milne, G. W. A., Miller, J. A. and Hoover, J. R.: The NCI Drug Information System. IV. Inventory and Shipping Modules. J. Chem. Inf. & Comp. Sci., 26, 179-185, (1986); Zehnacker, M. T., Brennan, R. H., Milne, G. W. A., and Miller, J. A.: The NCI Drug Information System. V. The DIS Biology Module. J. Chem. Inf. & Comp. Sci., 26, 186-193, (1986); Zehnacker, M. T., Brennan, R. H., Milne, G. W. A., Miller, J. A. and Hammel, M. J.: The NCI Drug Information System. VI. System Maintenance. J. Chem. Inf. & Comp. Sci., 26, 193-197, (1986);

6. The acronym "NSC" stands for National Service Center, a short form of Cancer Chemotherapy National Service Center, the early name for the program. The "NSC Number" is used by NCI as a Registry Number. Other Registry Numbers, such as the CAS Registry Number are not useful for NCI because CAS Registry Numbers can not be assigned to the confidential structures which constitute about half of the NCI database.

7. Hodes, L. "Computer-Aided Selection of Novel Antitumor Drugs for Animal Screening". ACS Symp. Ser. 1979, 112, 583-603. Hodes, L. "Selection of Molecular Fragments for Structure-Activity Studies in Antitumor Screening". J. Chem. Inf. & Comp. Sci., 1981, 21, 132-136. Hodes, L. "A Two-Component Approach to Predicting Antitumor Activity from Chemical Structure in Large Scale Screening". J. Med.

Chem., 1986, 29, 2207-2212. Paull, K., Hodes, L., and Simon, R. M. "Efficiency of Antitumor Screening Relative to Activity Criteria". J. Natl. Cancer Inst., 1986, 76, 1137-1142.

8. Much of the data discussed here may be found in: Hodes, L. J. "Computer-Aided Selection of Compounds for Antitumor Screening: Validation of a Statistical-Heuristic Method". J. Chem. Inf. & Comp. Sci. 1981, 21, 128-132.

9. Five of the original 400 compounds were dropped from consideration when it was discovered that their structures were incompletely described. About 2% of all compounds tested have incompletely defined structures but are tested in spite of this for a variety of reasons.

10. To be "highly active" against P388, a compound must show a reproducible %T/C greater than 175. Compounds whose %T/C values fall below 120 are considered "inactive" and the remainder, with %T/Cs lying between 120 and 175 are regarded as "moderately active".

11. As an example, in the oxazolidine shown here, the most characteristic

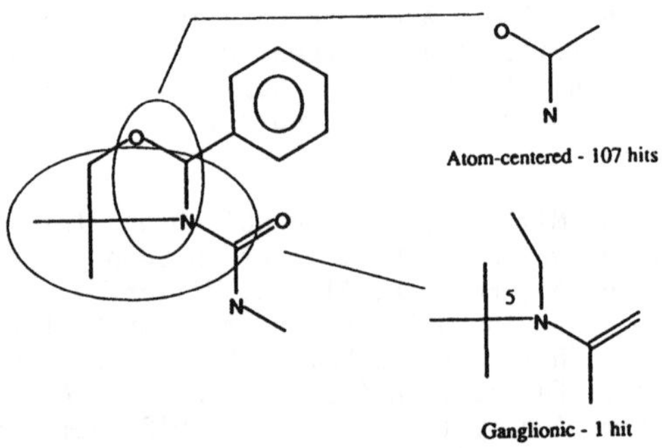

atom-centered fragment is found in 107 compounds that have been tested against P388. The corresponding ganglia augmented atom key is found in only one of these compounds.

12. Milne, G. W. A., Feldman, Alfred, Miller, J. A., Daly, G. P. and Hammel, M. J.: The NCI Drug Information System. II. The DIS Pre-Registry. J. Chem. Inf. & Comp. Sci., 26, 159-168, (1986).

13. In an attempt to create a "National Screening Center" as a service, NCI policy has always tested chemicals at no charge and has also declined, as a general matter, to purchase samples. Most of the compounds tested are not available for sale and even if they were, the Program could not afford to buy them. Testing data are supplied immediately to the supplier and, if so requested, the NCI will afford confidentiality to all such data.

14. In 1986, the decision was made to shift from animal, or in-vivo testing in the primary screen to in-vitro testing against human tumor cell lines. This decision renders the Hodes screen invalid at a stroke and is a good example of the susceptibility of such methods to external events.

Physico-Chemical Property Data Bank of the Prague Institute of Chemical Technology

Pavel Chuchvalec, Kvetoslav Ruzicka, Stanislav Labik, and Vlastimil Ruzicka, Jr.

Department of Physical Chemistry Institute of Chemical Technology, 166 28 Prague 6, Czechoslovakia

A knowledge of physico-chemical properties is a fundamental requirement of all design activities in the chemical industry. Experimental values comprise only a small part of the data required for further application. Estimation and prediction methods are in many cases the only way to get the information needed.

The development and modification of prediction methods as well as testing of different correlating relationships need accurate data for the assessment of the reliability of methods , tested. A database of critically evaluated data is therefore the necessary assumption for the successful activity in this region. Measurement and calculation of thermodynamic properties and phase equilibria of fluids represent traditional subjects of interest at our department. Large "personal" data files on different properties have been generated. In order to use the data more efficiently and to enable people from the industry a utilisation of this information, the physico-chemical property databank is being developed.

C. Jochum M. G. Hicks J. Sunkel (Eds.)
Physical Property Prediction
in Organic Chemistry

1. Compounds and properties on the database

The aim of our present work is to establish a set of critically evaluated data on the thermodynamic, thermophysical, transport and PVT properties of selected chemical compounds.

586 chemical compounds are on the database at present. The importance of a compound in the industrial chemical processes served as the main criterion for the selection that was carried out together with our partners from the industry and colleagues from technological departments of our Institute. Some homologous series were completed in order to achieve another objective - to prepare sets of data for development and testing of methods for prediction of physico-chemical properties. Comparatively a small number of inorganic compounds (such as the most common gases and solvents) were included in the database. Selected compounds cover the following groups: -inorganic compounds (34 substances) -compounds of carbon and hydrogen (147) -compounds of carbon,hydrogen and halogen (55) -compounds of carbon hydrogen and nitrogen (68) -compounds of carbon hydrogen and oxygen (175) -compounds of carbon hydrogen and sulphur (44) -other compounds (63).

The following physico-chemical properties are stored: -molar mass, gyration radius, refractive index, dipole moment, permitivity -critical temperature, critical pressure, critical volume -acentric factor -density of liquid -normal boiling point temperature -melting point temperature, enthalpy of fusion -enthalpy of formation, $(G_T-H^o_o)/T$ and $(H_T-H^o_o)/T$ functions of ideal gas -pure component vapour pressure, enthalpy of vaporization -molar heat capacity of ideal gas and of liquid -second virial coefficient -dynamic viscosity of gas and of liquid -thermal conductivity of gas and of liquid -surface tension. Parameters of equations enabling prediction of PVT and of thermodynamic properties for pure compounds and their mixtures are also included. The database contains both the "one-point" properties, i.e. values of a particular physico-chemical property under exactly specified state conditions, and parameters of suitable correlating and/or predicting equations. The emphasis was laid on the reliability of the data stored, on their self-consistency and on the completeness of data files.

2. Establishing of database

Establishing of the database realizes in three stages: 1. Collection of data on thermodynamic, transport and PVT properties for pure chemical compounds from primary literature sources and main compilations. 2. Assessment of data, testing of consistency, temperature correlation, and prediction of properties where experimental data are missing (if possible).

3. Storage of recommended values in the external memory of a computer and their check.

Recommended data on all properties are accompanied with information on the primary literature source. The quality of data is characterized by assigning the reliability classes (1 to 5) estimated on the basis of the accuracy of measurement claimed in the literature, subjective view of the compilers on the reliability of sources and correlation procedure used. Some statistical criteria (e.g. standard deviation) are presented, if possible. In case of correlating equations the ranges of validity are stored along with each set of correlation parameters.

A set of sotware tools for creation, modification and selection from the database files have been developed. The main part of the set is a menu-based program for retrieving physico-chemical properties and constants of correlating equations. The proposed form of the program permits a comfortable selection of a compound ı a property, and incorporates large on-screen helps for easy usage with the minimal forward knowledge.

3. The selection of data

One of the primary requirements in establishing our database is to achieve the high level of completeness of data files. This means that not only experimental but also predicted data are stored. In the first stage we have focused our attention to the prediction of missing critical constants because their knowledge is required for the estimation of many other properties especially when methods based on the corresponding states theory (CST) are used.

Experimental values of critical constants were found for about 350 compounds of the given set. For more than 200 another substances critical properties were estimated. Fourteen different methods were used for that purpose employing the following procedure. The suitability of methods was evaluated always for the particular group of compounds. The method with the best performance for the group was then used for the prediction in the corresponding group. As the most successful of all procedures tested appeared the group contribution method of Ambrose. The reliability of the estimation was checked by the comparison of the critical compressibility factor calculated from predicted critical constants with values of this factor for other members of the given group. Every estimated value stored on the database is indicated by the negative sign of the numerical code for the reliability class and the literature citation refers to the relevant estimation method.

The test of the thermodynamic consistency of the data stored can be demonstrated on the case of vapour pressure equations. Files of data for vapour pressure calculation can be divided into two groups. The first group covers constants of the Antoine equation (data for atmospheric and subatmospheric region). The constants were selected mostly on the basis of critical compilation from different data collections. For a limited number of compounds parameters determined from experimental data measured at our department were used. Constants of the Antoine equation are stored for 478 compounds of the given set at present. The rest of about 100 compounds includes substances unstable at temperatures close to their melting point and inorganic compounds which are at normal conditions in the gaseous state. The consistency checks were carried out by comparing normal boiling point temperature calculated from the Antoine equation with independently prepared files of critically evaluated normal boiling point temperatures. Another check consisted in the comparison of critically evaluated calorimetric heats of vaporization with the values obtained from the Antoine equation. The accuracy of selected constants of the Antoine equation was expressed by assigning the following reliability classes:

0.......it is not possible to determine the reliability
1.......reliability is better than 0.02 K
2.......reliability is better than 0.1 K
3.......reliability is better than 1 K
4.......reliability is better than 5 K
5.......reliability is better than 10 K.

The second group of vapour pressure data includes constants of the Wagner equation for calculation in a large temperature range. The file was checked by the same procedure as in the case of the Antoine equation.

Special attention was devoted to the selection of suitable forms of correlating equations. Let us take the temperature dependency of molar heat capacity of compounds in the state of ideal gas as an example. We are using the following equation in the database

$$C_p^{go} = A + B.(C/T)^2 . \exp(-C/T)/[1-\exp(-C/T)]^2 + D.(E/T)^2 . \exp(-E/T)/[1-\exp(-E/T)]^2$$

This function is better than a polynomial expression especially when extrapolated outside the original fit. In contrast to a polynomial expression the proposed equation produces finite limits for both extremes of temperature. The expression is analytically integrable and makes the correlation possible in a large temperature limit (200 to 3000 K). We have successfully applied this equation for calculation of chemical equilibria at high temperatures.

4. Conclusion

Reliable data for pure compounds is an essential prerequisite for a prediction of mixture properties. Our databank is being prepared to permit the calculation of mixture properties for the single-phase region (gaseous state, liquid state, supercritical state) as well as for the region of coexistence of liquid and gas phases. The data files of the pure compounds will support these calculations by providing all the information needed.

Molecular Orbital and Force-Field Calculations for Structure and Energy Predictions

Timothy Clark

Institut für Organische Chemie der Friedrich-Alexander-Universität
Erlangen-Nürnberg, Henkestrasse 42, 8520 Erlangen,
Federal Republic of Germany

This article aims to provide a short review of the methods available for the calculation of molecular structures and energies and other properties. The programs employed all use a guessed initial geometry as starting point for the calculation and then optimize the structure in order to find the minimum that can be found by moving down in energy from the starting point. This leads to the first problem with such methods, especially for very large molecules. There are at present very few programs that can investigate a series of possible structures in order to identify as many minima as possible and to be able to find the global (most stable) minimum within any degree of certainty. Even when this is possible (at present only for molecular mechanics calculations) the cost in computer time can be very large. For the other methods, the chemist must have enough imagination to be able to predict all the possibilities open to the molecule in order to find the global minimum.

Nevertheless, the classical situation that "Molecular Mechanics isn't parametrized, semi-empirical is too unreliable and *ab initio* is too expensive" is rapidly changing. The major cause for this change, which will eventually become a revolution, are the new generations of computer that are becoming available. The VAX generation of minicomputers is now being replaced by more modern super-minis that not only provide a major speed advantage, but also provide the programmer with significant amounts of fast memory that change the face of computational chemistry. At present, the Convex C-120 in Erlangen (a dedicated machine used only for structure and energy calculations) runs with 1.2 GigaBytes of scratch disk in order to be able to store the integrals used in an *ab initio* calculation. The calculations are

C. Jochum M. G. Hicks J. Sunkel (Eds.)
Physical Property Prediction
in Organic Chemistry

input/output limited on many fast computers because of the need to continuously read the integrals into core in an iterative self-consistent-field calculation. However, GigaByte main memory machines are no longer illusory, and will very soon become commonplace in the super-mini class. Current programs are not written to be able to use the capabilities of such machines. Gaussian 82[1], the most commonly used *ab initio* program, typically runs in less than 2% of the 32 MegaByte main memory on the Erlangen Convex. On the other hand, input/output rates of more than 2 MegaBytes/second are not uncommon. In 80% of routine calculations, this i/o load could all be shifted to main memory on a machine with 128 Megabytes of main memory, and the problem could be effectively vectorized as a side-effect. Once programs are rewritten along these lines, *in-core ab initio* programs that run up to 5x faster than the current Gaussian82 on vector machines like the Convex should be practicable. Similarly, the most modern semi-empirical programs, such as VAMP,[2] run within about 20-25 Megabytes for 100 atoms and are highly vectorized, so that full geometry optimizations without symmetry on molecules of this size have become routine on the C-120 in Erlangen. One of the highest priorities at present, however, is to rewrite the standard semiempirical, *ab initio*, and even molecular mechanics (modelling) programs to make use of the facilities and performance offered by the new generation of machines. The current programs are often 20 years old and were written for machines with very limited core and i/o facilities, so that they often spend their time trying to make a modern vector computer behave as if it were a 20-year-old scalar machine. Computational chemistry programs have a very bad reputation as i/o intensive and poorly vectorizable. There is, however, no reason for this *as long as sufficient fast memory is available*. Once the problems can be run in-core, the vectorization problems also tend to disappear and the performance increases accordingly.

The three main methods that are used for structure and energy prediction are Molecular Mechanics (Force-Field, or Modelling), semiempirical molecular orbital (MO) calculations such as MINDO/3, MNDO and AM1, and *ab initio* MO theory. Table 1 shows a comparison of the cpu-times and results of calculations using the three methods for propane. The *ab initio* runs use two different basis sets: 3-21G, a simple split-valence basis set, and 6-31G*, a polarization basis set that is becoming standard for geometry optimizations.

Table 1: Cpu-Times and Performance of Common Calculational Methods.[a]

	Force Field	Semi-empirical	ab initio		exp.
	MM2	MNDO	3-21G	6-31G*	
cpu-sec(Cyber 845)	0.8	11	550	4,700	
r_{CC} (Å)	1.534	1.530	1.541	1.528	1.526
CCC angle (°)	111.7	115.4	111.6	112.7	112.4
ΔH_f^0 (kcal mol^{-1})	-24.8	-24.9	-25.5[b]	-25.2[b]	-25.0

[a] Data taken from reference 3.

[b] Calculated using an atom-equivalent scheme based on the total energy (M. R. Ibrahim and P. v. R. Schleyer, *J.Comput.Chem.*, *1985*, **6**, 157).

The difference in cpu-time between the force-field calculation and the 6-31G* optimization is a factor of almost 6000. However, each of the methods has its advantages and disadvantages and even different ranges of applicability, so that there is often little choice for a particular problem.

Molecular Mechanics

Molecular mechanics calculations (also known as force-field calculations or as molecular modelling) are based on a very simple mechanical model of molecules.[4] Basically, the forces such as bond-stretching, angle-bending, or torsion are modelled by simple quadratic potentials, and the van der Waal's forces by a 6/12 potential. This makes the calculations very economical in terms of cpu-time, and indeed they are now more often performed on PC's than on mainframe computers. Such calculations can be extremely accurate as long as there are sufficient data for the parametrization. For alkanes, for instance, MM2 heats of formation are probably more accurate than experiment in most cases, and most certainly easier to determine. This is, however a special case, and the parametrization may not be as reliable for more "exotic" types of molecule. There are two major disadvantages to molecular mechanics calculations. The first is that the calculations are necessarily limited to compound types for which adequate experimental data are available for a reliable parametrization, and the second is that any extrapolation outside the range of the parametrization set (for instance to extremely strained molecules or to transition states) is likely to be unreliable. However, for many applications, the accuracy of the force-field predictions is not of paramount importance (e.g. in molecular modelling) and the speed of the method often outweighs its disadvantages. The problem of finding the global minimum becomes especially critical for the very large molecules that are often calculated by molecular mechanics.

Semi-Empirical MO Theory

The most common semi-empirical methods, MINDO/3,[5] MNDO,[6] and AM1[7] were all developed by Michael Dewar and his group in Austin, Texas. The declared aim is to provide cheap and easily applicable molecular orbital methods that can be used by experimentalists as a type of "computer spectrometer" in order to be able to determine the structure, energy and properties of the molecule in question. This is a very attractive concept because it avoids the need to make the compound - often the most expensive step in such determinations. Programs such as AMPAC,[8] MOPAC[9] or VAMP[2] are capable of optimizing molecules containing 100 or more atoms within reasonable cpu-times on modern super-minis, and make many experimental systems calculationally accessible. There is often no longer any need to replace ^{t}butyl groups by hydrogen or THF by water in order to be able to carry out the calculation. The only possible problem is the accuracy of the results, which have often been criticised by *ab initio* specialists. The major shortcomings[3] of MINDO/3 and MNDO were their inability to treat the rotation barriers in pi-conjugated systems properly, considerable errors for highly branched hydrocarbons, and a complete inability to reproduce hydrogen bonds. The latter defect is particularly unfortunate as these methods would otherwise be well suited to treating biological systems. The newest development, AM1, does not suffer as badly from these faults, although there is not yet a large enough body of data to judge its reliability. Table 2 shows some typical errors for standard molecules.

Table 2: MNDO and AM1 errors in Heats of Formation (kcal mol^{-1}).[a]

Compound(s)	AM1	MNDO
Hydrocarbons(average)	5.1	5.9
C,N,O,H Compounds (average)	5.9	6.6
Largest error (80 compounds)	26.2[b]	-49.6[c]
Cyclopentane	-10.5	-12.0
Cyclohexane	-9.0	-5.3
Allene	0.6	-1.6
Acetone	2.7	2.5
DMF	8.9	8.8
Hydrogen Bonds	yes	NO
π-rotation barriers	OK	too low

[a] Data taken from reference 7.

[b] Bicyclobutane [c] Cubane

At present, the calculations only contain *d*-orbitals for chromium, and so results for hypervalent molecules, and especially for sulfoxides and sulfones, which are often present in organic molcules of interest, are in error. Table 3 shows the Elements for which MNDO (italics and bold) and AM1 (bold only) parameters are currently available.

Table 3 Elements for which MNDO and AM1 Parameters are available.

H						
Li	*Be*	**B**	**C**	**N**	**O**	**F**
		Al	**Si**	*P*	*S*	**Cl**
					Cr	
			Ge			**Br**
			Sn			**I**
	Hg		*Pb*			

Ab Initio Molecular Orbital theory

Ab initio molecular orbital theory is, in principle at least, capable of providing results as good as, or better than, experiment. Calculations of such quality are, however, seldom reported and are only practicable for very small molecules. The vast majority of *ab initio* calculations reported are of much lower quality, with correspondingly larger errors. There is nothing inherently superior in *ab initio* calculations at lower levels of theory, but the mistake is often made that minimal basis set *ab initio* with incomplete geometry optimization are reported for systems that could have been treated very well by one of the semi-empirical methods. The major limitation of *ab initio* calculations is that they require very large amounts of cpu-time and disk if they are to be performed at high levels on large molecules. The practical limit for 6-31G* calculations with full geometry optimization and subsequent energy calculations using a fourth order Møller-Plesset correction for electron correlation is between 4 and 6 heavy (non-hydrogen) atoms for most well-equipped research groups. Heats of Formation are not given directly by *ab initio* calculations, but can be calculated using a variety of additive schemes. Some typical results are shown in Table 4.

Table 4: Errors in Ab Initio Heats of Formation (kcal mol^{-1}).[a]

Compound(s)	3-21G	6-31G*
29 Hydrocarbons (average)	2.7	1.3
Cyclopropene	+14.5	+3.6
Bicyclobutane	+19.5	+3.8
Benzene	-8.7	-7.7
27 C,H,N,O Compounds	1.3	1.2
Ketene	1.0	3.0
Cyanogen	1.7	7.1

[a] Data taken from M. R. Ibrahim and P. v. R. Schleyer, *J.Comput.Chem., 1985*, **6**, 157 and based on Hartree-Fock total energies. See also "Ab Initio Molecular Orbital Theory", W. J. Hehre, L. Radom, P. v. R. Schleyer and J. A. Pople, Wiley, New York, 1986.

Although there is still a considerable number of "problem molecules", such as benzene, for which the errors are quite large, and although the 3-21G basis set gives poor results for strained molecules, this approach does seem to show promise, especially if electron correlation is included in the calculations.

Outlook

None of the methods discussed are universally applicable and reliable. However, current progress is very rapid and we can expect both semi-empirical and *ab initio* methods to improve in accuracy and range of application dramatically in the next decade. This will be largely due to the opportunities offered by the new generations of computers. In short, computational chemistry has long been a science waiting for someone to build the right computer, and should now take advantage of the facilities being offered.

References

1. J. S. Binkley, R. A. Whiteside, K. Raghavachari, R. Seeger, D. J. DeFrees, H. B. Schlegel, M. J. Frisch, J. A. Pople and L. R. Kahn, "Gaussian 82", Carnegie-Mellon University, Pittsburgh, 1982.

2. VAMP (T.Clark, unpublished) is a vectorized semi-empirical program that retains the features and user-interface of AMPAC and MOPAC. It is currently available for Convex Computers.

3. "A Handbook of Computational Chemistry", T. Clark, Wiley, New York, 1985.

4. "Molecular Mechanics", U. Burkert and N. L. Allinger, ACS Monograph 177, American Chemical Society, Washington, D.C., 1982.

5. R. C. Bingham, M.J.S. Dewar and D. H. Lo, *J.Am.Chem.Soc. 1975*, **97**, 1285, 1294, 1302, 1307, 1311.

6. M. J. S. Dewar and W. Thiel, *J.Am.Chem.Soc.*,B 1977, **99**, 4899.

7. M. J. S. Dewar, E. G. Zoebisch, E. F. Healy and J. J. P. Stewart, *J.Am.Chem.Soc.,1985*, **107**, 3902.

8. The Dewar Group, QCPE Program No. 455.

9. J. J. P. Stewart, QCPE Program No. 536.

Estimation of Thermodynamic Properties of Organic Compounds in the Gas, Liquid, and Solid Phases at 298.15 K

Eugene S. Domalski and Elizabeth D. Hearing

Chemical Thermodynamics Division, Center for Chemical Physics, National Bureau of Standards, Gaithersburg, MD 20899, USA

Abstract

About one hundred years ago, the relationship between the boiling point of a substance and its heat of vaporization at the boiling point (Trouton's Rule) was reported in the chemical literature. Since that time, much progress has been made in the correlation of physical and thermodynamic properties. Physical organic chemists in particular have been active in the development of correlation and estimation schemes linking the molecular structures of organic compounds and various thermodynamic properties. Some correlation and estimation schemes are more successful than others. Of the estimation schemes developed, the one put together by S.W. Benson and co-workers has had more universal acceptance because of its overall simplicity, ease of application, and general good agreement between estimated and experimental values. This scheme assigns individual group energy values for molecular fragments which are additive, account for nearest neighbor interactions, and give special consideration to corrections for steric strain or stereoisomerism. The primary focus of this scheme has been on organic molecules in the gas phase, although some applications to the liquid and solid phases have been reported.

Our recent development of the Benson approach to the estimation of thermodynamic properties (enthalpy of formation, heat capacity, and entropy) at 298.15 K has focused upon the liquid and solid phases. The gas phase has been included also for the sake of continuity and internal consistency of the calculation of thermodynamic properties among the three phases. Studies showing the predictive capability of this scheme toward estimating the thermodynamic properties of hydrocarbons and organic compounds containing the elements: CHO, CHN, and CHNO have already been carried out. Work is in progress to complete the scope of the predictive capability to cover organic compounds containing the elements: sulfur, phosphorus, fluorine, chlorine, bromine, iodine, and metals.

This paper provides a general overview of the NBS program in the estimation of the thermodynamic properties of organic compounds using the Benson approach and presents discussions of selected topics, such as: (1) relationships of the enthalpy of formation, heat capacity, and entropy at 298.15 K to other thermodynamic properties (Gibbs energy of formation, equilibrium constants, enthalpies and entropies of transition) which extend predictive capabilities, (2) descriptions of the Benson notation, group values and their application to the estimation of thermodynamic properties, and (3) explanations of unique solutions to the estimation of two selected

C. Jochum M. G. Hicks J. Sunkel (Eds.)
Physical Property Prediction
in Organic Chemistry

classes of organic compounds.

Introduction

The development of numerical methods which correlate the molecular structure of a compound with its measured physical and thermodynamic properties is a normal consequence of interpreting experimental results for such measured properties. Such efforts are reasonable since only a limited number of substances can be subjected to experimental scrutiny. Extension of the experimental data to comparisons with estimate values for compounds should be pursued and the application carried to compounds for which experimental data are lacking for one reason or another.

During the 1950's, S.W. Benson and co-workers have developed an estimation method for the calculation of the enthalpy of formation, heat capacity, and entropy of organic compounds in the gas phase between 298 and 1500 K. The approach used is attractive because of many desirable features which include simple additivity, use of a simple notation scheme, second order character which means the identification of nearest-neighbor interactions, ease of application, and overall satisfactory agreement between the estimate thermodynamic property and the its experimentally determined value. Some studies by these researchers were devoted to the estimation of thermodynamic properties of organic compounds in the liquid and solid phases, but primary attention was given to the gas phase and its relationship to homogeneous gas phase kinetic processes.

The purpose of this paper is to describe the activities underway at the National Bureau of Standards which pursue the development of estimation procedures used to calculate the thermodynamic properties of organic compounds. The main effort is directed toward the extension of the estimation methods developed by Benson, et al. for the gas phase to the liquid and solid phases at 298.15 K. The activities have been arranged to proceed in three stages; first, estimation of the enthalpy of formation, heat capacity, and entropy at 298.15 K for hydrocarbon compounds, second, estimation methods for CHO, CHN, and CHNO compounds, and third, estimation methods for organic compounds containing S, P, halogens, and metals. At this time, the first stage of work is finished and a paper on hydrocarbon compounds is expected to appear in the Journal of Physical and Chemical Reference Data later this year. Also, work on CHO, CHN, and CHNO compounds has been completed and a manuscript is in preparation. Work is going on to complete the last stage of this three part estimation effort. Also, the groups of organic compounds in each the three stages will be available on diskettes which allow a user to interactively find out on his or her own personal computer which group contribution values were used to estimate a particular compound, how to estimate the thermodynamic properties for a compound from the user's choice of group values, or what the difference in magnitude is between an experimental and estimated property.

The thermodynamic data used as the data-base for the generation of group contribution values and for comparison to estimated values derived from the group values were obtained primarily from a variety of selected thermodynamic compilations which are either well established or are as current as possible. The primary compilations were: 53ROS/PIT, 69STU/WES, 70COX/PIL, 77PED/RYL, 84DOM/EVA, 86MAR, and 86PED/NAY. Upon occasion a thermodynamic value listed in a compilation is not the actual value reported in the journal article by an investigator, but an adjusted or smoothed value which differs a few tenth or hundredths of an energy unit from the original literature value. Fortunately, such adjustments or smoothing methods caused only very small changes. Some thermodynamic values used in the data-base come directly from journal articles because they have not been included in the compilations or have just recently appeared in the literature.

Relationships to Other Thermodynamic Properties at 298.15 K

Many relationships exist between the enthalpy of formation, heat capacity, and entropy at 298.15 K with other useful thermodynamic properties. Some of the properties and the mathematical relationships at 298.15 K are as follows:

Gibbs energy of formation, $\Delta_f G° = \Delta_f H° - \Delta_f S°(298.15)$

Equilibrium constant, $\ln K_f = -\Delta_f G°/R(298.15)$

Enthalpy of vaporization at 298.15 K, $\Delta H°_{vap} = \Delta_f H°(gas) - \Delta_f H°(liquid)$

Enthalpies of fusion +
$\sum$enthalpies of transition at 298.15 K, $\Delta H°(fusion) + \sum\Delta H°(transition) = \Delta_f H°(liquid) - \Delta_f H°(solid)$

Heat capacity $C_p° = C_v° + R$

Care must be taken to understand that the above properties are at 298.15 K and will be different at other temperatures. Depending upon the property, some differences will be larger than others. The temperature dependence of a thermodynamic property is important because it sometimes has been measured and reported at a temperatures other than 298.15 K. Data reduction of such a property to 298.15 K is necessary for the development of an estimation scheme, such as the one under discussion.

Three useful thermodynamic relationships as a function of temperature are:

(1) Heat capacity

$C_p° = a + bT + cT^2 + dT^3 + eT^4 + \ldots.$, where a, b, c, d, and e are constants for a particular compound.

(2) Equilibrium constant and enthalpy of reaction,

$\ln(K_2/K_1) = (\Delta H°/R)[(T_2 - T_1)/(T_1T_2)]$, and

(3) Correction of enthalpies of transition to 298.15 K,

$$\Delta H°(\text{transition}) \text{ at } 298.15 \text{ K} = \Delta H°(\text{transition}) \text{ at } T - \Delta Cp°(T - 298.15)$$

where $\Delta C_p°$ is the difference in the heat capacities between the two phases.

The Benson Notation

The group contribution values which have been calculated correspond to a particular molecular segment of an organic compound. For example,

$C-(H)_3(C)$	corresponds to	$-CH_3$	
$C-(H)_2(C)_2$	corresponds to	$-CH_2-$	
$C-(H)(C)_3$	corresponds to	$-CH<$	
$C-(C)_4$	corresponds to	$>C<$	
$CO-(C)_2$	corresponds to	C-CO-C	
O-(H)(C)	corresponds to	C-O-H	
N-(H)(C)(CO)	corresponds to	C-NH-CO	
$C_d-(H)_2$	corresponds to	$=CH_2$	(C_d is a doubly bonded carbon)
$C_B-(H)$	corresponds to	>C-H	(C_B is a benzene/aromatic carbon)
$C_{BF}-(C_{BF})$	corresponds to	$>C=C<$	(C_{BF} is a fused benzene/aromatic carbon)
$C-(H)_2(C)(CN)$	corresponds to	$-CH_2-CN$	(aliphatic nitrile group)
$C_B-(NO2)$	corresponds to	$>C_B-NO2$	(benzene/aromatic nitro group)

At this point, with a preliminary understanding of the notation used to identify segments or groups which comprise the molecular structure of an organic compound, a few sample calculations will serve to show how one can estimate the enthalpy of formation, heat capacity, or entropy at 298.15 K. The development of groups and group value followed a prescribed path from alkanes to alkenes, alkynes, aromatic hydrocarbons, cycloalkanes, alcohols, ethers, esters, acids, and so on. The kind of agreement one should expect between an estimate for a thermodynamic property is related to what can be expected in the level of precision and accuracy possible from a laboratory measurement. Most of the enthalpies of formation in the literature were derived from combustion bomb calorimetric experiments which in rare instance can yield uncertainties of +/- 0.4 kJ/mol, with +/- 1 to 3 kJ/mol being more common, and in some instances +/- 4 to 8 kJ/mol being the best that can be expected. Occasionally, improved precision and accuracy can be derived from calorimetric experiments which measure enthalpies of hydrogenation or reaction, rather than the combustion process. Heat capacity and entropy values are obtained primarily from low temperature adiabatic calorimetry. High quality precision measurements yield uncertainties in the +/- 0.5 to 1 J/mol-K range. Average precision measurements give uncertainties of +/- 3 to 6 J/mol-K, with some falling into the range above +/- 8 J/mol-K. One can argue about the expression of uncertainties and that a percentage of the values measured is more appropriate and a better indicator of precision, however, when the difference in an estimated and an experimental value is examined, the general guidelines for precisions of these properties is more helpful for this discussion based on units of kJ/mol or J/mol-K. In general, good agreement is present between an estimated and an experimental value when the difference for enthalpies of formation is 4.0 kJ/mol or less and that for heat capacities and entropies is 4.0 J/mol-K or less.

A limited list of group contribution values is provided for the purpose of demonstrating what they are like and how they can be applied to estimate the enthalpy of formation, heat capacity, and entropy at 298.15 K for any organic compound.

Some Additive Group Contribution Values at 298.15 K

Values for $\Delta_f H°$ are in kJ/mol and those for $C_p°$ and $S°$ are in J/mol-K.

	(gas phase)			(liquid phase)			(solid phase)		
Group	$\Delta_f H°$	$C_p°$	$S°$	$\Delta_f H°$	$C_p°$	$S°$	$\Delta_f H°$	$C_p°$	$S°$
C-$(H)_3$(C)	-42.26	25.73	127.32	-47.61	36.48	83.30	-46.74	67.45	56.69
C-$(H)_2(C)_2$	-20.63	22.89	39.16	-25.73	30.42	32.38	-29.41	21.92	23.01
C-(H)$(C)_3$	-1.17	20.08	-53.60	-4.77	21.38	-23.89	-5.98		
-CH_3 corr	-2.26			-2.18			-2.34		
C-$(C)_4$	19.20	16.53	-149.49	17.99	10.24	-98.65	12.47	-83.63	-33.19
-CH_3 corr	-4.56			-4.39			-4.35		

	(gas phase)	(liquid phase)	(solid phase)
Group	$\Delta_f H°$	$\Delta_f H°$	$\Delta_f H°$
C-$(H)_2$(C)(CO)	-21.84	-24.14	-14.20 (*)
CO-(C)(O)	-137.24	-149.37	-153.60

O-(CO)(H)	-254.30	-285.64	-295.80

(*) see discussion on corrections to $\Delta_f H°$ for carboxylic acids.

Example Calculation for n-Hexane in the Liquid Phase

n-Hexane, CH_3-CH_2-CH_2-CH_2-CH_2-CH_3

Enthalpy of Formation

No. of Groups	Group	Group Value	Sum(kJ/mol)
2	C-$(H)_3$(C)	-47.61	-95.22
4	C-$(H)_2(C)_2$	-25.73	-102.92
			-198.14

Compare to experimental value: -198.66 kJ/mol (44PRO/ROS, 69GOO/SMI, 86PED/NAY)

Heat Capacity

No. of Groups	Group	Group Value	Sum(J/mol-K)
2	C-$(H)_3$(C)	36.48	72.96
4	C-$(H)_2(C)_2$	30.42	60.84
			194.64

Compare with experimental value: 194.97 J/mol-K (46DOU/HUF, 84DOM/EVA)

Entropy

No. of Groups	Group	Group Value	Sum(J/mol-K)
2	C-$(H)_3$(C)	83.30	166.60
4	C-$(H)_2(C)_2$	32.38	64.76
			296.12

Compare with experimental value: 296.06 J/mol-K (46DOU/HUF, 84DOM/EVA)

Example Calculation for Naphthalene in the Solid Phase

Naphthalene, $C_{10}H_8$

Enthalpy of Formation

No. of Groups	Group	Group Value	Sum(kJ/mol)
8	C_B-(H)	6.53	52.24
2	C_{BF}-(C_{BF})	14.10	28.20

			80.44

Compare with experimental value: 77.90 kJ/mol (86PED/NAY)

Heat Capacity

No. of Groups	Group	Group Value	Sum(J/mol-K)
8	C_B-(H)	20.13	161.04
2	C_{BF}-(C_{BF})	2.30	4.60

			165.64

Compare with experimental value: 165.69 J/mol-K (57MCC/FIN, 84DOM/EVA)

Entropy

No. of Groups	Group	Group Value	Sum(J/mol-K)
8	C_B-(H)	22.75	182.00
2	C_{BF}-(C_{BF})	-6.00	-12.00

			170.00

Compare with experimental value: 167.40 J/mol-K (57MCC/FIN, 84DOM/EVA)

Unique Solutions to the Estimation of Thermodynamic Properties

---

Methyl Repulsion Corrections for Tertiary and Quaternary Carbon Atoms

During the evaluation of data on the enthalpies of formation of substituted alkanes and alkenes, an approach was developed which corrects for the repulsion of interactions of hydrogen atoms on methyl groups attached to tertiary or quaternary carbon atoms. This technique differs from the approach Benson and co-workers (58BEN/BUS, 69BEN/CRU, 76BEN) have adopted in which a correction is applied for gauche isomers. A group value is assigned to tertiary or quaternary carbon atom along with a group value for the number of methyl groups attached to the tertiary or quaternary carbon atom. For example, ΔfH° for 2-methylpropane requires 3 methyl repulsion correction values, 2-methylbutane requires 2 methyl repulsion correction values, 3-methylpentane requires 1 methyl repulsion value, 3-ethylpentane does not need any correction. The following examples deal with the application of the methyl repulsion correction to the calculation of enthalpies of formation for gaseous hydrocarbons with quaternary carbon atoms.

2,2-Dimethylpropane, C_5H_{12}

No. of Groups	Group	Group Value	Sum(kJ/mol)
4	$C-(H)_3(C)$	-42.26	-169.04
1	$C-(C)_4$	19.20	19.20
4	Quat. CH_3 corr.	-4.56	-18.24
			-168.08

Compare with experimental value: -167.99 kJ/mol (70GOO, 86PED/NAY)

Compare with estimated value from 76BEN: -168.62 kJ/mol

2,2-Dimethylbutane, C_6H_{14}

No. of Groups	Group	Group Value	Sum(kJ/mol)
4	$C-(H)_3(C)$	-42.26	-169.04
1	$C-(H)_2(C)_2$	-20.63	-20.63
1	$C-(C)_4$	19.20	19.20
3	Quat. CH_3 corr.	-4.56	-13.68
			-184.15

Compare with experimental value: -186.10 kJ/mol (41PRO/ROS, 47OSB/GIN, 86PED/NAY)

Compare with estimated value from 76BEN: -182.55 kJ/mol

3,3-Dimethylpentane, C_7H_{16}

No. of Groups	Group	Group Value	Sum(kJ/mol)
4	C-$(H)_3(C)$	-42.26	-169.04
2	C-$(H)_2(C)_2$	-20.63	-41.26
1	C-$(C)_4$	19.20	19.20
2	Quat. CH_3 corr.	-4.56	-9.12
			-200.22

Compare with experimental value: -201.17 kJ/mol (41PRO/ROS 2, 47OSB/GIN, 86PED/NAY)

Compare with estimated value from 76BEN: -196.48 kJ/mol

3-Methyl-3-ethylpentane, C_8H_{18}

No. of Groups	Group	Group Value	Sum(kJ/mol)
4	C-$(H)_3(C)$	-42.26	-169.04
3	C-$(H)_2(C)_2$	-20.63	-61.89
1	C-$(C)_4$	19.20	19.20
1	Quat. CH_3 corr.	-4.56	-4.56
			-216.29

Compare with experimental value: -214.85 kJ/mol (45PRO/ROS, 47OSB/GIN, 86PED/NAY)

Compare with estimated value from 76BEN: -210.41 kJ/mol

3,3-Diethylpentane, C_9H_{20}

No. of Groups	Group	Group Value	Sum(kJ/mol)
4	C-$(H)_3$(C)	-42.26	-169.04
4	C-$(H)_2(C)_2$	-20.63	-82.52
1	C-$(C)_4$	19.20	19.20

			-232.36

Compare with experimental value: -232.34 kJ/mol (47JOH/PRO, 61LAB/GRE, 86PED/NAY)

Compare with estimated value from 76BEN: -224.35 kJ/mol

Similar methyl repulsion corrections are needed for calculating the enthalpies of formation of hydrocarbon compounds which contain tertiary and quaternary carbon atoms in liquid and solid phase and are applied in the same manner as shown above. A comparable correction did not emerge from an evaluation of data on the heat capacities and entropies of corresponding hydrocarbons.

Enthalpies of Formation of Aliphatic Acids in the Solid Phase

Development of the group value for C-$(H)_2$(C)(CO) (which corresponds to C-CH_2-CO) needed for the estimation of aliphatic carboxylic acids in the solid phase requires special attention. For example, let us examine the following carboxylic acids and their enthalpies of formation:

acid	structure	ΔfH°(solid), kJ/mol, expt'l
ethanoic acid	CH_3-CO-O-H	-496.06 +/- 0.5
propanoic acid	CH_3-CH_2-CO-O-H	-521.40 +/- 0.5
butanoic acid	CH_3-CH_2-CH_2-CO-O-H	-546.40 +/- 0.8
pentanoic acid	CH_3-CH_2-CH_2-CH_2-CO-O-H	-573.60 +/- 1.0
hexanoic acid	CH_3-CH_2-CH_2-CH_2-CH_2-CO-O-H	-599.20 +/- 2.0
heptanoic acid	CH_3-$(CH_2)_4$-CH_2-CO-O-H	-627.70 +/- 1.8
octanoic acid	CH_3-$(CH_2)_5$-CH_2-CO-O-H	-657.40 +/- 1.4

decanoic acid	CH_3-$(CH_2)_7$-CH_2-CO-O-H	-713.70 +/- 1.0
dodecanoic acid	CH_3-$(CH_2)_9$-CH_2-CO-O-H	-774.60 +/- 1.0
hexadecanoic acid	CH_3-$(CH_2)_{13}$-CH_2-CO-O-H	-891.50 +/- 1.7
eicosanoic acid	CH_3-$(CH_2)_{17}$-CH_2-CO-O-H	-1011.9 +/- 1.6

Decanoic through eicosanoic acid are solids at 298.15 K and exhibit an average -CH_2-increment for the enthalpy of formation satisfactorily comparable to that in n-alkanes. Values of ΔfH° for ethanoic through nonanoic acid in the solid phase were derived by combining data for ΔfH° for the liquid with the appropriate ΔH°(fusion). Experimental values for $\Delta_f H°$ and ΔH°(fusion) for the carboxylic acids were taken from 86PED/NAY and 84DOM/EVA, respectively. The average -CH_2- increment for hexanoic through nonanoic acid is the same as one finds with the n-alkanes, however, a correction must be applied to the group value C-$(H)_2$(C)(CO) for propanoic, butanoic, and pentanoic acids so that reasonable agreement is obtained between the experimental and estimated values. This correction is not required for the gas and liquid phases, but applies only to the solid phase. The estimation of ΔfH° for ethanoic acid does not involve a group value for C-$(H)_2(C)_2$ or C-$(H)_2$(C)(CO) and in a certain respect because of this is unique.

For estimating ΔfH° (in kJ/mol) for propanoic, butanoic, and pentanoic acids use:

C-$(H)_2$(C)(CO) = -14.20 + (6-N)(-3.70), where N is the number of carbon atoms in the carboxylic acid and is limited to having N equal to 3, 4, 5, or 6.

For estimating ΔfH° (in kJ/mol) for hexanoic acid and acids with larger numbers of carbon atoms use:

C-$(H)_2$(C)(CO) = -14.20.

Example Calculations for Enthalpies of Formation of some Carboxylic Acids in the Solid Phase at 298.15 K

Propanoic Acid

No. of Groups	Group	Group Value	Sum(kJ/mol)
1	C-$(H)_3$(C)	-46.74	-46.74
1	C-$(H)_2$(C)(CO)	-25.30	-25.30
1	CO-(C)(O)	-153.60	-153.60
1	O-(CO)(H)	-295.80	-295.80

			-521.44

Butanoic Acid

No. of Groups	Group	Group Value	Sum(kJ/mol)
1	$C-(H)_3(C)$	-46.74	-46.74
1	$C-(H)_2(C)_2$	-29.41	-29.41
1	$C-(H)_2(C)(CO)$	-21.60	-21.60
1	CO-(C)(O)	-153.60	-153.60
1	O-(CO)(H)	-295.80	-295.80
			-547.15

Pentanoic Acid

No. of Groups	Group	Group Value	Sum(kJ/mol)
1	$C-(H)_3(C)$	-46.74	-46.74
2	$C-(H)_2(C)_2$	-29.41	-58.82
1	$C-(H)_2(C)(CO)$	-17.90	-17.90
1	CO-(C)(O)	-153.60	-153.60
1	O-(CO)(H)	-295.80	-295.80
			-572.86

Hexanoic Acid

No. of Groups	Group	Group Value	Sum(kJ/mol)
1	$C-(H)_3(C)$	-46.74	-46.74
3	$C-(H)_2(C)_2$	-29.41	-88.23
1	$C-(H)_2(C)(CO)$	-14.20	-14.20
1	CO-(C)(O)	-153.60	-153.60
1	O-(CO)(H)	-295.80	-295.80
			-598.57

It is likely that a decrease in polar interactions induced by an increase in the paraffinic character of the solid carboxylic acids is being reflected in the correction to the C-$(H)_2$(C)(CO) group. The following table shows the kind of agreement that is possible between the experimental and estimated values for the enthalpies of formation of carboxylic acids in the solid phase.

acid	ΔfH°, kJ/mol, expt'l	ΔfH°, kJ/mol, calc	difference
ethanoic acid	-496.06	-496.14	0.08
propanoic acid	-521.40	-521.44	0.04
butanoic acid	-546.40	-547.15	0.75
pentanoic acid	-573.60	-572.86	-0.74
hexanoic acid	-599.20	-598.57	-0.63
heptanoic acid	-627.70	-627.98	0.28
octanoic acid	-657.40	-657.39	-0.01
decanoic acid	-713.75	-716.21	2.46
dodecanoic acid	-774.60	-775.03	0.43
hexadecanoic acid	-891.50	-892.67	1.17
eicosanoic acid	-1011.9	-1010.3	-1.6

Summary and Conclusions

In summary, a series of group contribution values have been developed which permit the estimation of the enthalpy of formation heat capacity and entropy at 298.15 K for compounds containing the elements carbon, hydrogen, oxygen, and nitrogen in the gas, liquid, and solid phases. The limitations which are imposed upon the precision of the experimental data reported for ΔfH°, Cp°, and S° for these compounds by calorimetric or related techniques can be accommodate by the Benson second order group additivity approach to the estimation of these properties in an acceptable and satisfactory manner.

Also, the work which has been carried out and which will appear in J. Phys. & Chem. Ref. Data demonstrates that this approach can provide valid and useful estimates in the liquid and solid phases. This paper, in a brief way, discusses the basic concepts of the group additivity method developed many years ago by Benson and co-workers and gives some examples of how estimates are calculated for enthalpies of formation, heat capacities, and entropies. It also provides some detail regarding the accommodation of unique energy corrections in the calculation of enthalpies of formation at 298.15 K for two selected classes of organic compounds, namely, hydrocarbons containing tertiary and quaternary carbon atoms in the gas, liquid, and solid phases, and carboxylic acids in the solid phase. The estimation scheme described here for calculating the thermodynamic properties of organic compounds in the gas, liquid, and solid phases operates, on the average, at about the current level of capability in experimental calorimetry. However, large differences between experimental and estimated thermodynamic values are likely for highly branched, highly polar, or strained organic compounds.

References

41PRO/ROS Prosen, E.J., and Rossini, F.D., J. Research Nat. Bur. Stds. 27, 289-310 (1941).

41PRO/ROS 2 Prosen, E.J., and Rossini, F.D., J. Research Nat. Bur. Stds. 27, 519-528 (1941).

44PRO/ROS Prosen, E.J., and Rossini, F.D., J. Research Nat. Bur. Stds. 33, 255-272 (1944).

45PRO/ROS Prosen, E.J., and Rossini, F.D., J. Research Nat. Bur. Stds. 34, 163-174 (1945).

46 DOU/HUF Douslin, D.R., and Huffman, H.M., J. Am. Chem. Soc. 68, 1704-1708 (1946).

47JOH/PRO Johnson, W.H., Prosen, E.J., and Rossini, F.D., J. Research Nat. Bur. Stds. 38, 419-422 (1947).

47OSB/GIN Osborne, N.S., and Ginnings, D.C., J. Research Nat. Bur. Stds. 39, 453-477 (1947).

53ROS/PIT Rossini, f.D., Pitzer, K.S., Arnett, R.L., Braun, R.M., and Pimentel, G.C., "Selected Values of Physical and Thermodynamic Properties of Hydrocarbons and Related Compounds", (Carnegie Press, Pittsburgh, PA 1953).

57MCC/FIN McCullough, J.P., Finke, H.L., Messerly, J.F., Kincheloe, T.C., and Waddington, G., J. Phys. Chem. 61, 1105-1116 (1957).

58BEN/BUS Benson, S.W., and Buss, J.H., J. Chem. Phys. 62, 271-280 (1958).

61LAB/GRE Labbauf, A., Greenshields, J.B., and Rossini, F.D., J. Chem. Eng. Data 6, 261-263 (1961).

69BEN/CRU Benson, S.W., Cruickshank, F.R., Golden, D.M., Haugen, G.R., O'Neal, H.E., Rodgers, A.S., Shaw, R., and Walsh, R., Chem. Rev. 69, 269-324 (1969).

69GOO/SMI Good, W.D., and Smith, N.K., J. Chem. Eng. Data 14, 102-106 (1969).

69STU/WES Stull, D.R., Westrum, E.F., Jr., and Sinke, G.C., "The Chemical Thermodynamic Properties of Organic Compounds" (J. Wiley & Sons, Inc., New York, 1969).

70COX/PIL Cox, J.D., and Pilcher, G., "Thermochemistry of Organic and Organometallic Compounds", (Academic Press, London, 1970).

70GOO Good, W.D., J. Chem. Thermodynam. 2, 237-244 (1970).

76BEN Benson, S.W., "Thermochemical Kinetics", Second Edition, (J. Wiley & Sons, Inc., New York, 1976).

77PED/RYL Pedley, J.B., and Rylance, J., "Sussex - Computer Analysed Thermochemical Data: Organic and Organometallic Compounds", (University of Sussex, School of Molecular Sciences, Falmer, Brighton, U.K., 1977).

84DOM/EVA Domalski, E.S., Evans, W.H., and Hearing, E.D., J. Phys. & Chem. Ref. Data 13, Supplement No. 1, 286 pp., (1984).

86PED/NAY Pedley, J.B., Naylor, R.D., and Kirby, S.P., "Thermochemical Data of Organic Compounds", Second Edition, (Chapman and Hall, London and New York, 1986).

88MAR Marsh, K.N., et al., "Selected Values of Hydrocarbons and Related Compounds", Thermodynamics Research Center, Texas A & M University, College Station, TX (loose-leaf data sheets, extant 1988).

Empirical Methods for the Calculation of Physicochemical Data of Organic Compounds

Johann Gasteiger

Institute of Organic Chemistry, Technical University of Munich, 8046 Garching, Federal Republic of Germany

Abstract:

Methods have been developed that allow the calculation of energetic and electronic effects in organic molecules. The values thus obtained can be used for the prediction of a variety of physical and chemical data of organic compounds.

Introduction:

The results presented here are an offspring of our work on the development of EROS (Elaboration of Reactions for Organic Synthesis), a program system for the prediction of chemical reactions and for the design of organic syntheses [1,2]. At the outset, we decided not to draw the reactions from a data base of known reactions stored in the computer but to generate reactions as formal bond and electron shifting processes. This allows to obtain not only known but also novel reactions.

The task is then to develop methods that automatically determine reactive bonds and give an estimate for their relative reactivity. To achieve that we embarked on a program to develop procedures for the quantification of chemical effects like charge distribution, inductive, resonance, and polarizability effect, as well as bond dissoziation energies.

In order to be able to calculate large data sets and sizeable molecules rigorous quantum mechanical methods had to be rejected. Rather, rapid empirical methods were developed for the calculation of the various chemical effects. Because of the empirical nature of these methods the significance of the calculated values has to be established. This was accomplished by comparing the results for a large selection of organic molecules with physical and chemical data of these compounds.

C. Jochum M. G. Hicks J. Sunkel (Eds.)
Physical Property Prediction
in Organic Chemistry

The Basis of the Models

The empirical methods that we have developed for the calculation of energetic and electronic effects in organic compounds all rest on two premises:

1. A structural formula is a valuable model for the representation of an organic molecule.

2. The influence of atoms on site-specific effects decrease with the number of bonds between the location of the effect and the atom considered.

Accepting these two assumptions, one is faced with the task of having to represent the atoms and bonds of a structural formula with appropriate parameters to calculate a specific effect. Furthermore, for site-dependent effects, the network of bonds has to be dealt with by the mathematical form of the procedure that combines the atomic parameters.

Four basic approaches to the solution of this issue have been explored. The algorithms developed for these methods are briefly explained in the following. Representative results obtained in these calculations are given to show the scope of the methods.

1. Additivity Scheme:
Heats of Formation; Heats of Reaction; Bond Dissociation Energies

It has long been known that many molecular properties can be estimated by summation of parameters assigned to the substructures of a molecule. A hierarchy of approximations can be established that depends on the size of the substructures [3]:

atomic parameters	zero-order approximation
bond parameters	first-order approximation
group parameters	second-order approximation

The number of parameters that has to be considered in an additivity scheme increases with the order of approximation chosen.

We developed a procedure for the calculation of heats of formation that works with parameters for bonds (1,2-interactions) and also considers next-nearest neighbors (1,3-interactions). [4] It can be shown that this approach is equivalent to a group additivity scheme.

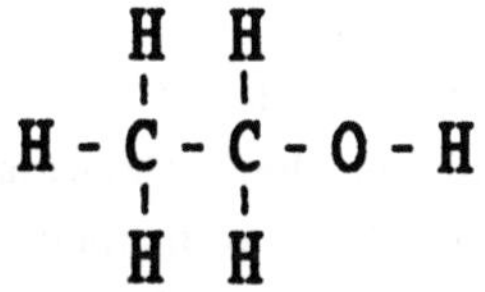

ΔH_f^o (exp) -56.24 kcal/mol

1,2 - interactions

1 C - C
5 C - H
1 C - O
1 O - H

ΔH_f^o (calc) -52.59

1,2 - + 1,3 - interactions

1 C - C
5 C - H
1 C - O
1 O - H
+ 1 C - C - O

ΔH_f^o (calc) - 56.19

Fig. 1. Calculation of the heat of formation of ethanol by consideration of 1,2-interactions (bond parameters) as well as of 1,2- and 1,3-interactions.

Fig. 1 shows how the estimate of the heat of formation of ethanol by a bond additivity scheme can be improved by inclusion of a parameter for a three-atom fragment. The parameters for 1,2- and 1,3-interactions have been obtained through multi-linear regression analyses on experimental heats of formation.[5)] A large variety of classes of compounds has been parameterized. Ring strain energies and aromatic delocalization energies are automatically taken care of.[6)] Table 1 gives some representative examples to assess the quality of the estimation.

compound	heat of formation (gas; 298 K; in kcal/mol) exp.	calc.
propane	-24.83	-25.28
cyclopropane	12.73	12.90
butanol	-48.94	-49.52
ethyl acetate	-106.34	-106.81
1,1,2-trichloroethane	-35.50	-36.28
allyl iodide	22.80	22.46
leucine	-116.34	-118.39
piperidine	-11.76	-11.11
aniline	20.81	18.71
trifluorotoluene	-143.20	-140.20
indene	44.60	45.38

Table 1. Experimental [7)] and calculated heats of formation

Furthermore, heats of reaction can easily be obtained from calculations of the heats of formation of starting materials and products of a reaction. The determination of parameters for free radicals allows the estimation of bond dissoziation energies. The example of Fig. 2 shows how even small variations in the dissociation energies of C-C and C-H bonds can thus be represented.

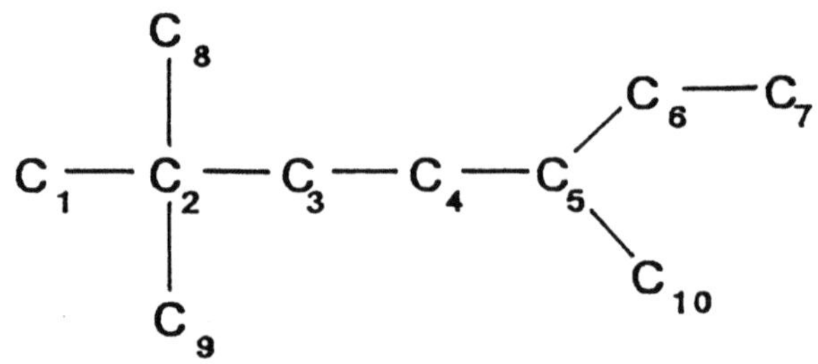

	calc	exp
C_1-H	98.7	98
C_3-H	94.9	95
C_5-H	92.5	92

	calc	exp
C_3-C_4	82.6	81.8
C_4-C_5	79.9	80.4
C_3-C_2	78.4	77.7
C_6-C_7	86.5	84.8
C_5-C_{10}	83.8	84.2
C_1-C_2	82.4	81.8

Fig. 2. Experimental [8] and calculated bond dissoziation energies in 2,2,5-trimethylheptane

2. An Iterative Walk through the Constitution of a Molecule: C-1s ESCA-Shifts, ^{1}H NMR Shifts, Dipole Moments

Quite a different look at a structural formula is taken in our method for the calculation of partial atomic charges in σ-bonded molecules [9]. The task is there to calculate a site-specific property, the partial charge on a specific atom. Thus, we have to account for the second assumption, mentioned above, that more remote atoms have lesser influence.

The basis of the model is the electronegativity concept. Values for the electronegativities of the orbitals of an atom in a given hybridization state can be obtained from valence bond ionization potentials and electron affinities. [10] The electronegativity of an orbital is dependent on the occupation number of that orbital, or, equivalently, on the charge in this orbital.

A polynomial of degree two is taken for this dependence, the coefficients a, b, and c can be determined from the ionization potentials and electron affinites in the neutral, positive, and negative state. On bond formation charge is shifted between the atoms of a bond. The amount of charge shift is dependent on the elctronegativity difference of the two atoms. The overall charge on an atom is obtained from the individual charge shifts in the bonds. The mutual dependence of electronegativity on charge and the charge shift on electronegativity is handled by an iterative prozedure, that is graphically represented in Fig. 3.

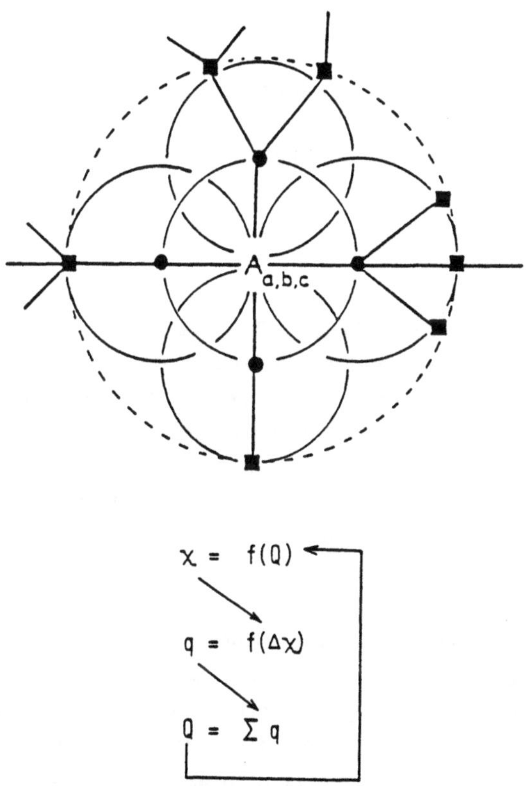

Fig. 3. Graphical representation of the method for partial equalization of orbital electronegativity (PEOE) [9]

Only the direct neighbors of an atom are considered in each iteration. However, this iterative nature has the effect that the influence of successive neighbor spheres is implicitly accounted for with each new iteration. After six iterations the charge shifts become so small that the procedure is stopped. Thus, the influence of atoms that are up to six bonds away from the atom considered is accounted for.

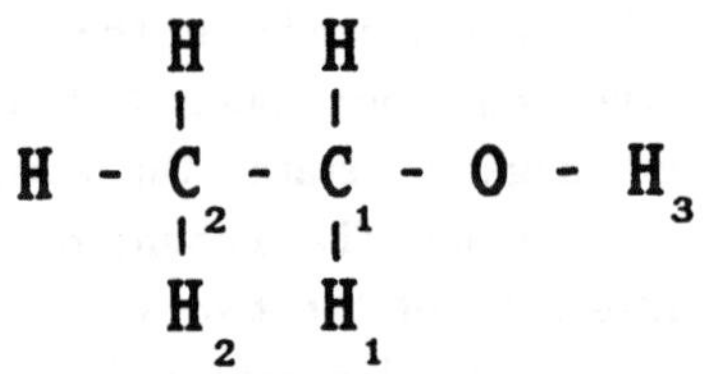

atom	q_σ	χ_σ
O	- 0.396	9.29
C_1	0.042	8.36
C_2	- 0.041	7.60
H_1	0.055	7.51
H_2	0.025	7.33
H_3	0.210	8.45

Fig. 4. σ-charge (q_σ; in electron units) and σ-electronegativites (χ_σ; in eV) obtained for ethanol

Fig. 4 shows the results for ethanol. The charge values obtained are an indication of the type of atom and its molecular environment. Thus, the two carbon atoms and the three sets of hydrogen atoms are each differentiated due to their different chemical environment. Because of the charge dependence of electronegativity, also the residual electronegativity values, χ_σ, of each atom are quite sensitive to the molecular structure. These electronegativity values are a useful quantitative measure of the inductive effect (vide infra).

The notion of partial charges on the atoms of a molecule has been of immense usefulness in chemistry. However, it is an artificial concept, values of partial charges are not directly measurable. On the other hand, several physical properties have been related to partial atomic charges. These physical properties have to be taken as a measure to assess the quality of any method for the calculation of partial charges.

It has been established that shifts in the core electron binding energies as measured by ESCA are rather directly depending on the valence electron distribution. We have therefore correlated the partial charges on the carbon atoms of a variety of organic compounds to the C-1s ESCA shifts. Fig. 5 shows that this correlation is quite a good one. [9] This establishes the physical significance of the partial charges calculated by our method.

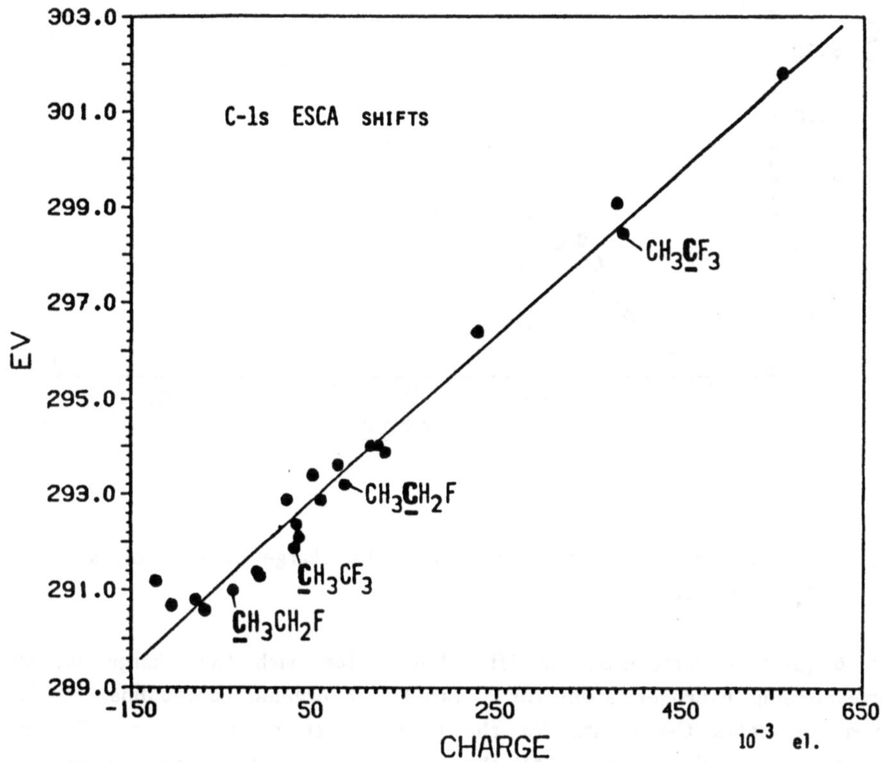

Fig. 5. Correlation of C-1s ESCA shifts with the partial charge on the carbon atom for a variety of organic compounds .

With this firm basis, the use of the partial atomic charges for correlating or predicting other physical data was investigated.

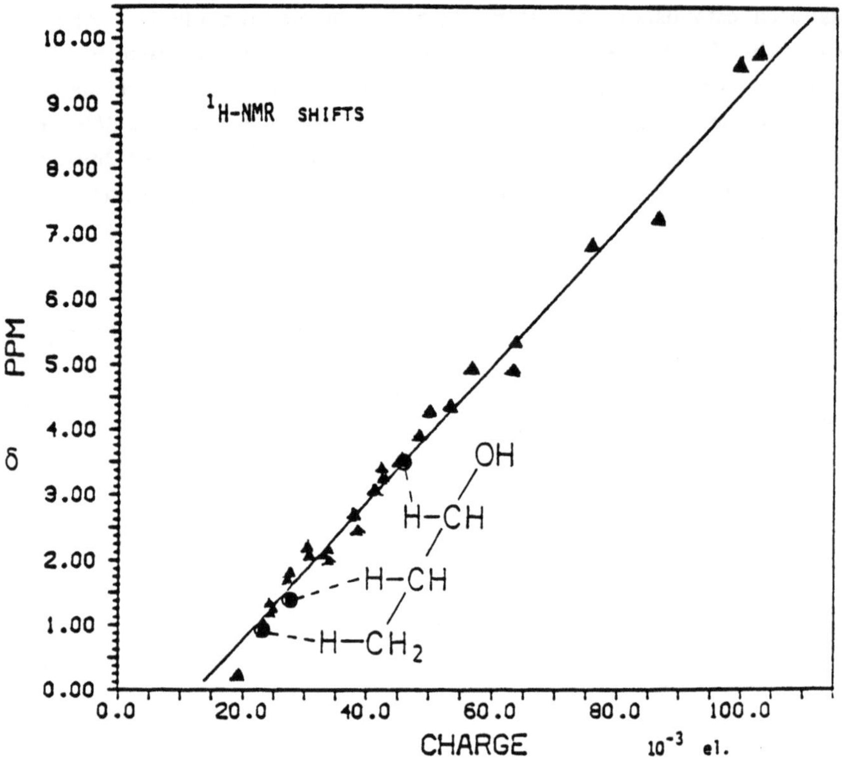

Fig. 6. Correlation of ^1H NMR-shifts with the charge on the corresponding hydrogen atom. [11]

Fig. 6 gives a correlation of ^1H NMR shifts with the charge on the corresponding hydrogen atom for a variety of organic compounds. [11] The figure highlights the points for the hydrogen atoms on the α-, β-, and γ-carbon atom of 1-propanol. All three points lie on the overall correlation line indicating that the magnitude of the inductive effect of the oxygen atom and its attenuation over two, three, and four bonds is quantitatively reproduced.

In another study, dipole moments of various organic compounds were calculated from the partial charges and experimental bond lengths. [12] Table 2A gives some representative examples.

As expected, the calculated values were consistently smaller than the experimental ones. This has to be attributed to the extra moment of free electron pairs. Proper allowance for that was made for halogen compounds. Those values are contained in Table 2B.

compound	dipole moment (in D) exp.	calc.
A) H_2O	1.85	1.15
HCl	1.08	0.89
H_2S	0.97	0.87
CH_3OH	1.70	1.55
CH_3NH_2	1.30	1.15
$(CH_3)_2NH$	1.01	0.83
B) CH_3CH_2F	1.96	1.99
CH_3CH_2Cl	1.75-2.09	2.01
CH_3CH_2Br	1.96-2.02	2.07
CH_3CH_2I	1.9	1.89

Table 2. Experimental and calculated values of dipole moments. Values in A) without, in B) with inclusion of free electron pair moments.

3. Beyond a Single Valence Bond Structure: ^{13}C NMR Shifts, C-1s ESCA Shifts, Dipole Moments

The reader might have realized that the PEOE method tacitly assumes that a chemical compound is well represented by a single valence bond structure. In conjugated systems this is no longer the case. Here, an extension of the method has been developed. First, the σ-charge distribution is calculated and then the relaxation of the π-electrons under the influence of the σ-charges is considered. To this effect, all those valence bond structures of the conjugated system are generated that have a maximum of two opposite charges. The π-charges are obtained by assigning weights to these valence bond structures that depend both on topological considerations and on π-electronegativities. [13]

Good correlations have been obtained between the total charges (σ and π) and C-1s ESCA shifts in fluoroalkenes as well as with ^{13}C NMR shifts in mono-substituted benzenes (Fig. 7). [13]

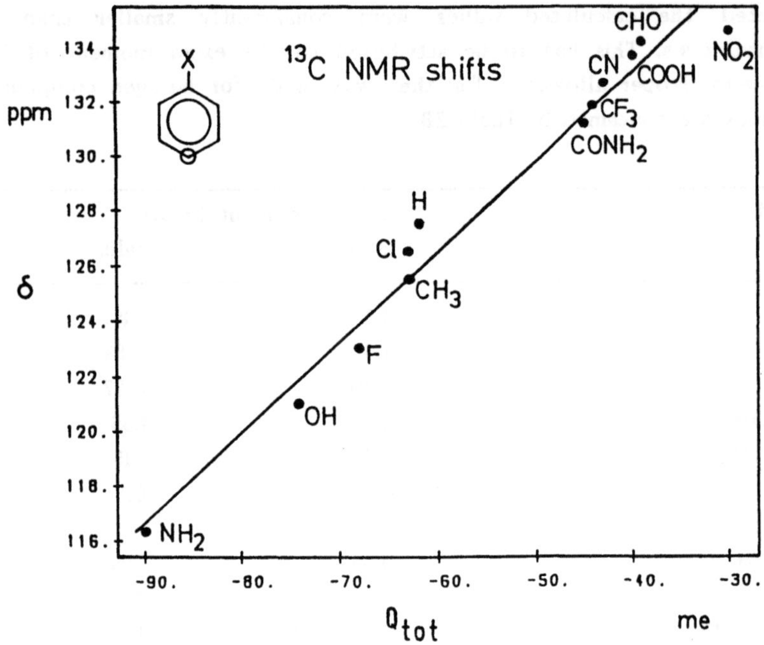

Fig. 7. Correlation between ^{13}C NMR shifts and total charge for the para-carbon of mono-substituted benzenes. [13]

In addition, good correspondence between experimental dipole moments and those calculated from the σ- and π-charges has been obtained. [13] **Three examples are given in Fig. 8.**

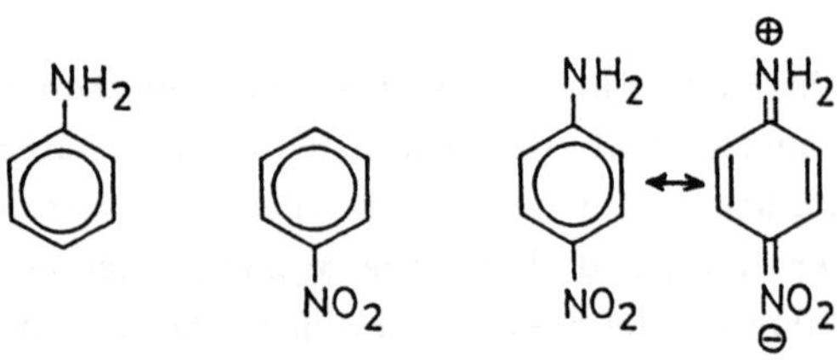

μ	exp.	1.15	4.16	6.91
(D)	calc.	1.18	4.11	6.83

Fig. 8. Experimental and calculated dipole moments for some substituted benzenes. [13]

It is particularly noteworthy, that the dipole moment of p-nitroaniline is larger than the sum of its components. This is attributed to the extra contribution of through-conjugation, an effect reproduced by our method.

4. An Attennation Model: Mean Molecular Polarizability, Auger Shifts, Proton Affinities

Yet another way of looking at a structural formula and accounting for the diminishing influence of more remote atoms on a localized effect was explored.

Mean molecular polarizability, $\bar{\alpha}$, can reasonably well be reproduced by an additivity scheme working with parameters, $\bar{\alpha}_i$, for atoms in different hybridization states. [14] Mean molecular polarizability is, however, not directly useful to represent the stabilizing effect of a polarizable polyatomic molecule on a charge introduced into this molecule. To this end we have defined an effective polarizability, α_d, and have given an algorithm for its calculation [15].

Fig. 9. graphically illustrates this method for the calculation of the effective polarizability of the nitrogen atom in 2-aminopropane as needed, e.g. when introducing a positive charge through protonation of the amino group.

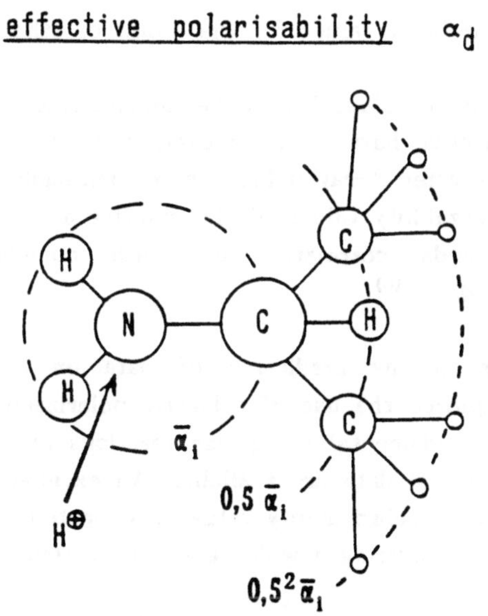

Fig. 9. Outline of the method for the calculation of effective polarizability. [16]

Parameters, $\bar{\alpha}_i$, are used that are typical for an atom in a particular hybridization state. These parameters are added in a way that accounts for the number of bonds between the site of the effect and the atom being considered. This has the result, as with the charge calculation, that the values obtained for an atom reflect its molecular environment. In Fig. 10 the results for each atom of ethanol are given.

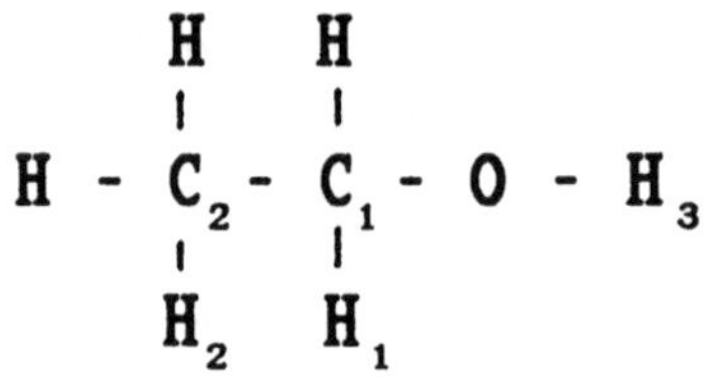

	α_d (in Å^3)
O	3.321
C_1	4.336
C_2	4.100
H_1	2.893
H_2	2.775
H_3	2.186

Fig. 10. Effective polarizability values for the atoms of ethanol.

The merit of the calculated values has to be demonstrated with experimental data. Polarizability effects have been invoked for the relaxation of the electrons in a molecule after X-ray induced core ionization of a molecule. In fact, the effective polarizability values of the chlorine atom for a series of 14 organochlorine compounds correlate well with combined ESCA-Auger photoelectron spectral data. [16]

Our main interest is in the prediction of data on chemical reactivity. Therefore, we have explored the use of effective polarizability values in this area, concentrating on fundamental gas phase reactions as there larger series of high quality experimental data are available. An excellent correlation was obtained between effective polarizability values and proton affinity data for a series of 49 alkyl amines covering a wide selection of structural changes (Fig. 11).

proton affinity (PA) of amines

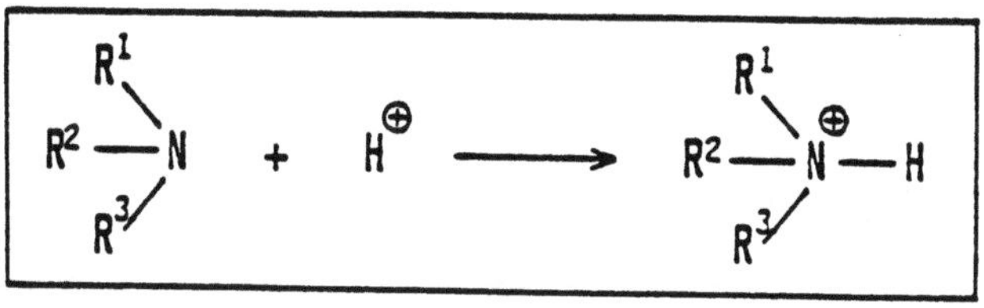

PA = - ΔH_r

alkyl amines (49 compounds)

e.g.

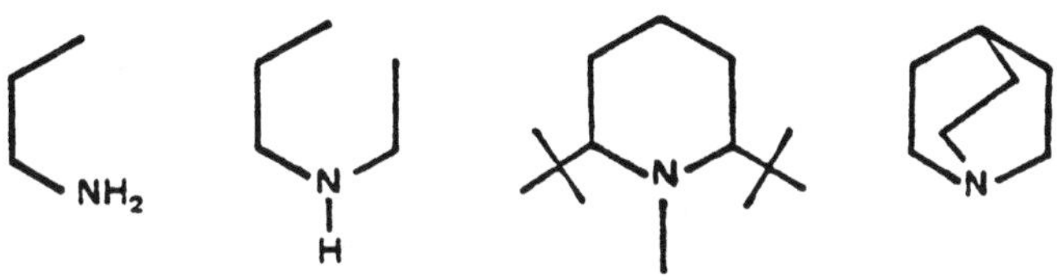

$$PA = 209.2 + 2.7\ \alpha_d$$

Fig. 11. Correlation between effective polarizability values and proton affinities of alkyl amines. [16]

This result merits some comments and reflections on how it was achieved. The parameters used in the attenuation model for calculating effective polarizability have been derived from data on mean molecular polarizability, $\bar{\alpha}$, through multi-linear regression analyses. For many important compounds no $\bar{\alpha}$-values could be found in standard reference tables. For a lot of those cases a look into Beilstein was helpful as there data on the refractive index, n_D, and the density, d, of these compounds could be found. This allowed, together with the value of the molecular weight, MW, the calculation of mean molecular polarizabilities, $\bar{\alpha}$, through the well-known Lorenz-Lorentz equation.

Thus, in effect, a way has been developed for these alkyl amines from such elementary data as refractive index, molecular weight, and density to gas phase reactivity data, i.e. proton affinity.

This path involves a series of well-defined equations and simple empirical models (Fig. 12.)

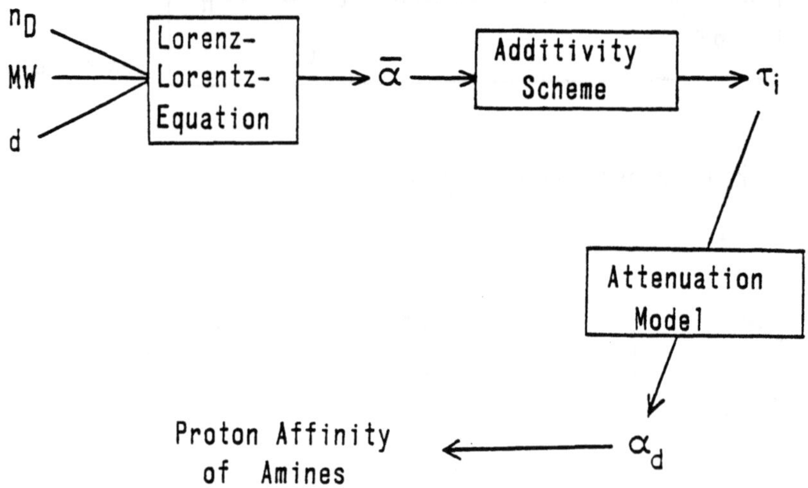

Fig. 12. The series of equations and models that lead from data on refractive index, n_D, molecular weight, MW, and density, d, to proton affinities.

It is clear, that online access to the data of Beilstein would have greatly facilitated the development of this path and its results. Furthermore, it opens the prospect to similar findings.

The Complexity of the World: **Multi-Parameter Applications**

In the preceding examples, the physical or chemical data have all directly been calculated from the values of a single chemical effects. This must be viewed more as the exception than the rule. Most physical or chemical data of organic compounds are under the simultaneous influence of several effects, they depend on more than one parameter.

As a case in point, let us look at the ^{13}C NMR shifts of the various halomethanes. Viewing these shifts against the charge on the carbon atom shows quite a vexing behaviour in the fluoro-, chloro-, bromo-, and iodo-compounds (Fig. 13).

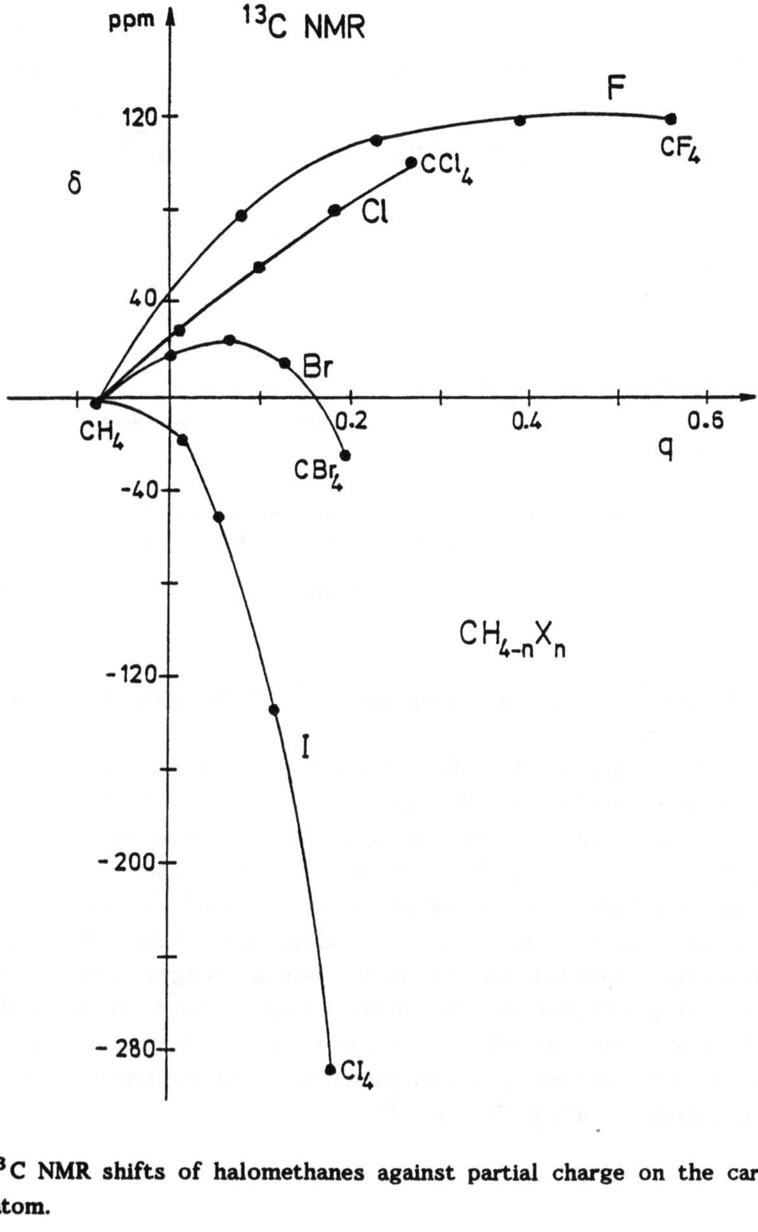

Fig. 13. ^{13}C NMR shifts of halomethanes against partial charge on the carbon atom.

However, a good correlation could be obtained, when the square of the mean molecular polarizability, $\bar{\alpha}$, was used as a second parameter (Fig. 14). [17] This equation could cover the entire range of ^{13}C NMR shifts form CH_3F to CI_4.

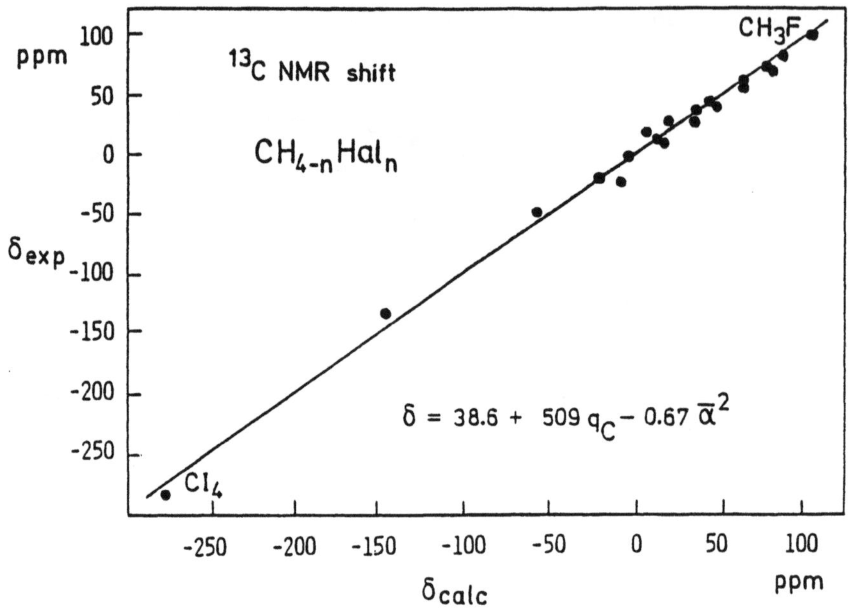

Fig. 14. An equation for the prediction of ^{13}C NMR shifts in halomethanes.[17)]

Use of the values obtained from the above empirical methods will have a wide range of applications. In line with our main interest we have largely concentrated on data on chemical reactivity. Extending the study on proton affinities of amines (Fig. 11) to those compounds that also bear strongly electron withdrawing groups asked for inclusion of the inductive effect. It was found that the values of the residual electronegativities, χ, that are obtained simultaneously with the partial atomic charges in the PEOE method are a good quantitative measure of the inductive effect. Thus, all the proton affinities of substituted alkyl amines available at the time of the study (80 data) could be handled by a two parameter equation based on the inductive and polarizability effect (Fig. 15).[18)]

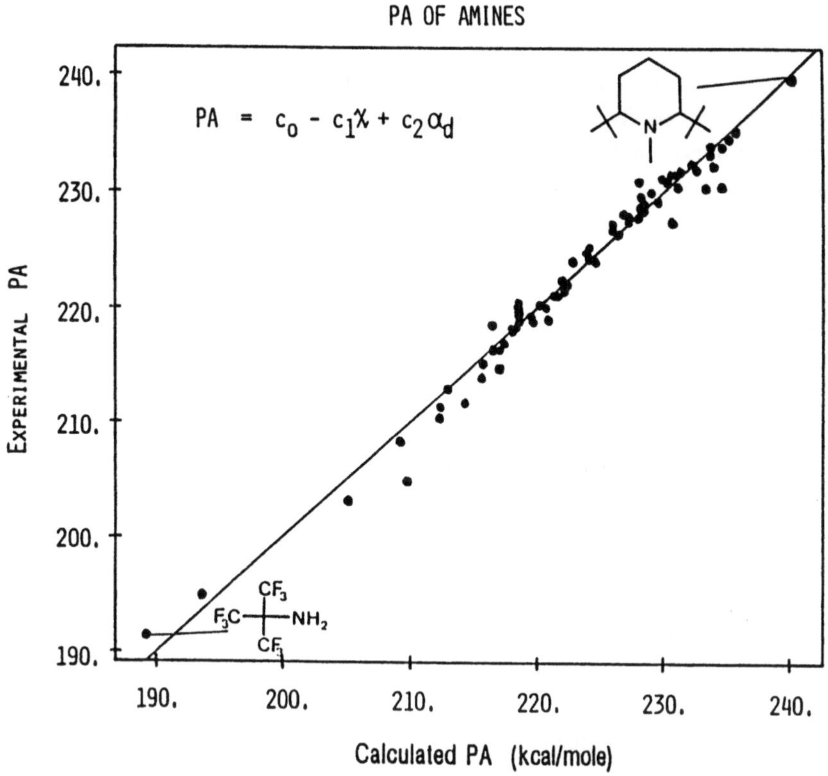

Fig. 15. Experimental and calculated proton affinities of amines (χ, residual electronegativity; α_d, effective polarizability). [18]

Similar results have been obtained for the proton affinities of alcohols and ethers as well as of thiols and thioethers, [19] **for the proton affinities of carbonyl compounds and gas phase acidities of alcohols.** [20] **These studies have been extended to solution data as was demonstrated with pK_a values of alcohols.** [21]

Conclusion

Quantitative values for important chemical effects have been made available through empirical methods. These procedures are very rapid, allowing the calculation of large molecules and big data sets. The properties that can be calculated include

atomic properties
σ- and π-partial charges
σ- and π-residual electronegativities
effective polarisabilities

bond properties
difference in σ- and π-partial charges
difference in σ- and π-residual electronegativities
bond polarities
resonance effect
bond polarizabilities
bond dissociation energies

molecular properties
mean molecular polarizability
heat of formation
strain and delocalization energy

These parameters are useful for the calculation of a wide variety of physical and chemical data through single or multi-parameter equations.

These parameters can thus be used in a factual data base for the prediction of unknown data as well as for checking data before storing them.

Acknowledgements

The work reported here rests on the endeavour of many dedicated persons. Their names are found in the references. I extend my thanks to them. Support of this work by Deutsche Forschungsgemeinschaft, Imperial Chemical Industries, plc, England, Sumitomo Chemical Co., Ltd, Japan and Tecnofarmaci, S.p.A.., Italy is gratefully acknowledged.

References

1) J. Gasteiger, C. Jochum
Topics Curr. Chem. 74, 93-126 (1978)

2) J. Gasteiger, M. G. Hutchings, B. Christoph, L. Gann, C. Hiller, P. Löw, M. Marsili, H. Saller, K. Yuki
Topics Curr. Chem., 137, 19-73 (1987)

3) S. W. Benson, J. H. Buss
J. Chem. Phys. 29, 546 (1958)

4) J. Gasteiger
Tetrahedron 35, 1419-1426 (1979)

5) J. Gasteiger, P. Jacob, U. Strauß
Tetrahedron 35, 139-146 (1979)

6) J. Gasteiger, O. Dammer
Tetrahedron 34, 2939-2945 (1978)

7) J. D. Cox, G. Pilcher
Thermochemistry of Organic and Organometallic Compounds, Academic Press, New York, 1970

8) D. F. Mc Millen, D. M. Golden
Ann. Rev. Phys. Chem. 33, 493 (1982)

9) J. Gasteiger, M. Marsili
Tetrahedron 36, 3219-3228 (1980)

10) J. Hinze, H. H. Jaffe
J. Am. Chem. Soc. 84, 540 (1962)

11) J. Gasteiger, M. Marsili
Org. Magn. Resonance 15, 353-360 (1981)

12) J. Gasteiger, M. D. Guillen
J. Chem. Research (S) 1983, 304-305; (M) 1983, 2611-2624

13) J. Gasteiger, H. Saller
Angew. Chem. 97, 699-701 (1985)
Angew. Chem. Intern Ed. Engl. 24, 687-689 (1985)

14) Y. K. Kang, M. S. Ihon
Theoret. Chim. Acta 61, 41 (1982)

15) J. Gasteiger, P. Löw, unpublished.
This procedure is different from the originally published method.[16] The modification has been made to repair a deficiency of the original method in accounting for the effect of atoms that are more than four bonds away from the site being considered. Apart from that the results are very similar.

16) J. Gasteiger, M. G. Hutchings
J. Chem. Soc. Perkin 2, 1984, 559-564

17) J. Gasteiger, I. Suryanarayana
Magn. Reson. Chem. 23, 156-157 (1985)

18) M. G. Hutchings, J. Gasteiger
Tetrahedron Lett. 24, 2541-2544 (1983)

19) J. Gasteiger, M. G. Hutchings
J. Amer. Chem. Soc. 106, 6489-6495 (1984)

20) M. G. Hutchings, J. Gasteiger
J. Chem. Soc. Perkin 2, 1986, 447-454

21) M. G. Hutchings, J. Gasteiger
J. Chem. Soc. Perkin 2, 1986, 455-462

Statistical Thermodynamics: Current Perspectives and Limitations of Fluid Property Estimation

K. Lucas

GHS Duisburg, 4100 Duisburg, Federal Republic of Germany

1. Introduction

Statistical thermodynamics provides a simple formal connection between the thermodynamics of a system, as represented by its free energy A, and the molecular properties of a system, as represented by its canonical partition function Q [1]:

$$A^{(T,V,\{N_j\})} = - kT \ln Q(T,V,\{N_j\}) \quad , \qquad (1)$$

where k is Boltzmann's constant, T is the thermodynamic temperature, V the volume, $\{N_j\}$ the total amount of molecule numbers of the various components, and

$$Q = \sum_i e^{-E_i/kT} \quad , \qquad (2)$$

Here E_i, the molecular energy of the system in its quantum state i, is the key quantity. It contains all the molecular information like molecular properties of single molecules, intermolecular forces etc. Since it is impossible to solve the Schrödinger equation for E_i and perform the summation of equ. (2) for the general case, the applications of statistical thermodynamics originate from various approximations to

C. Jochum M. G. Hicks J. Sunkel (Eds.)
Physical Property Prediction
in Organic Chemistry

equ. (2). It is convenient to look at these approximations separately for the various regions of the state diagram.

2. The Ideal Gas

The ideal gas is a particularly important model system. Its molecules have no volume and exert no intermolecular forces upon each other. Thus the molecular situation of an ideal gas consisting of Argon atoms, Hydrogen Chloride molecules and Chloride molecules resembles the one shown in fig. 1. Any molecular interactions as shown in the upper right are extremely unprobable. The thermodynamic functions of the ideal gas

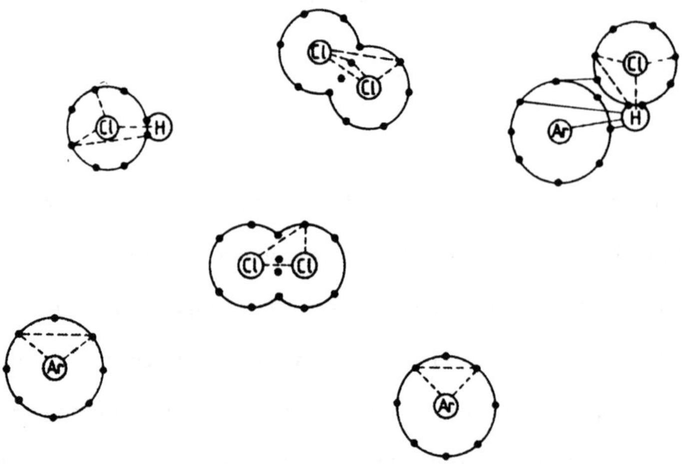

Figure 1 Molecular situation of an ideal gas

describe correctly some aspects of real gas behaviour in the limit of low density. Due to the independence of the single molecules the canonical partition function can be written as

$$Q^{id} = q^N \frac{1}{N!} \quad , \tag{3}$$

where

$$q = \sum_i e^{-\varepsilon_i/kT}$$

is the molecular partition function defined in terms of the quantum energy values ε_i of a single molecule. Since even a single molecule is a complicated system for quantum mechanical treatment, the total energy of a molecule is split into various parts which are assumed to be independent.

$$\varepsilon = \varepsilon_{tr} + \varepsilon_{ntr} = \varepsilon_{tr} + \varepsilon_r + \varepsilon_v + \varepsilon_{ir} + \varepsilon_{el} \tag{4}$$

Here, ε_{tr} represents the translational energy of a molecule which is entirely kinetic. The associated molecular partition function is

$$q_{tr} = \left(\frac{2\pi\, m\, kT}{h^2}\right)^{3/2} V \tag{5}$$

which leads to the famous ideal gas equation of state:

$$p = nkT \quad , \tag{6}$$

where $n = N/V$ is the number density.

The energy ε_{ntr} summarizes the non-translational contributions to the energy of the individual molecules, which do not influence the equation of state. They can be split to a high degree of accuracy into a contribution of rotation (ε_r), vibration (ε_v), internal rotation (ε_{ir}) and electronic energy (ε_{el}). At very high temperatures and for very high accuracy it may be necessary to consider coupling effects as a correction to the simple split of equ. (4). Rotational and vibrational contributions are easily treated by the rigid rotator/harmonic oscillator approximation. To evaluate these contributions for a particular molecule we need its structure and its vibration frequencies. It is a standard procedure to calculate the principal moments of inertia I_A, I_B and I_C of a known molecular structure. If we have them the molecular partition function of rotation is easily calculated from

$$q_r = \frac{\pi^{1/2}}{\sigma_r} \sqrt{\frac{T^3}{\Theta_{r_A} \Theta_{r_B} \Theta_{r_C}}} \qquad (7)$$

where

$$\Theta_{r_A} = \frac{h^2}{8\pi^2 I_A k}$$

is one of the characteristic temperatures of rotation with similar definitions for Θ_{r_B} and Θ_{r_C}, h is Planck's constant and σ_r the symmetry number of rotation, which is equal to the number of rotational arrangements of a molecule that represents the same quantum state. The fundamental vibration frequencies $\nu_{o,j}$ of a molecule can be used to define characteristic temperatures of vibration, $\Theta_{v,j} = h\nu_{o,j}/k$, in terms of which the molecular partition function of vibration reads:

$$q_V = \sum_j \frac{1}{1 - e^{-\Theta_{v,j}/T}} \tag{8}$$

The electronic energy can in most cases be included in the zero point of energy. A particularly important contribution for the larger organic molecules is the contribution of internal rotation. Let us consider a molecule like propanol(2) (C_3H_8O) as shown in fig. 2. Due to its 12 atoms this molecule

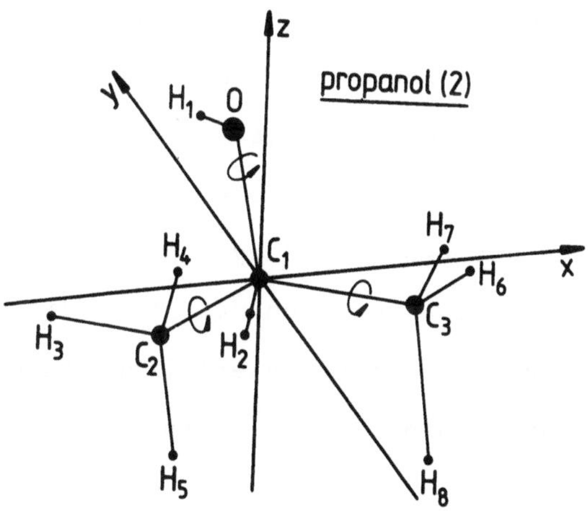

Figure 2 Internal rotation in the propanol(2) molecule

will have 36 degrees of freedom. We have 3 translational, 3 rotational and a rest of 30 internal degrees of freedom connected to the non-rigidity of the propanol(2) molecule. Looking at the molecular structure we see that 3 internal rotations will have to be considered, two CH_3-rotators and one OH-rotator. The molecular partition function of free internal rotation is given as

$$q_{ir} = \frac{1}{\sigma_{ir}} \sqrt{\left(\frac{T}{\Theta_{ir}}\right)} \quad , \qquad (9)$$

with $\Theta_{ir} = h^2/8\pi^2 \, k \, I_{ir}$ as the characteristic temperature of internal rotation, σ_{ir} the symmetry number of internal rotation and

$$I_{ir} = I_1 \left[1 - I_1 \left(\frac{\cos^2\alpha}{I_A} + \frac{\cos^2\beta}{I_B} + \frac{\cos^2\gamma}{I_C}\right)\right] \quad , \qquad (10)$$

where I_1 is the moment of inertia of the rotational group, and α, β, γ are the angles between the rotational axis and the principal axes of inertia of the molecule. Equ. (10) is easily evaluated for a given molecular structure. Generally, internal rotation is not free but restricted by a potential barrier and the molecular partition function must be modified accordingly. Since three modes of internal rotation have to be taken into account, the number of vibrational modes reduces to 27.

The properties of organic systems in the ideal gas state can be calculated with remarkable accuracy from statistical thermodynamics if the pertinent molecular properties are available. Such molecular properties are normally found from spectroscopic investigations. When molecules become more complicated, it becomes impossible to make proper assignments of vibrational fundamental frequencies. It is then more promissing to assume a model for the intramolecular force field and determine its force constants from measured frequencies of other molecules. Calculations of this kind have been performed for many molecules and calculated frequencies are stored in various references [2]. Structural data of the molecules are often available, but barriers for internal rotation must either be estimated from data on other molecules or from heat capacities or entropies. It can be expected that today's ab-initio calculations will provide us with the pertinent molecular data for very complicated organic molecules in the near future [3]. Summarizing it can be said that early applications of statistical thermodynamics to the ideal gas thermodynamic functions have been extended from simple anorganic molecules to quite complicated organic systems and that this process is expected to continue in the future to more and more complicated molecules.

3. Intermolecular Forces

If we apply some sufficiently high pressure to the ideal gas, we observe that its properties change in a way that is not consistent with the laws of an ideal gas. The ideal gas has become a real gas. Fig. 3 shows the molecular situation of such a real gas. In addition to the various forms of molecular energy associated with the individual molecules which are already present in the ideal gas calculations, a further type

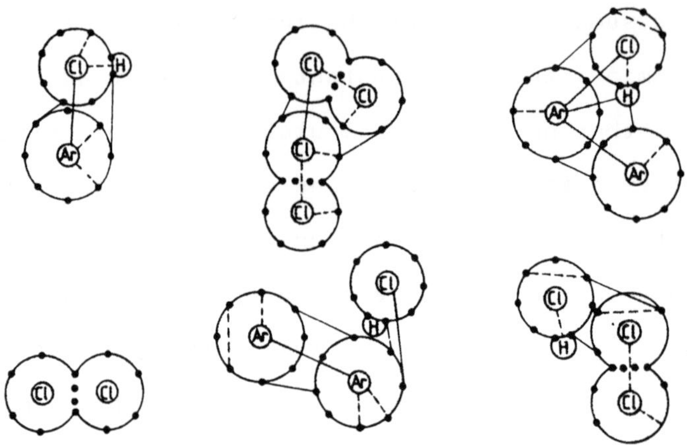

Figure 3 Molecular situation in a real gas

of molecular energy now becomes evident. It is symbolized by tie-lines between different molecules in fig. 3 and is associated with the intermolecular forces between them. In the typical range of real gases only two or at most three molecules approach each other simultaneously with sufficient probability such that the associated intermolecular energy becomes important. If we can assume that the molecules are rigid, i.e. that the vibration frequencies are independent of density, we can describe the intermolecular potential energy U by a function that only depends on external molecular configuration coordinates like distances of molecular centers and orientations of the molecules, i.e. $U = U(\bar{r}^N \omega^N)$. The precise form of this intermolecular energy function is crucial for the quality of the calculated thermophysical properties of the gas. This becomes even more so when by further increasing the density we reach the state of a liquid or a dense gas. The molecular situation is then as shown in fig. 4. The molecules are now closely packed together, and every molecule feels the intermolecular forces of many others.

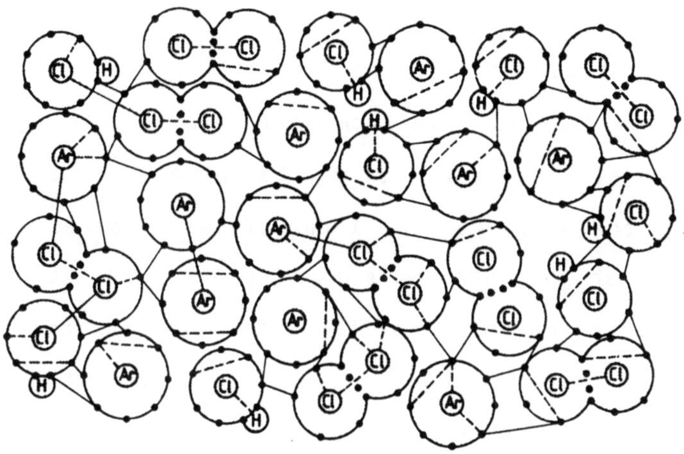

Figure 4 Molecular situation in a liquid

How can we model the intermolecular forces between molecules properly? Quantum mechanical perturbation theory gives us a good deal of information at long range. We find that the intermolecular energy function can be formulated in terms of pair and three-body potentials containing molecular properties like multipole moments, polarizabilities etc. For example, the interaction between two dipoles is found to be

$$\phi^{\mu\mu}_{\alpha_i\beta_j} = -\frac{\mu_\alpha \mu_\beta}{r^3_{\alpha_i\beta_j}} [2\cos\vartheta_{\alpha_i} \cos\vartheta_{\beta_j} - \sin\vartheta_{\alpha_i} \sin\vartheta_{\beta_j} \cos\varphi_{\alpha_i\beta_j}] \tag{11}$$

The full pair potential at long range for non-linear molecules is a rather complicated expression. Frequently there is a lack of information about molecular constants like quadrupole moment and polarizability tensor. Still, even an incomplete model of the intermolecular forces is better than no information at all. Further, there is the expectation that these parameters may soon be available from ab-initio calculations [3]. At short range, we must find a way to introduce molecular shape into the model for the intermolecular forces. mation at all. At short range, we must find a way to introduce molecular shape into the model for the intermolecular forces. This can be done by placing centers of repulsion into the molecule, usually close to the atomic sites. A r^{-12}-power law can be assumed to hold between these repulsive sites. I refer to the superposition of the quantum mechanical long range terms with the site-site short range terms as to the SSR-MPA potential model. Such a model contains three parameters that must be fitted to experimental data in addition to a number of molecular constants that can be taken from independent sources.

4. Real Gases

When intermolecular forces are present the evaluation of the canonical partition function becomes much more complicated. For real gases we can expand the thermodynamic functions around the ideal gas state. The equation of state, i.e. the virial equation, then has the form:

$$\frac{p}{nkT} = 1 + B(T,\{x_i\})n + C(T,\{x_i\})n^2 + \ldots \tag{12}$$

For rigid molecules the second virial coefficient B is related to the pair potential by

$$B_{\alpha\beta} = - \frac{2\pi}{\int d\omega_1 d\omega_2} \int_{\omega_1} \int_{\omega_2} \int_0^\infty (e^{-\phi_{\alpha\beta}(r\omega_1\omega_2]/kT} - 1)\, d\omega_1\, d\omega_2\, r^2\, dr \tag{13}$$

and

$$B = \sum_\alpha \sum_\beta x_\alpha\, x_\beta\, B_{\alpha\beta} \tag{14}$$

Similar, yet much more complicated expressions hold for the third virial coefficient. Other properties of real gases can be calculated from the virial coefficients and are thus also probes of the intermolecular pair potential. In order to evaluate the second virial coefficient B for a non-linear molecule, we need an expression for the pair potential $\phi_{\alpha\beta}$ and then must perform a six-fold numerical integration. This can be done with standard computers without much difficulty. Fig. 5 shows some results on organic refrigerants R12, R22 and R23 [4]. The procedure to produce these results has been

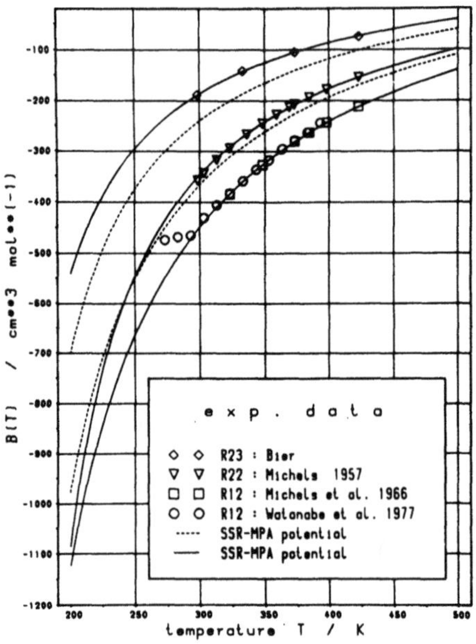

Figure 5 Virial coefficient B(T) in the systems R12, R22 and R23. The dashed lines denote the predicted interaction coefficients.

to fit the three adjustable parameters of the SSR-MPA potential model to data of the second virial coefficient and the Joule-Thomson coefficient in a limited temperature range [1]. The data are of course accurately reproduced and the extrapolations made to lower and higher temperatures are considered to be reliable. Then, without a further use of data, the unlike interaction virial coefficients have been predicted by assuming universal combination rules for the potential parameters. Very few data are available for the second virial coefficients of mixtures, and this is the usual situation. One of the main objectives of such calculations is to perform reliable extrapolations from pure fluids to mixtures. Typically such extrapolations are difficult for molecules that differ considerably in their intermolecular force fields. It is also possible to give estimates of the transport properties. Although the kinetic theory of polyatomic gases is not in a well-developed state yet, it turns out that by slightly adjusting the potential parameters for the pure components rather accurate predictions for the gas mixture can be made. At the moment, there remain considerable difficulties in the calculation of the third virial coefficient since a 12-fold integration has to be performed there which leads to the limits of even high-speed computers.

5. Liquids

For liquids an expansion of the thermodynamic properties in terms of density is not possible and other perturbation expansions have been studied. The canonical partition function reads in the semi-classical approximation for rigid molecules:

$$Q = \frac{1}{N!} Q^{id} Z \quad , \qquad (15)$$

where Q^{id} represents the contribution considered before for the ideal gas state and

$$Z = \frac{1}{\int d\omega^N} \int e^{-U(\bar{r}^N \omega^N)/kT} \, d\bar{r}^N \, d\omega^N \qquad (16)$$

is the configuration integral. Quantum effects can be included by additional terms, but normally do not contribute much for organic molecules, except for say CH_4 and C_2H_6. The configuration integral has such a high dimension that a direct numerical solution is out of question. However, an essentially exact calculation is possible by computer simulations. In these computer simulations a limited number of molecules in a box is considered, between 100 and 1000. Two methods are available, molecular dynamics and Monte Carlo. We consider here only Monte Carlo simulations, since these are particularly suited for the thermodynamic functions, pressure and internal energy. The expression to be considered for the configurational part of the internal energy is:

$$U^c = \frac{\int U(\bar{r}^N \omega^N) \; \exp(-U(\bar{r}^N \omega^N)/kT) \; d\bar{r}^N \; d\omega^N}{\int \exp(-U(\bar{r}^N \omega^N)/kT) \; d\bar{r}^N \; d\omega^N} \qquad (17)$$

The Monte Carlo method selects various configurations, evaluates the intermolecular energy function for them and adds the results up to the value of the integral. The crucial step is the generation of configurations to be evaluated. In the theory of Markov chains it is shown that it is possible to generate such configurations with a given normalized probability distribution. In the present case, configurations are generated according to $\exp(-U/kT)/\int \exp(-U/kT)\, d\Gamma$. In order to keep the computing time in sensible limits, a limited number of molecules is looked at and particular measures are taken so that this small ensemble gives realistic estimates of the true molecular system. In the liquid range, small inaccuarcies of the intermolecular energy function have dramatic effects on the thermodynamic functions. At the same time extensive fitting of potential parameters is not possible due to excessive demands for computer time for such calculations. Thus, only for very few systems calculations of this kind have been carried out with good technical accuarcy. Typical applications have been the noble gases and, more recently, some linear molecules like CO_2 [5]. It seems to be possible to generate data over the whole region of state starting from very few experimental data to fit the potential parameters. Non-linear organic molecules have not been treated by computer simulations over the whole density range, although calculations in limited regions have been performed with simple model potentials. For mixtures only very limited calculations on simple model systems have been published. Still it is to be expected that these computer simulations will play an increasingly important role in the generation of data for liquids from potential models in the near future.

Particularly for mixtures a perturbation expansion around a spherical reference system has proved to be rather successful. Phase diagrams in fluid mixtures are very much influenced by polar properties of the molecules and these can well be introduced by a pertubation expansion of the kind

$$A = A^{ref} + A^{(1)} + A^{(2)} + \ldots \tag{18}$$

with

$$\phi = \phi^{ref} + \lambda\phi^{p} \tag{19}$$

and

$$A^{ref} = A\,(\phi^{ref}) \quad ,$$

which is considered to be known, and

$$A^{(\nu)} = \left(\frac{\partial^{\nu} A}{\partial \lambda^{\nu}}\right)_{\lambda=0} \quad ,$$

which can easily be calculated from the known properties of the reference system. In fig. 6 we show the results for the vapor-liquid equilibrium of the system C_2H_6 + CO_2, as calculated from a pure Argon equation of state as a reference and including essentially the quadrupoles of the pure components. The agreement of the molecular model with the data is quite good, contrary to the simple Argon corresponding states model without molecular parameters. Calculations of this kind have been performed for many systems and are generally more reliable than calculations with purely empirical methods.

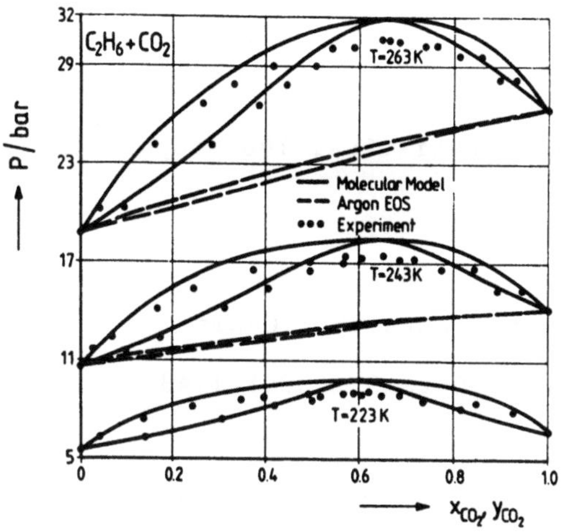

Figure 6 Vapor-liquid equilibrium in the system $C_2H_6 + CO_2$

All these applications of statistical thermodynamics are limited to rigid molecules for which we can derive sensible models of the intermolecular forces. It is expected that such calculations will be extended to more complicated rigid molecules in the future. It is not clear at the moment, whether statistical thermodynamics based on explicit laws for the intermolecular forces can also contribute to flexible molecules in a similar way. Today rather more empirical methods are applied to them.

6. References

[1] K. Lucas: Applied Statistical Thermodynamics (in German) Springer 1986

[2] Shimanouchi, T.; Matsuura, H.; Ogawa, Y.; Harada, F.: J. Phys. Chem. Ref. Data 9 (1980) 1149

[3] Hehre, W.J.; Radom, L.; Schleyer, P.V.R.; Pople, J.A.: Ab Inition Molecular Orbital Theory. John Wiley & Sons, New York 1986

[4] Ameling, W.; Ripke, M.; Lucas, K.: Int. J. Thermophysics, to be published

[5] Luckas, M.; Lucas, K.: Fluid Phase Equilibria, submitted

One and Multidimensional Numerical Interpolation Methods

Peter Jochum

Softron GmbH, Rudolf-Diesel-Strasse 1, 8032 Gräfelfing, Federal Republic of Germany

1. General Interpolation Problem

Given the interpolation points (x_i, f_i) ("knots") we search for coefficients a_i satisfying the <u>interpolation condition</u> [6]:

(1) $I(a_o, \ldots, a_n ; x_i) = f_i$, $i = 0, \ldots, m$ $(m = n)$, where

$x_i \in \mathbf{R}^N$ and $f_i \in \mathbf{R}$ ($\mathbf{R}$ = real space).

1.1 Classification

The interpolation problem is characterized by different properties which determine its degree of sophistication:

1.1.1 One versus Multidimensional

$N = 1$: 1-dimensional: x is a simple real variable. Its range is an interval. $I(a_o, \ldots, a_n ; x)$ is a curve.

$N > 1$: Multidimens.: x is an N-dimensional real vector ranging in a polygonal domain. $I(a_o, \ldots, a_n ; x)$ is a hyper surface. The number of coefficients is <u>independent</u> of the dimension!

1.1.2 Linear or Nonlinear

This property describes whether or not the mapping:

(2) $a_o, \ldots, a_n \;\; \text{-->} \;\; I(a_o, \ldots, a_n ; x)$

is linear. It does <u>not</u> specify whether the function $I(\ldots ; x)$ is linear!

C. Jochum M. G. Hicks J. Sunkel (Eds.)
Physical Property Prediction
in Organic Chemistry

1.1.3 Overdetermined Interpolation Condition (m > n)

In Practice m >> n (by a factor of 5 to 10). The interpolation condition (1) cannot be satisfied any more. It is replaced by an approximation condition ("fit"):

$$(3) \quad g(a_o,\ldots,a_n) = \sum_{i=0}^{m} (I(a_o,\ldots,a_n; x_i) - f_i)^2 \rightarrow \underset{a \in \mathbf{R}^N}{\text{Minimum !}}$$

1.2 Origin

The "origin" specifies whether the interpolation function I(...) is derived from "heuristic considerations" or from a systematic "physical model". The approach considerably determines the stability of results versus measurement or systematic errors. If possible a physical model is superior because in general fewer parameters must be introduced to fit the data.

Examples:

1.2.1 Heuristic Approach

RIA (Radio Immuno Assay):	Smoothing splines or "logit-log" functions for calibration.
Chromatography:	"Skewed exponential fit".
Image analysis:	Fourier transform

1.2.2 Systematic Approach

Beer's law for the determination of conc.:	A linear polynomial.
Mass action law:	A rational or square root function.

2. Examples from Different Classes

2.1 Linear 1-Dimensional Interpolation

An analyst wants to compensate for compressibility in HPLC. As solvent he uses a mixture of (a) water and (b) methanol. When running a gradient (e.g. %B = 0...100%) the compressibility C changes by an order of magnitude. To compensate he must know C for all %B-values. This is done by a heuristic approach:

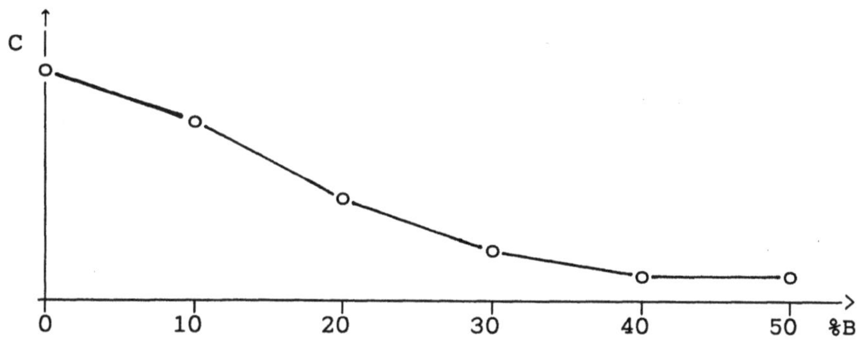

Notation: $x := \%B$, $f := C$, $n = 5$, $N = 1$

2.1.1 Polynomial Interpolation

Find $a_o, \ldots, a_n$ such that:

$$(4) \quad f(x) = \sum_{i=0}^{n} a_i x^i \quad \text{and} \quad f(i * 10) = C_i, \; i=0,\ldots,n.$$

The dependence upon the a_i is linear but the function $f(x)$ is a polynomial of degree 5 and, therefore, nonlinear!

Solution: Explicit by the formulas of NEWTON or LAGRANGE.

2.1.2 Spline Interpolation

A spline is an interval-wise given function which on each interval coincides with a polynomial of a certain degree (the polynomials are generally different on each interval). Like polynomials, splines depend linearly upon their coefficients.

<u>Linear (polygonal) splines</u>:

(4) $s(x) = \sum_{i=0}^{n} a_i \, l^i(x)$

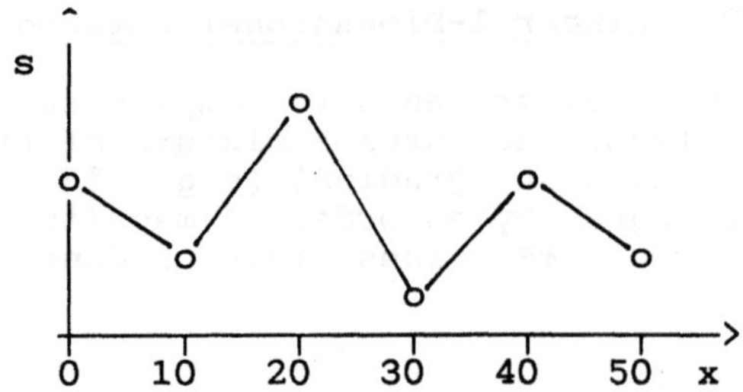

(5) $l_i(x)$:

"Basis function"

(Lagrange)

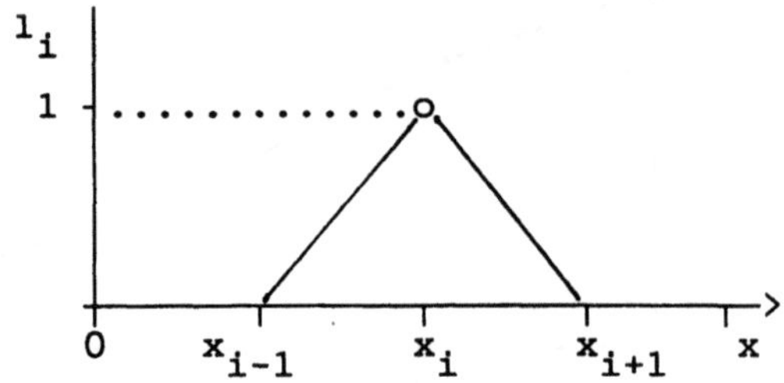

Solution: Explicit (piecewise linear)
Evaluation: Finding interval + 1 Operation per point.

<u>Cubic splines</u>:

(6) $s(x) = \sum_{i=0}^{n} a_i \, b^i(x)$

s is twice differentiable,

$s \,|[x_{i-1}, x_i] \in P_3$.

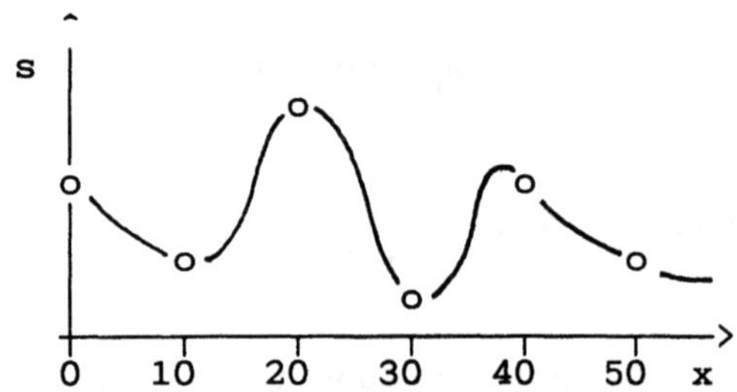

(7) $b_i(x)$:

"Basis functions",

each covering 4 intervals.

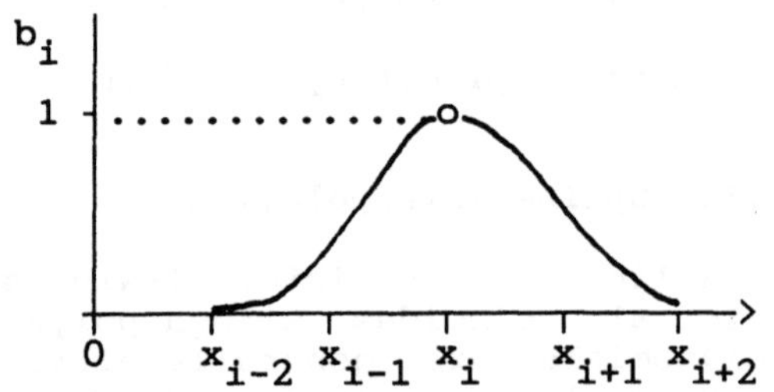

<u>Solution</u>: Linear tridiagonal system of n linear equations. <u>All</u> knots must be known <u>before</u> the spline can be computed (contrary to linear splines).

Evaluation: Finding interval + 3 Operations per point.

2.2 Nonlinear Interpolation Problems

2.2.1 Nonlinear 1-Dimensional Problem

Consider the chemical reaction [2]:

$$(8) \qquad P + Q \underset{k'}{\overset{k}{\longleftrightarrow}} PQ$$

where:
p = total concentration of P
q = total concentration of Q
B = concentration of the product PQ ("Bound")
k = forward velocity
k'= backward velocity

Mass action law:

$$(9) \qquad \frac{(p - B)(q - B)}{B} = K = \frac{k}{k'} = \text{Dissociation const.}$$

Solving for B:

$$(10) \qquad B(K,q;p) = \tfrac{1}{2}\{K + p + q - [(K + p + q)^2 - 4pq]^{\frac{1}{2}}\}$$

Given (measuring) $B_o = B(K,q;\ p_o)$ and $B_1 = B(K,q;\ p_1)$ we are looking for the parameters K and q.

<u>Solution</u>: B depends <u>nonlinearly</u> on K and q. But K and q can still be computed through a linear system of equations (what is unusual!): Let $c_i = B_i/(p_i - B_i)$:

$$(11) \qquad q - c_o K = c_o, \quad q - c_1 K = c_1 \qquad (\text{"Scatchard"})$$

Evaluation: 4 operations and a square root.

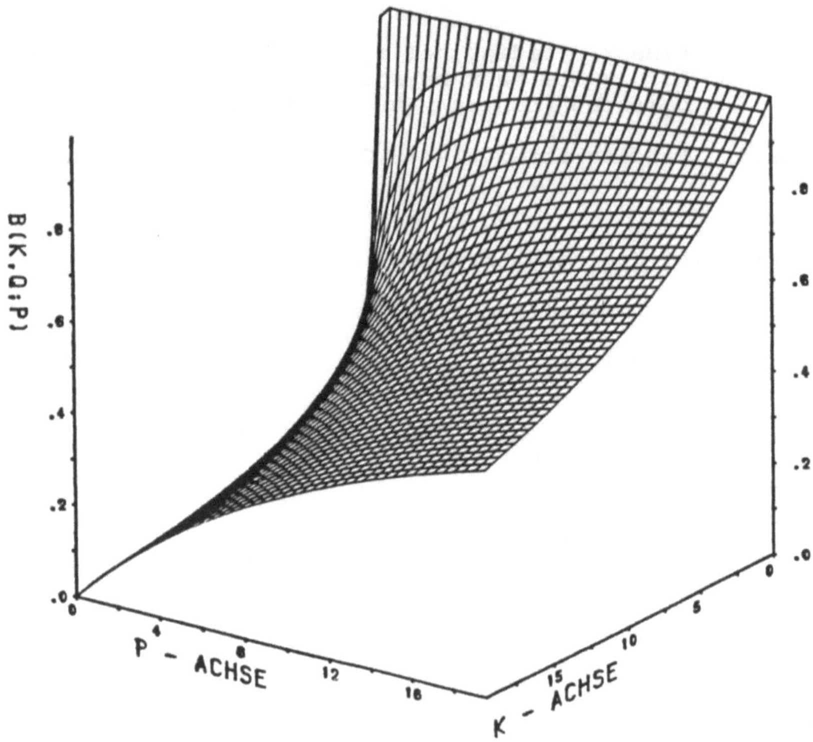

Fig. 1: B as a function of K and p (q fixed) [2]

2.2.2 Nonlinear Overdetermined 1-Dimensional Problem

We consider the same problem as before but measure at $m > 2$ points:

(12) Given: $(p_o, B_o), \ldots , (p_m, B_m)$

Again we are searching K and q. Since the problem is now overdetermined, we must solve problem (3) which in this notation is:

$$(13) \quad F(K, q) = \sum_{i=0}^{m} (B(K,q; p_i) - B_i)^2 \rightarrow \underset{K,q \in R}{\text{Minimum !}}$$

Solution: Not only the distance function (Euclidian norm) is nonlinear but also B as a function of K and q. Thus a 2-dimensional (!) nonlinear least squares problem must be solved in order to determine K and q. Note: The interpolation problem is only 1-dimensional because there is only the single independent variable p!
The numerical effort is by 3 orders of magnitude higher than for m = 2.

Evaluation: 4 operations and a square root.

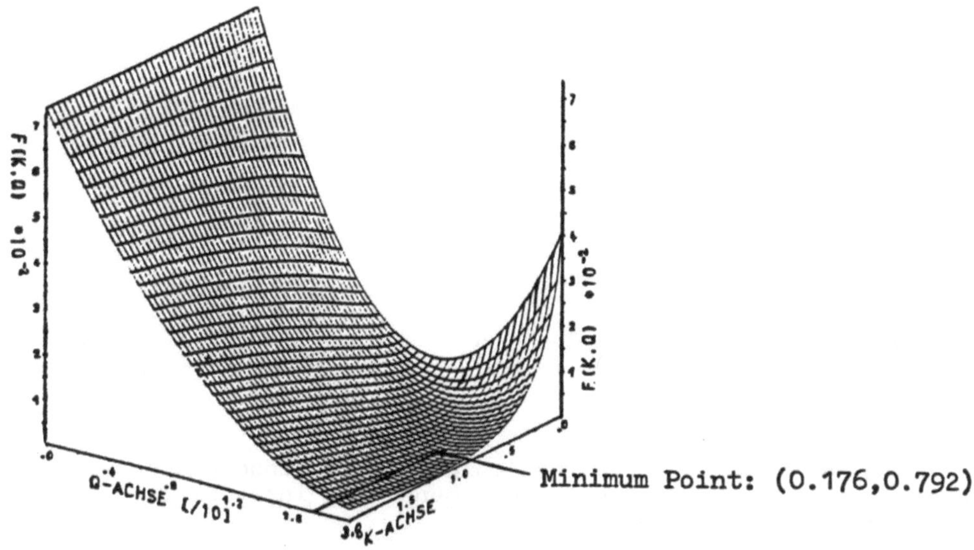

Fig. 2: The function F(K, q) to be minimized

One can prove that F(K,q) is a convex function but a very <u>ill conditioned</u> one: The surface increases with very little slope if both K and q tend versus infinity. Thus the determination of the minimum point is difficult and highly sensitive with respect to measurement errors.

2.2.3 Multidimensional Interpolation Problem

Lets choose an example from chromatography. A typical HPLC detector measures absorbance versus time at a fixed wavelength. More up-to-date detectors are able to vary their wavelength rapidly to record one or, even better, several spectra during a run. Unfortunately time does not stop during such a spectrum scan. Thus the following distribution of data points typically occurs:

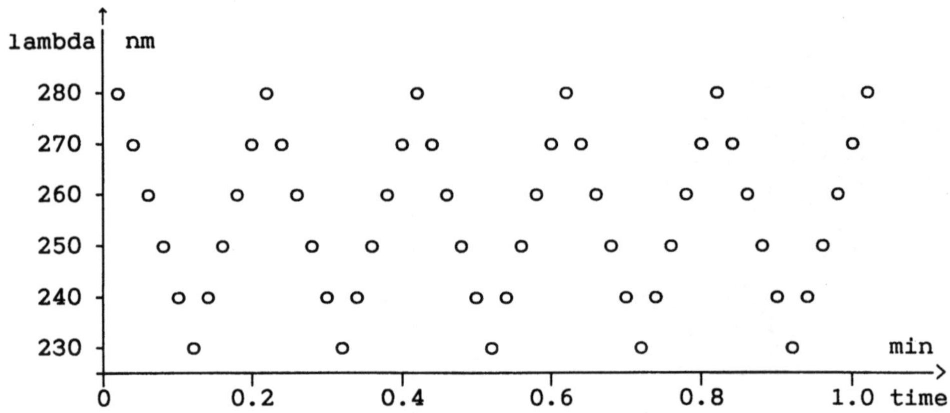

If you want to retrieve the 'pure' and continuous spectra from such an irregular data set two-dimensional interpolation on an irregular grid is needed.

Solution: 1. 2-dim. splines ("finite elements"): Very complicated because of the continuity conditions at the trapezoidal boundaries. The construction and implementation of appropriate "basis" functions as in the 1-dimensional case ("Lagrange functions") is much more involved. This is not so difficult on a rectangular grid.
But: Convergence is very good!

2. 2-dim. polynomials: Equally simple as in the 1-dimensional case. The simplest (but worst) way is to choose 2-dim. monomials as a basis and solve a linear system of equations.
But: Convergence is poor! The method is not even feasible for more than 100 data points.

3. Avoid the problem! Take a "diode array" as detector! If this is impossible, try to produce a rectangular data set.

3. Error Considerations

3.1 Classification

By the word error we denote any deviation of the interpolated (or approximated) value from the "real" physical fact. There are several classes of error sources which should be thoroughly distinguished:

3.1.1 Systematic Errors

They arise from a wrong mathematical model. Examples are:

- o Too many terms neglected in a physically derived model equation (e.g. mass action law above).
- o Linear assumptions in nonlinear problems (e.g. in a partial differential equation).
- o Inappropriate heuristic function space (e.g. polynomials instead of splines).

Their influence must be estimated by physical or chemical rather than mathematical considerations.

3.1.2 Numerical Effects ("Artefacts")

They are caused by intrinsic properties of the chosen interpolation space. In many cases their impact can be estimated by mathematical analysis a priori. The mathematical error estimate usually consists of a worst case study and a subsequent derivation of deterministic error bounds.

3.1.3 Data Perturbation

This type of consideration measures the influence of measurement deviations (usually random) at the 'knots' (known data) on the quality of the interpolation results.

3.2 Typical Behavior of Different Function Spaces

We have only room to give a rough overview of the error behavior of the two most important interpolation spaces, polynomials and cubic splines.

3.2.1 Systematic Errors

Lets assume we measured "bell shaped" data but did not recognize that they could be modelled by a "Gaussian exponential". Instead we try splines and polynomials:

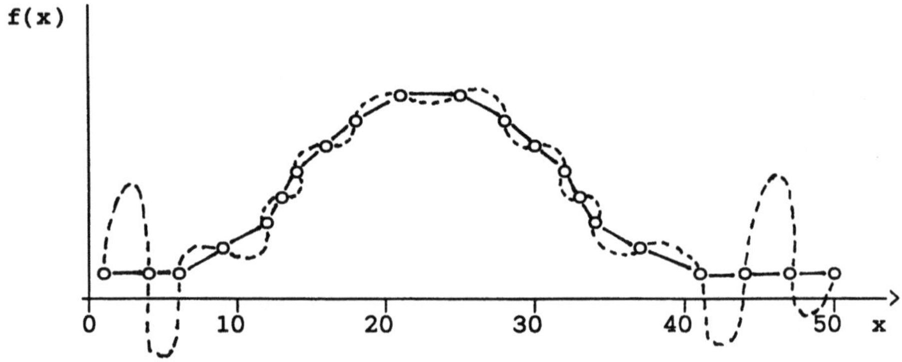

Splines have much better approximation characteristics: They have the famous minimum curvature property:

Among all twice differentiable functions which interpolate a given data set the cubic spline has the smallest mean curvature (integral over the squared second derivative).

3.2.2 Artefacts

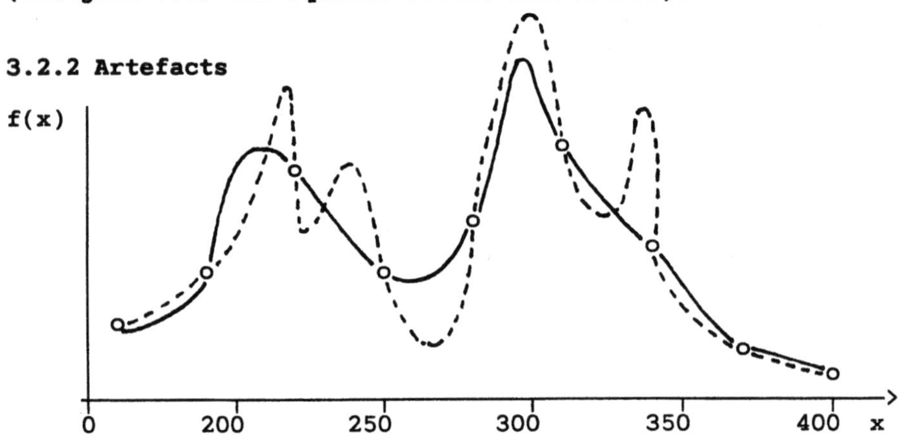

We interpolated a spectrum given by 9 data points. With a polynomial of degree 8 we receive "ghost peaks" not contained in the original continuous spectrum. A spline smoothly fits the data.

3.2.3 Noise

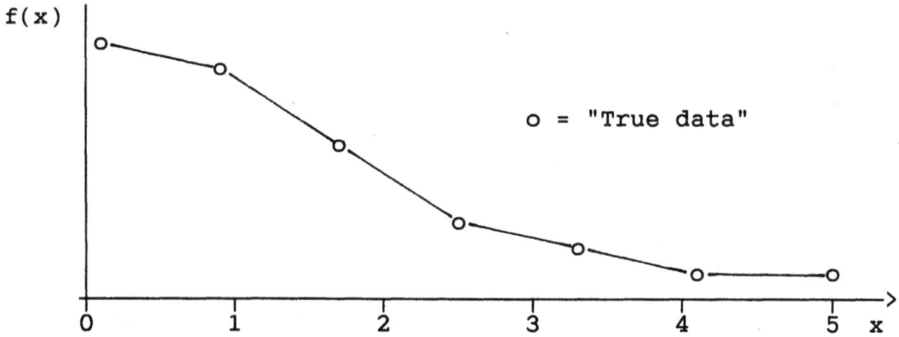

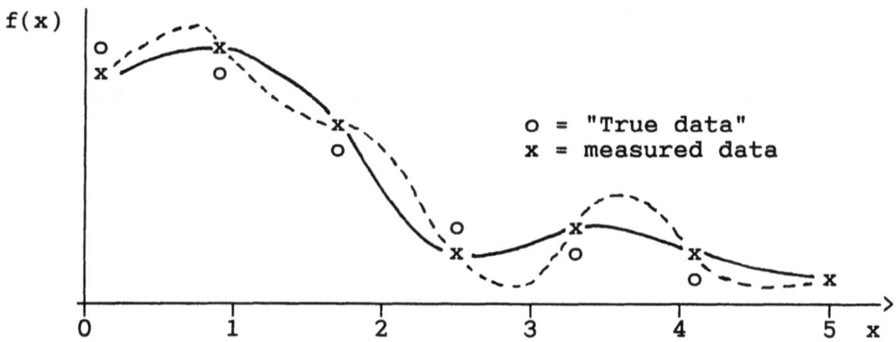

The spline is relatively stable versus noise whereas the polynomial produces artefacts again. A very common way to interpolate data of this type is to take a function class which has only one inflection point (e.g. a rational function).

4. Current and Future Trends

In many 'real life' cases the situation is too complex to be described by a meaningful physical model (such as the simple mass action law or the heat equation). But frequently other mathematical properties are known from either physical measurements or theoretical predictions. Among these properties the most important are:

- o Nonnegativity of physical data
- o Monotonicity of f (nonnegative or nonpositive slope)
- o Convexity ("pot shaped" function, nonnegative second deriv.)

These properties can be taken into account when choosing the appropriate interpolation space. They are not a substitution for a physical model but their effect is regularization [4]. This means that data perturbations have less effect on the interpolation accuracy. But nothing comes for free. Instead of a relatively simple interpolation or approximation task the following considerably more involved problem must be solved:

$$(14)\quad g(a_o,\dots,a_n) = \sum_{i=0}^{m} \left(I(a_o,\dots,a_n;\, x_i) - f_i \right)^2 \rightarrow \text{Minimum !}$$

$$\text{where}\quad a \in \{\, a \in \mathbf{R}^N \mid I(a;\, x) \text{ has certain properties } \}.$$

But the effect is often worth the burden! Examples can be found in [1], [3] and [5].

5. References

1. Hämmerlin, G., Hoffmann, K.H., Improperly Posed Problems and Their Numerical Treatment. ISNM 63, Birkhäuser Basel **1983**

2. Jawny,J., Jochum,P., Eiermann,W.; J. Steroid Biochem. **1984** 20, No. 2, 595

3. Jochum, P., Schrott, E.L., Anal. Chimica Acta **1984**, 157, 211

4. Lawson, C.L., Hanson, J.H.; Solving Least Squares Problems, Prentice-Hall, Englewood Cliffs, New Jersey **1974**

5. Sharaf, M.A., Kowalski, B.R., Anal. Chem. **1982**, 54, 1291

6. Stoer,J.; Einführung in die Numerische Mathematik I, 4th Edition. Springer Berlin **1983**

A Fuzzy Approach to Predicting Chemical Data from Incomplete, Uncertain and Verbal Compound Features

Matthias Otto[1] and Hans Bandemer[2]

Department of Chemistry[1] and Department of Mathematics[2],
Bergakademie Freiberg, Akademiestrasse 6, 9200 Freiberg,
German Democratic Republic

Estimation of missing data in an incomplete data base is possible by means of physico-chemical (deterministic) models or by using inherent information in the data base under study. The use of deterministic modelling techniques is limited to those objects (compounds) that fit within all the necessary assumptions and therefore this method will be applicable to a subset of compounds only or several models have to be created in combination with the data base.
The second method, i.e. data estimation based on inherent data base information, avoids specific mechanistic models and represents the more general estimation method.
There are only few proposals to solve this problem because methods are needed that can deal with incomplete data.
If the data base consists of numerical data only, e.g. the SIMCA-algorithm [1] could be used. With this method missing data can be estimated by filling in the missing values as the corresponding variable averages, estimating the principal component model with these data, and replacing the old "missing" data with new data predicted by the principal component model. This

C. Jochum M. G. Hicks J. Sunkel (Eds.)
Physical Property Prediction
in Organic Chemistry

procedure is repeated until convergence is obtained.

In case of a data base that also contains verbal compound features (linguistic variables) and highly uncertain numerical data conventional statistical or numerical methods cannot be applied. Therefore, we propose here a method that is based on the theory of fuzzy sets [2.3]. No physico-chemical model is needed since the information from the data base is used. In addition, the method enables linguistic variables to be handled, e.g. the color of dyes, and it makes use of the similarity between objects, such as organic compounds. and between features, such as physical data of those compounds.

THEORY

The data base

A common data base consists of n objects (compounds) that are characterized by q features each. For each feature a scale is introduced, say $X^{(j)}$, according to which the objects are valued. The datum x_{ij} representing the value of the j-th feature for the i-th object can be a number. Then it will be named <u>crisp</u>; x_{ij} can be a vague datum, e.g. a fuzzy number, or it can be a value of a linguistic variable (see below). In the latter two cases the datum x_{ij} is called <u>fuzzy</u>, specified by a membership function

$$m_{ij}\left(x^{(j)} \right) \; ; \qquad x^{(j)} \in X^{(j)} \qquad (1)$$

denoted by M_{ij} and where $X^{(j)}$ is the corresponding universe of discourse.

Then the data are arranged in an n times q matrix X as follows:

$$
\begin{array}{lllll}
x_{11} & x_{12} & x_{13} & \ldots & x_{1q} \\
x_{21} & x_{22} & ?_{23} & \ldots & x_{2q} \\
x_{31} & ?_{32} & x_{33} & \ldots & x_{3q} \\
\vdots & & & & \\
x_{n1} & x_{n2} & x_{n3} & \ldots & x_{nq}
\end{array}
\qquad (2)
$$

The indexed question marks point to the fact that the corresponding data are missing. Estimation of these missing data by interpolation will be our following subject.

<u>Specifying the data base</u>

As an example a data base is considered here, that contains among others organic compounds (indicators) absorbing light in the visible spectral range. The data are needed for designing fiber optic chemical sensors for measuring pH or other analytes [6]. Table I gives an idea of this data base. The data base comprises both observations, such as the molecular weight, the molar absorbance coefficient at maximum wavelength, ε_{max}, the negative logarithmic protolysis constant, pK, or the eigenvalues, λ_i, from the adjacency matrix, and verbal compound features, such as the color of the compounds and their solubility in water. The adjacency matrix (cf. [7]) was used for characterizing the chemical structure of the indicators. This matrix was specified

TABLE I Data Base Example

Name	Molecular weight	Color	ε_{max}	pK	lambda 1	lambda 2	Solubility in water
1 Bromocresol green	698.0	blue	35040	4.66	30.08	0	more or less high
2 Bromothymol blue	624.4	blue	23400	7.10	36.07	0	very low
3 Thymol blue	466.6	blue	4224	8.90	29.09	0	very low
4 Phenol red	354.4	red	37740	7.81	2.236	0	low
5 Bromocresol purple	540.2	purple	63650	6.12	22.11	0	more or less low
6 Bromophenol blue	670.0	?	67840	3.85	16.16	0	low
7 m-Cresol red	382.4	purple	9560	8.3	14.18	0	low
8 Cresol red	382.4	purple	24378	8.25	14.18	0	low
9 Xylenol blue	410.5	blue	16000	8.8	28.09	0	low
10 Chlorophenol red	423.3	orange	23280	5.6	12.21	0	very low
11 Bromocresol green	698.0	yellow	16370	4.66	30.08	0.99	more or less low
12 Bromothymol blue	624.4	yellow	16990	7.10	36.07	0.99	very low
13 Thymol blue	466.6	yellow	2007	8.90	28.09	0.99	very low
14 Phenol red	354.4	yellow	16640	7.81	2.288	0.99	low
15 Benzene	78.11	colorless	204	-	-	-	low
16 Toluene	92.14	colorless	225	-	-	-	low
17 Chlorobenzene	112.56	colorless	190	-	-	-	low
18 Bromobenzene	157.01	colorless	192	-	-	-	low
19 Phenol	94.11	colorless	269	9.62	-	-	more or less low
20 Aniline	93.13	colorless	1430	4.78	-	-	more or less low

by the assumptions that the indicators of interest are of the sulphophthalein type with the general structures of:

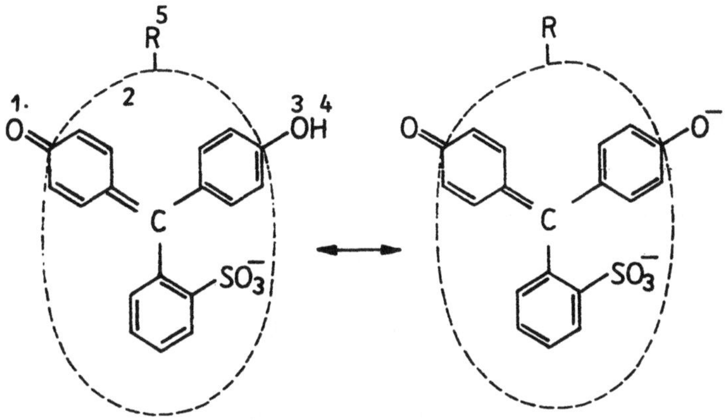

This structure can be represented simply by a fragment representing the sulphophtalein substructure and the terminal atoms =0 and -0^- or -0H . Substituents R are introduced as an additional unit (no. 5), a "superatom", so that in case of the protonated sulphophthalein the adjacency matrix is obtained as given in Table II.

The connection between the fragments 1 and 2 is characterized by representing the double bond, and the substituents are specified as the sum of their spectroscopic moments s_i [8], e.g. for the indicator bromocresol green one obtains for the 4 Br-atoms and 2 CH_3-atoms $R = \sum s_i = 4*4+2*7 = 30$.

TABLE II. Adjacency matrix for the sulphophthalein structure

atom #	atom # 1	2	3	4	5
1	0	2	0	0	0
2	2	0	1	0	R
3	0	1	0	1	0
4	0	0	1	0	0
5	0	R	0	0	0

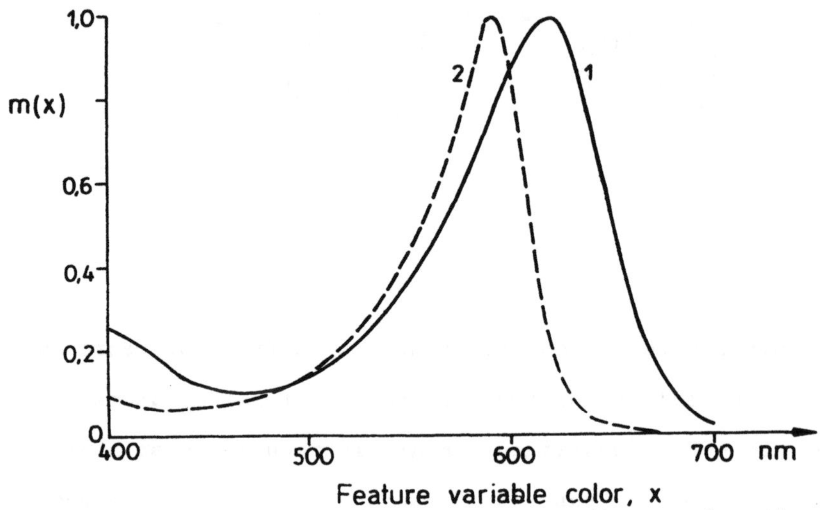

Fig. 1. Visible spectrum of indicators in the wavelength range between 400 and 700 nm renormed to the interval [0,1] and used as membership function for the feature "color". 1 - Bromocresol green, 2 - Bromophenol blue.

For representing the spectrum of the graphs the eigenvalues of the adjacency matrix were computed as given in Table I, column 6 and 7.

To compare the verbal compound features, such as color or solubility, a membership function m(x) is assigned to the corresponding feature variable X, as suggested in (1), suppressing the upper index for the moment. In case of characterizing the color of the indicator compound the visible spectrum is taken as the membership function renormed to the interval [0,1].

Fig. 1 gives examples for the compounds bromocresol green (curve 1) and bromophenol blue (curve 2).

The solubility in water can be expressed verbally as being "high", "low", "very low" etc. These linguistic variables are specified by truth values for which the universe is the unit interval [0,1]. As the general membership function we use for "high" solubility (cf. ref. 9, p.160):

$$m(x) = \begin{cases} 0 & \text{for } x<0 \\ 2x^2 & \text{for } 0 \le x \le 0.5 \\ 1-2(x-1)^2 & \text{for } 0.5 \le x \le 1 \\ 1 & \text{for } x > 1 \end{cases} \qquad (3)$$

To specify other labels for characterizing the solubility common suggestions for modifying membership functions are applied (Fig. 2) [9]:

very high	$m_v(x) = m(x)^2$	
more or less high	$m_{mh}(x) = m(x)^{1/2}$	
low (not high)	$m_l(x) = 1 - m(x)$	(4)
very low	$m_{vl}(x) = [1 - m(x)]^2$	
more or less low	$m_{ml}(x) = [1 - m(x)]^{1/2}$	
medium	$m_{me} = \min\,[m(x), 1-m(x)]$	

The chosen modifying functions applied on m(x) for obtaining the different labels of the linguistic variables are only a set of proposals, but they are rather frequently used.

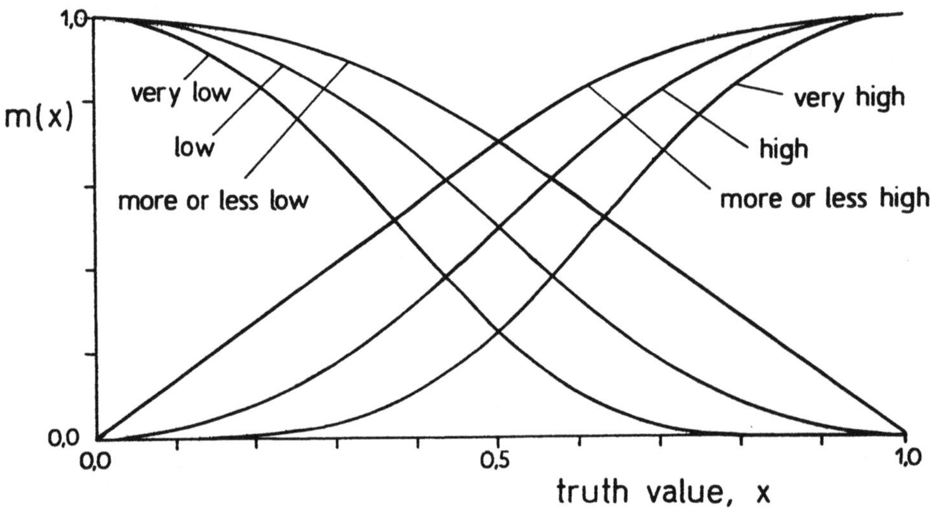

Fig. 2. Membership functions for characterizing "solubility in water".

Estimation by interpolation

The aim in every interpolation problem is to find neighbouring data from which one can start the interpolation. Usually this demands the introduction of a distance measure in the whole feature space, i.e. $X^{(1)} \times \ldots \times X^{(q)}$). This is a crucial task when the scales in the $X^{(j)}$ differ essentially and it is a new problem when the data are fuzzy.

In our context we will overcome this difficulties by introducing a suitable similarity relation R_j in each of the feature universes $X^{(j)}$. These relations will then be aggregated to a fuzzy relation

$$R = R_1 \circ R_2 \circ \dots \circ R_q \qquad (5)$$

which will indicate the neighbours of each object, respectively, given the gap $?_{i_o j_o}$ which is to be filled in by interpolation.

The neighbours of the i_o-th object are inspected for neighbouring features of the j_o-th feature, from which the needed interpolation can be obtained by interpolation of an approximate linear functional relationship linking these features with the j_o-th one.

Search for neighbouring objects

If the data in the j-th feature are fuzzy, i.e. M_{ij} with $m_{ij}(x^{(j)})$ according to (1), then the degree of similarity $r_j(i_1, i_2)$ of two objects i_1 and i_2 with respect to the j-th feature can be given by

$$r_j(i_1, i_2) = \text{card}\,(M_{i_1 j} \cap M_{i_2 j})/\text{card}\,(M_{i_1 j} \cup M_{i_2 j}) \tag{6}$$

defining the fuzzy similarity relation R_j (cf. ref. 3).

In formula (6) the power of the intersection $M_{i_1 j} \cap M_{i_2 j}$ given by the membership function

$$\underline{m}_{i_1 i_2}(x^{(j)}) = \min\,\{m_{i_1 j}(x^{(j)}),\ m_{i_2 j}(x^{(j)})\} \tag{7}$$

is compared with the power of the union $M_{i_1 j} \cup M_{i_2 j}$ given by the corresponding membership function

$$\overline{m}_{i_1 i_2}(x^{(j)}) = \max\,\{m_{i_1 j}(x^{(j)}),\ m_{i_2 j}(x^{(j)})\} \tag{8}$$

The quotient is one, if the two sets $M_{i_1 j}$ and $M_{i_2 j}$ are identical, and it is zero if they are disjoint.

The power is measured, as usually, by

$$\text{card}\ M = \int_X m(x)\,dx \qquad \text{or} \qquad \text{card}\ M = \sum_{x \in X} m(x) \tag{9}$$

in dependence on whether the universe X is continuous or discrete, respectively.

If the data in the j-th feature are crisp, then the fuzzy similarity relation can be specified directly by choosing a two-dimensional fuzzy set characterizing the degree to which two points, say u and v, in $X^{(j)}$ should be considered approximately equal, e.g.

$$m_j(u,v) = [\, 1 - c_j \,|\, u - v \,|^{p_j} \,]^+ \qquad (10)$$

with suitably chosen constants c_j and p_j being positive numbers. (The plus suffix denotes truncation of the argument to zero if $[1-c_j |u-v|^{p_j}] < 0$.)

This membership function gives the value one iff u = v and it decreases to zero with increasing differences between u and v. The degree of similarity of two crisp data $x_{i_1 j}$ and $x_{i_2 j}$ is then obtained by:

$$r_j(i_1,i_2) = m_j(x_{i_1 j}, x_{i_2 j}) \qquad (11)$$

Having specified the q fuzzy similarity relations R_j in this manner, they are to be aggregated. For this purpose we choose the weighted mean, i.e.

$$r(i_1,i_2) = m_R(i_1,i_2) = \sum_{j=1}^{q} g_j m_j(i_1,i_2) \qquad (12)$$

$$\text{with } g_j \in (0,1) \quad \text{and} \quad \sum_{j=1}^{q} g_j = 1$$

where g_j can indicate the importance the experimenter attributes to the j-th feature. In the case of indifference we may choose $g_j = 1/q$ for all j as it is done in the present work.

In order to select neighbours of our i_o-th object where the j_o-th feature value is missing, we specify a threshold r_o for the degree of similarity, $r(i_o,i)$, depending on the actual data set. All objects with a degree larger than this r_o will be considered neighbours of the i_o-th object.

This selection will form the partial data base for the next step - the search for neighbouring features.

Search for neighbouring features

We consider all pairs (jo,j) and specify two-dimensional fuzzy observations by

$$m_i(x^{(jo)},x^{(j)}) = \min \{ m_{ijo}(x^{(jo)}), m_{ij}(x^{(j)}) \} \tag{13}$$

(We restrict ourselves to the only interesting case that at least one of the features involved in (13) is fuzzy-valued).

The fuzzy observations are aggregated with respect to i, e.g. by

$$m(x^{(jo)},x^{(j)}) = \max_i m_i(x^{(jo)},x^{(j)}) \tag{14a}$$

or

$$m(x^{(jo)},x^{(j)}) = \sum_{i \in I} m_i(x^{(jo)},x^{(j)})/n_I \tag{14b}$$

where I is the set of all i taken into account ("the neighbourhood of i_o", see above) and n_I is the number of elements in I.

Then the aggregation in (14) is approximated by a parameter-fuzzy functional relationship, say

$$x^{(jo)} = a_{oj} + a_{1j} x^{(j)} \tag{15}$$

in the simplest case.

The parameters (a_{oj}, a_{1j}) are estimated using a method proposed in [5], which consists in summing up for each value of (a_{oj}, a_{1j}) the membership values of m as defined in (14) along the graph of the functional relationship with these values, e.g.

$$m_E(a_{oj}, a_{1j}) = \int_X m(a_{oj} + a_{1j} x^{(j)}, x^{(j)})\, dx^{(j)} \tag{16}$$

These values are the corresponding sum $m_E(f_j)$ for local approximations given by

$$f_j(x^{(j)}) = \arg\max_{x^{(jo)}} m(x^{(jo)}, x^{(j)}) \tag{17}$$

hence resulting in

$$m_c(a_{oj}, a_{1j}; f_j) = m_E(a_{oj}, a_{1j}) / m_E(f_j) \tag{18}$$

where $m_c(a_{oj}, a_{1j}; f_j)$ reflects the degree of approximation of the fuzzy functional relationship by the chosen approximation function with the parameter values (a_{oj}, a_{1j}). The membership function m_c according to (18) can be regarded as defining a fuzzy set A_j over the corresponding parameter set.

The "best approximation" is obtained by that feature j* showing a maximum degree of approximation, i.e.

$$m(a_{oj*},a_{1j*};f_{j*}) = \max_{j=jo} m_c(a_{oj},a_{1j};f_j) \qquad (19)$$

Now, we will choose j* (or a j with a nearby-value of m_c for convenience) to interpolate the missing value by a fuzzy datum (cf. [5]):

$$m_{i_o j_o}(x^{(jo)}) = \sup_{\{(a_{oj*},a_{1j*}):x^{(jo)}=a_{oj*}+a_{1j*}x^{(j*)}\}} m_c(a_{oj*},a_{1j*};f_{j*}) \qquad (20)$$

RESULTS AND DISCUSSION

The performance of the method is demonstrated for estimating the color of the indicator bromophenol blue (object no.6). Thus, the objects with the most similar features have to be evaluated with the fuzzy search method according to equs. (6) to (12). For characterizing the comparability of the feature observations the membership function of equ. (10) was applied with p=2 and the constants $c=10^{-6}$ for the molecular weight; $c = 10^{-10}$ for ε_{max}; $c = 10^{-6}$ for the pK-value; c =0.001 for lambda1 and c=1 for lambda2. The membership function for the linguistic variable "solubility" was specified as described above (equ. (3) and (4)). The result for the similarity between the feature objects and the feature of object no. 6 (bromophenol blue) is given in Table III.

TABLE III. The similarity of object features compared to the features of object no. 6 (bromophenol blue) calculated with equ. (7) to (11) and the fuzzy relation according to equ. (12).

Compound	Molecular weight	ε_{max}	pK	lambda1	lambda2	Solubility in water	Fuzzy relation
1	0.999	0.892	0.993	0.806	1	0.289	0.830
2	0.998	0.874	0.894	0.604	1	0.766	0.856
3	0.959	0.595	0.745	0.833	1	0.766	0.816
4	0.900	0.909	0.843	0.806	1	1	0.910
5	0.983	0.998	0.948	0.965	1	0.792	0.948
6	1	1	1	1	1	1	1
7	0.917	0.660	0.802	0.996	1	1	0.896
8	0.917	0.811	0.806	0.996	1	1	0.885
9	0.932	0.731	0.755	0.858	1	1	0.865
10	0.939	0.801	0.970	0.985	1	0.766	0.876
11	0.999	0.735	0.993	0.806	0	0.289	0.560
12	0.998	0.741	0.894	0.604	0	0.766	0.579
13	0.959	0.567	0.745	0.858	0	0.766	0.600
14	0.900	0.738	0.843	0.808	0.236	1	0.654
15	0.649	0.543	0	0	0	1	0.275
16	0.666	0.542	0	0	0	1	0.278
17	0.689	0.542	0	0	0	1	0.282
18	0.737	0.543	0	0	0	1	0.289
19	0.668	0.543	0.667	0	0	0.792	0.355
20	0.667	0.559	0.991	0	0	0.792	0.408

In the last column of Table III the fuzzy relation of objects was computed according to equ. (12).
The following conclusions can be drawn from the values for the fuzzy relations: values between 0.816 and 0.948 are assigned to indicator compounds that are of the alkaline form of the sulphophthalein type (deprotonated, without atom no. 4).
The protonated indicators (objects 10 to 14) are similar to the bromophenol blue features with values ranging between 0.560 and 0.654. The difference is mainly due to the appearance of a second non-zero eigenvalue (lambda 2) reasoned by the altered component structure, i.e.there is an additional H-atom.
The colorless, UV-absorbing compounds revealed values between 0.275 and 0.408 showing the lowest degree of similarity with the bromophenol blue indicator.
A suitable threshold for selecting the most similar objects with respect to the features of bromophenol blue would be $r_o = 0.7$, i.e. the compounds 1 to 5 and 7 to 10.
With this subset of compounds now the features that are best related with the feature to be estimated, i.e. color, are evaluated by the formalism as given in equations (13) to (19). (Aggregation was performed according to equ. (14b).)
The feature variables molecular weight, ε_{max}, pK, lambda1 and lambda2 are taken as being crisp because no further information was available with the present data. This also simplifies computations since then for the fuzzy functional relationship of variables $(x^{(jo)}, x^{(j)})$ only the membership function of the color

(the indicator spectrum), $x^{(jo)}$, has to be taken into account.

In the present case the closest relationship was found between color and the first eigenvalue of the adjacence matrix with a degree of approximation of $m_c(a_{oj*}, a_{1j*}; f_{j*}) = 0.751$ and the parameters $a_{oj*} = 90$ and and $a_{1j*} = 5.1$.

With this relationship the missing value can be estimated from the given lambda1 - value of 16.16 using the equation for fuzzy predictions (equ.(20)). The estimated membership function is given in Fig. 3, curve 2.

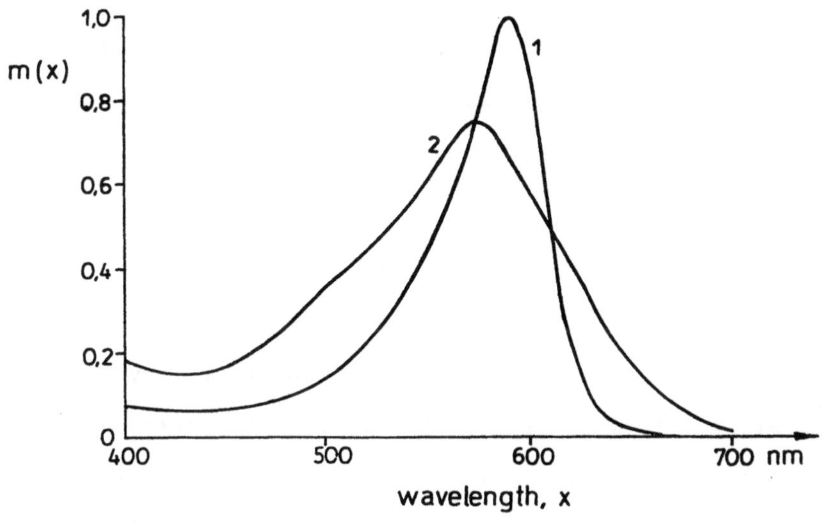

Fig. 3. The spectrum of bromophenol blue (1) compared with the estimated spectrum obtained as a membership function of the fuzzy prediction (2) from the first eigenvalue of the adjacency matrix.

Compared to the known indicator spectra a color of purple can be derived as the most possible color for bromophenol blue in its alkaline form. (The spectra comparison can be carried out by means of a previously described method for comparing fuzzy functions [10]). Often, however, the retrieval of the whole spectrum will be still more important than just the evaluation of the color of the compound.

In Fig. 3 the estimated spectrum (curve 2) is compared to the renormed measured absorbance spectrum (curve 1). The differences, especially in the hight of the two spectra must be attributed to the existing vague relationship between color and the first eigenvalue.

If, for example, the wavelength of maximum absorbance is plotted against lambda1 one obtains deviations in the range of 10 to 20 nm that correspond with the fuzziness of the found predictions in comparison with the original bromophenol blue spectrum.

Therefore, the comparability of the estimated spectrum with the real spectrum can be taken as being satisfactory.

CONCLUSION

The proposed fuzzy method of data estimation by interpolation in a data base has been used here for predicting the spectrum (color) of indicator compounds. As introduced by the general theory of the fuzzy interpolation method its applicability is, of course, not restricted to estimating colors of dyes. Uncertain and imprecise physical data, data ranges rather than single feature values, e.g. the range of property change of an indicator, or other linguistic variables, such as taste, smell or more complex properties, could be handled with the proposed method likewise. Even the chemical compound name could be specified as a linguistic variable and similar compounds could be evaluated on the basis of such a definition.

These features of the fuzzy interpolation method extent common methods of interpolation to a great deal since known methods of multidimensional data interpolation can only deal with crisp numerical data.

REFERENCES

1 S. Wold, M. Sjöström, in; "Chemometrics: Theory and Applications", ACS Symposium Series 52 (B.R. Kowalski, ed.), Washington DC 1977.

2 D. Dubois and H. Prade, Fuzzy Sets and Systems: Theory and Application, Academic Press, New York, 1980.

3 H. Bandemer and M. Otto, Mikrochimica Acta [Wien] 1986II, 93-124.

4 M. Otto and H. Bandemer, Anal. Chim. Acta 184(1986)21-31.

5 M. Otto and H. Bandemer, Chemometrics Intelligent Laboratory Systems 1(1986) 71-78.

6 M. Arnold (Ed.), Talanta 35(2) (1988) 75-159.

7 I. Ugi, P.D. Gillespie, Angew. Chem. 83(1971) 980.

8. R. Borsdorf and M. Scholz, Spektroskopische Methoden in der organischen Chemie, WTB-Reihe, Band 21 (1974), p. 172.

9 L.A. Zadeh, A theory of approximate reasoning, in Machine Intelligence 9 (J. H. Hayes, D. Michie, L.I. Mikulich, Eds.), Wiley, New York, 1979.

10 M. Otto and H. Bandemer, Anal. Chim. Acta 191 (1987) 193.

Ranking and Clustering of Chemical Structure Databases

Peter Willett

Department of Information Studies, University of Sheffield, Western Bank,
Sheffield S10 2TN, United Kingdom

Abstract This paper summarises an extended research programme to investigate the use of fragment-based measures of inter-molecular similarity in chemical information systems, with particular reference to structure-property correlation. Comparative studies are reported of structural similarity measures and of clustering methods for chemical structure databases. The methods are most appropriate when very sparse data matrices are available; in such cases, a very fast nearest neighbour searching algorithm can be used for the calculation of the requisite similarities.

1 Introduction

Many methods have been described for the prediction of property values in previously untested molecules, these methods generally assuming that several compounds are available for which the appropriate property data are already available. This data can be used to set up some predictive mechanism by which the activity of new compounds can be estimated: examples of this general approach include the use of quantum mechanics, regression analysis and linear discriminant functions. Early work on property prediction focussed on chemical and physical properties, but there is now substantial interest in the prediction of biological activities. This paper discusses techniques for structure-activity relationship studies where structural features are used as the basis of the predictive mechanism.

The prediction of property values from structural data involves the implicit assumption that structurally similar molecules will exhibit comparable activity behaviour. Accordingly, several researchers have studied property prediction methods

C. Jochum M. G. Hicks J. Sunkel (Eds.)
Physical Property Prediction
in Organic Chemistry

which are based on explicit measures of inter-molecular structural similarity; the term *similarity* will be used throughout this paper, but it should be noted that the actual measure used may be one of similarity, of dissimilarity or of distance. Several factors need to be taken into account in calculating inter-molecular similarity measures [27], these including

- The *structural features* which are used to characterise the molecules which are to be compared and whose similarities are to be calculated
- The *weighting scheme* which is used to differentiate between features on the grounds of how important they are in determining similarity
- The *similarity coefficient* which is used to obtain a numeric quantification of the degree of similarity between a pair of structures

The first of these factors has been addressed by several workers [5]. Thus, Adamson and Bush have discussed the use of fragment substructure occurrence data [2] [3], Randic and Wilkins and Carhart *et al.* the use of inter-atomic path lengths [9] [17] [23], Gabanyi *et al.* the use of topological molecular transforms based upon distance matrices [13] and Broto *et al.* related work using the mathematical theory of autocorrelation functions [8].

This paper adopts the definition of inter-molecular similarity used by Adamson and Bush in which the degree of structural similarity between a pair of compounds is expressed in terms of the substructural fragments that are common, and that are not common, to the two molecules that are being compared. Section 2 discusses a comparison of the effectiveness of a range of ranking measures for property prediction using a number of small datasets for which both structure and property data are available. One of the main characteristics of similarity-based methods is their applicability to very large files of compounds, since they do not suffer from the dimensionality problems which can bedevil large-scale applications of parametric prediction methods [21], and do not require the extensive computation associated with quantum mechanical approaches. Section 3 accordingly presents an algorithm which allows the implementation of similarity matching on databases containing tens or thousands of molecules. Section 4 then discusses the use of cluster analysis methods for structure-property correlation and compares the ranking and clustering approaches. The paper closes with a brief summary of other sorts of similarity measure that could be used for property prediction.

2 Ranking Of Chemical Structure Databases

As noted in the introduction, the determination of inter-molecular similarity to rank chemical structures necessitates the specification of a weighting scheme and of a similarity coefficient. There is an extensive literature associated with both types of function [11] [20] but there are no clear guidelines as to what measures are likely to be most appropriate for a particular application area. We have accordingly carried out an extended comparison of some 36 similarity measures which use fragment substructure occurrence data [28].

The evaluation of the weighting schemes and similarity coefficients involved 16 sets of molecules selected from the structure-activity and structure-property literatures and containing between 20 and 129 molecules. These datasets span a wide range of chemical, physical and biological properties, and include both homogeneous and heterogeneous sets of compounds, it being hoped that the use of such a wide range of types of compound and property would ensure that the results were of some generality and not conditioned by the characteristics of a particular dataset.

The compounds were characterised by sets of substructural descriptors, specifically the *augmented atom* fragment which was generated for each and every non-hydrogen atom in a molecule [4] [27] For a data set containing N compounds characterised by M fragments, an $N \times M$ data matrix, D, was generated where $D[I, J]$ contained the number of occurrences of the J-th fragment in the I-th compound. This matrix was then used to generate a similar matrix, W, where $W[I, J]$ was the weight of the J-th fragment in the I-th compound using one of six weighting schemes. Each such data matrix was used to generate a symmetric $N \times N$ similarity matrix, S, the elements of which, $S[K, L]$, corresponded to the similarity between molecules K and L using one of six similarity coefficients. The effectiveness of each similarity matrix, S, for property prediction was assessed using a 'leave-one-out'approach. For each molecule, Q, the similarity matrix S was scanned to identify the *nearest neighbour*, or *best match*, i.e., that molecule which was most similar to Q. The predicted property value for compound Q was then set equal to the observed property value for this nearest neighbour compound and the overall measure of agreement between the N observed and predicted property values determined by means of the product moment correlation coefficient.

Several types of statistical information can be extracted readily from machine-readable representations of chemical structures to form the basis for a fragment weighting scheme.

- The simplest such scheme involves a consideration of the number of occurrences of each particular fragment type within a compound, with frequently occurring fragments being given a greater weight than those that occur less frequently. Thus, a pair of molecules that had several occurrences of a given fragment in common would be considered to be more similar to each other than if they had only a single occurrence in common.

- The second type of weighting takes into account the sizes of the molecules that are being compared, where by size is meant the number of fragments assigned to a compound. Thus, a fragment in a small molecule would be assigned a greater weight than the same fragment in a larger compound, and fragments common to a pair of small molecules would be regarded as a better indicator of structural similarity than the sharing of a common fragment by two large structures.

- The final type of weighting scheme involves the inverse frequency of occurrence of a fragment in the data set, with infrequently occurring fragments being assigned greater weights than those that occur throughout the dataset. Thus, a pair of compounds that both possessed some rare substructural feature would be regarded as being much more strongly related to each other than if they both possessed a feature such as a benzenoid ring.

Each of these schemes seems to be intuitively quite reasonable, and there seems to be no obvious *a priori* reason why any one of them should be preferred to the others; the weighting schemes tested reflected these three types of consideration.

There are four main types of similarity coefficient

- *Distance coefficients* are extensively used owing to their simple geometric interpretation. The most common is simply the Euclidean distance between a pair of points in a multi-dimensional space, the dimensions of which are defined by the variables used for the characterisation of the objects which are being compared. Other distance coefficients are discussed by Sneath and Sokal [20].

- *Association coefficients* are the most widely used type of coefficient when two-state, i.e., binary variables are involved. Many such coefficients have been described; an example is the Tanimoto coefficient which is calculated as

$$C/(A + B - C)$$

  for a pair of objects containing A and B attributes, C of which are in common. Such coefficients can be generalised to non-binary data if required [19].

- *Correlation coefficients* measure the degree of statistical correlation between a pair of objects in just the same way as a correlation coefficient measures the strength of the relationship between sets of values for a pair of variables in regression analysis.

- *Probabilistic coefficients* take account of the distribution of frequencies of occurrence of variables in a dataset; these coefficients are not considered further here in view of the unsatisfactory results reported by Adamson and Bush [3].

Correlation coefficients between the observed and the predicted property values were calculated as described above using each combination of similarity coefficient and weighting function [28]. An inspection of the resulting coefficients shows that, in general, fair correlations are obtained, typically in the range 0.6-0.8, though the extent of this correlation naturally varied from one dataset to another. In some cases, the correlations were generally low, but even here there were usually a few similarity measures that gave more acceptable results. A statistical analysis of the correlation coefficients demonstrated that

- The weighting schemes which took account of the number of occurrences of a fragment in a molecule gave slightly better correlations than the schemes based upon binary data matrices; however, the use of weighting schemes based on the frequencies of occurrence of fragments within a dataset, or the number of fragments within a molecule, did not seem to be of general utility.

- The distance coefficients gave rather poorer correlations than the other types of similarity coefficient which were tested, although the differences did not seem to be large. In subsequent work, Willett *et al.* [29] suggested that the Tanimoto coefficient mentioned previously seemed to be the most suitable coefficient for structure handling applications.

Perhaps the most important, general result from the experiments is that the use of simple, fragment-based, similarity measures enables the identification of structures that are closely related in activity to a specified query molecule. Thus, the ranking of a set of structures to identify the most similar may be sufficient to lead to several others in a machine-readable file that might be expected to exhibit comparable activities.

3 Implementation Of Ranking Methods

In the work described above, the nearest neighbour search for the molecule most similar to the chosen query molecule, Q, was effected by a *serial search* in which Q was compared with each molecule in the dataset in turn to identify the common fragments; the resulting similarity was then compared with that of the current nearest neighbour. This procedure can be described algorithmically as shown below, where it is assumed that the screen records for each of the molecules in a database are stored as a *bit map*, B, in which the bit $B[I, J]$ is set to TRUE if the J-th fragment has been assigned to the I-th molecule. The bit map can be regarded either as a *serial file*, in which the bit strings are inspected in sequence one after the other, or as an *inverted file*, in which access is available to all of the structures that contain a specific fragment screen. When used in the serial form, the nearest neighbour search is as follows:

1. Set I to 1, and NN and $MAXCOEFF$ to 0.

2. Compare the bit string representing the query molecule, Q, with that representing the I-th molecule. Identify the bits, i.e., fragment substructures, in common and then calculate the resulting similarity coefficient, $COEFF$. If $COEFF$ is greater than $MAXCOEFF$, then replace NN and $MAXCOEFF$ by I and $COEFF$ respectively.

3. Increment I by 1. Go to Step 2 if there are still structures remaining to be matched against Q.

4. Retrieve the appropriate property value for the NN-th molecule and return this as the calculated value for Q

It is assumed here that only the single nearest neighbour is required; simple modifications enable the algorithm to be used for the retrieval of K nearest neighbours ($K > 1$). Although simple in concept, this algorithm is clearly far too time-consuming for interactive nearest neighbour searching if the database is at all large, even if the program code is heavily optimised [9]. Alternatively, if the bit map is considered as an inverted file, it is possible to make use of an algorithm which is based upon the *addition* of the columns in the bit map that correspond to the fragments in the query (rather than to their intersection or union as in conventional substructure searching) and upon the fact that the matrix is very sparse, with only a small fraction of the elements containing a non-zero value.

The addition results in a vector, the elements of which contain the number of fragments in common between the query structure and each of the structures in the file. Given this information, it is trivial to evaluate the corresponding similarity coefficient, e.g., the Tanimoto coefficient, and hence to rank the compounds in order of decreasing similarity with the query structure so as to identify the nearest neighbour. The procedure is sufficiently fast in operation to allow for interactive nearest neighbour searching of large files of structures, with response times comparable to those for conventional inverted file substructure searching. The algorithm is as follows:

1. Initialise the elements of an N-element array, $COMMON$, to zero. Set J to 1.
2. Identify the column of the bit map which corresponds to the J-th fragment assigned to Q and add each element of it, regarded as an integer value of 0 or 1, to the corresponding element of $COMMON$.
3. Increment J by 1. Go to Step 2 if there are still more fragments in Q to be processed.
4. Set I to 1 and NN and $MAXCOEFF$ to 0.
5. Calculate the similarity for the I-th molecule using the information in $COMMON[I]$. If the resulting coefficient, $COEFF$, is greater than $MAXCOEFF$, then replace NN and $MAXCOEFF$ by I and $COEFF$ respectively.
6. Increment I by 1. Go to Step 5 if there are still coefficients to be calculated.
7. Retrieve the appropriate property value for the NN-th molecule and return this as the calculated value for Q.

It should be noted that since this algorithm is to be applied to a bit map that contains only binary fragment data, the resulting inter-molecular similarities are likely to be a less accurate reflection of the similarities between file compounds and the chosen target structure than if full fragment occurrence data was available; the algorithm can be used with such information, but this clearly increases the storage requirement. The algorithm can also be used with a conventional inverted file where only the non-zero elements of the bit map are stored [16]; the important point is that rapid access must be available to all of the molecules containing a particular screen. Willett *et al.* [29] discuss the implementation of this algorithm at Pfizer Central Research, where it is now used on a routine basis for nearest neighbour searching in a database containing *circa* quarter of a million chemical structures. The algorithm is mainly used to provide a structural browsing capability and has also been used for the ranking of broadly defined, conventional substructure searches.

4 Clustering Of Chemical Structure Databases

Rather than matching the query molecule, Q, against a structure database to identify the K nearest neighbours, an alternative approach to property prediction involves the clustering of the database. Cluster analysis, or automatic classification, is the name given to a range of methods that can be used to detect groupings present within multivariate data sets. Cluster analysis methods were first used extensively for the study of biological species, but have recently been applied to the analysis of chemical databases [27]. An early example of such work is that of Adamson and Bush [2] who reported the use of the single linkage clustering method to classify the 20 naturally occurring amino acids, and evaluated the classifications by means of simulated property prediction experiments using the amino acid pK_a values; a similar approach was adopted in an evaluation of similarity measures for the clustering of 39 structural diverse compounds with local anaesthetic activity [3]. This work has formed the basis for our more recent studies [27].

Two factors need to be taken into account if clustering methods are to be used.

- The similarity measure which is used to calculate the inter-molecular similarities.
- The clustering method which uses the similarity data to identify the groups of similar molecules that are present in the dataset.

The first of these factors has been discussed in Section 2 above; the second involves consideration of the many different clustering methods which are available.

The single linkage method used by Adamson and Bush is an example of the *hierarchic agglomerative* clustering methods; these produce classifications in which small clusters of very similar molecules are nested within larger and larger clusters of less closely related molecules. The classifications are built up by a series of agglomerations in which small clusters, initially containing individual molecules, are fused together to form progressively larger clusters [1] [24]. Less commonly, hierarchic *divisive* methods may be used in which clusters are generated in a top-down manner, by progressively sub-dividing the single cluster which represents an entire dataset. Experimental studies suggest that these methods are generally inferior to the agglomerative ones, especially when structurally heterogeneous datasets need to be processed [18] [24].

Non-hierarchic clustering methods result in a dataset being partitioned into a set of (generally) non-overlapping groups having no hierarchical relationships between them. A systematic evaluation of all possible partitions is quite infeasible; accordingly, many different heuristics have been described to allow the identification of good, but possibly sub-optimal, partitions; such methods are generally much less demanding of computational resources than the hierarchic methods [25].

Willett [27] reports an extended study of the effectiveness of over thirty hierarchic and non-hierarchic methods for the clustering of chemical structure databases using a leave-one-out approach analogous to that discussed in Section 2 for the evaluation of similarity measures. Specifically, the following procedure was adopted:

1. Cluster the dataset using some particular clustering method; the datasets used were the same as those used for the ranking measure tests in Section 2.

2. For each molecule, identify the cluster which contains it.

3. Calculate the predicted property value for this molecule as the average of the observed values for the other members of that cluster.

4. Calculate the correlation between the sets of observed and predicted property values.

The comparison suggested that non-hierarchic methods gave results that were comparable with, or superior to, those given by the computationally more demanding hierarchic methods and identified a non-hierarchic method due to Jarvis and Patrick [14] as the most generally useful for this application area. The Jarvis-Patrick method involves the use of a nearest neighbour table, an $N \times K$ integer array, $NNTABLE$, containing the identifiers for the K nearest neighbours for each of the N molecules; $NNTABLE[I, J]$ contains the identifier for the J-th nearest neighbour of the I-th molecule (using the Tanimoto coefficient in most of our experiments). Two molecules, I_1 and I_2, are placed in the same cluster if I_1 is a nearest neighbour of I_2, I_2 is a nearest neighbour of I_1, and I_1 and I_2 share at least S nearest neighbours in common, where S is a user-defined parameter. The nearest neighbour table can be generated efficiently using the algorithm described in the previous section; the method has been used on a large scale for compound selection in biological screening programmes and for the clustering of substructure search output [27].

Having summarised the studies of ranking and clustering carried out in our laboratory over the last few years, the question arises as to which of them is likely to be the more useful for property prediction in realistic circumstances, rather than in the unrealistic leave-one-out environment considered here. We first consider how the two approaches could actually be used in practice.

Given an appropriate measure of structural similarity, the following procedure allows the calculation of some property value for a query molecule, Q, for which the appropriate data is not available:

- Q is used as the basis for a nearest neighbour search of the database to identify the K most similar molecules, where K ($K \geq 1$) is a user-defined parameter.

- The property value for Q is then calculated as the average of the property values for the K nearest neighbours.

This assumes that property data is available for all of the K nearest neighbours. If this is not the case, two approaches seem possible:

- The nearest neighbour search is restricted to that part of the database for which the requisite property data is available.

- If any of the nearest neighbours do not have the required data, then replace Q by this molecule, $NN(Q)$ and use this as the basis of a nearest neighbour search. In this way, a chain of compounds

$$Q, NN(Q), NN(NN(Q)), NN(NN(NN(Q))) \ldots$$

  is generated until some molecule is retrieved for which the data is available. This is then used as the value for Q.

Two approaches are also possible if a clustering approach is adopted. These are

- Identify the cluster containing Q and then calculate the required value as the average of the values for those molecules in that cluster for which this value is available

- Carry out a nearest neighbour search of the clustered file so as to identify the cluster, or clusters, which is most similar to Q and predict the value for Q on the basis of the molecules in that cluster

Extensive studies in an analogous application area, *viz* the prediction of relevance in documents characterised by keywords, have shown the searching approach to be much superior [30]; similar, unpublished results have been obtained in our laboratory with chemical datasets containing several hundreds of structures for which partition coefficients and heats of formation data were available. However, the clustering method used in this chemical comparison was the single linkage method, which has been shown subsequently to be noticeably inferior to other clustering methods; moreover, as with all of the experiments here, the property prediction involved the highly unrealistic leave-one-out method, which assumes that property data is available for all of the structures with the exception of the one for which a value is being calculated. Thus, while it is likely that the methods of Section 2 are to be preferred, much further, detailed work is required.

5 Conclusions

In conclusion, the reader should note that the definition of similarity used in this paper is a very restricted one, taking account as it does only of local topological relationships between pairs of molecules. Since topographic relationships are ignored, it is possible for structures to be identified as being identical to each other, i.e., having a very high calculated similarity and thus being comparable in activity, when this is not in fact the case: obvious examples of such behaviour are given by stereoisomers or different conformations of the same molecule. None the less, the results do show that even the crude definition of inter-molecular structural similarity that we have used is sufficient to identify at least some sort of relationship between structure and property.

More precise definitions of inter-molecular similarity are, of course, possible. One readily computable measure that might be used is the *maximal common substructure* (MCS), i.e., the largest substructure common to a pair of molecules. Although more difficult to compute than the simple, fragment-based schemes used in the work reported here, practical MCS algorithms have been developed to support systems for structure elucidation and chemical reaction indexing [10] [22] [26]; recent work by Johnson *et al.* suggests that these increased computational requirements result in substantial improvements in predictive performance [15]. A further line of investigation is the use of 3-D structural information, rather than the 2-D structure diagrams used in the work to date. The increasing availability of 3-D coordinate data, from X-ray crystallography and molecular mechanics programs, has already led to some work on the identification of MCSs in 3-D [6] [7] [12]; to date, however, no structure-property studies seem to have been reported.

Acknowledgements Thanks are due to David Bawden and Vivienne Winterman for their contributions to this work and to Pfizer Central Research and the Science and Engineering Research Council for funding.

References

[1] Adamson, G.W. and Bawden, D. Comparison of hierarchical cluster analysis techniques for the automatic classification of chemical structures. *Journal of Chemical Information and Computer Sciences* **21**, 204-209, 1981.

[2] Adamson, G.W. and Bush, J.A. A method for the automatic classification of chemical structures. *Information Storage and Retrieval* **9**, 561-568, 1973.

[3] Adamson, G.W. and Bush, J.A. A comparison of the performance of some similarity and dissimilarity measures in the automatic classification of chemical structures. *Journal of Chemical Information and Computer Sciences* **15**, 55-58, 1975.

[4] Ash, J.E., Chubb, P.A., Ward, S.E., Welford, S.M. and Willett, P. *Communication, Storage and Retrieval of Chemical Information* Ellis Horwood, Chichester, 1985.

[5] Bawden, D. Computerized chemical structure-handling techniques in structure-activity studies and molecular property prediction. *Journal of Chemical Information and Computer Sciences* **23**, 14-22, 1983.

[6] Brint, A.T. and Willett, P. Algorithms for the identification of three-dimensional maximal common substructures. *Journal of Chemical Information nd Computer Sciences* **27**, 152-158, 1987.

[7] Brint, A.T. and Willett, P. Identifying 3-D maximal common substructures using transputer networks. *Journal of Molecular Graphics* **5**, 200-207, 1987.

[8] Broto, P. Moreau, G. and Vandycke, C. Molecular structures: perception, autocorrelation descriptor and SAR studies. *European Journal of Medicinal Chemistry* **19**, 66-70, 1984.

[9] Carhart, R.E., Smith, D.H. and Venkataraghavan, R. Atom pairs as molecular features in structure-activity studies: definition and application. *Journal of Chemical Information and Computer Sciences* **25**, 64-73, 1985.

[10] Cone, M.M., Venkataraghavan, R. and McLafferty, F.W. Molecular structure comparison program for the identification of maximal common substructures. *Journal of the American Chemical Society* **99**, 7668-7671, 1977.

[11] Cormack, R.M. A review of classification. *Journal of the Royal Statistical Society* **134**, 321-367, 1971.

[12] Crandell, C.W. and Smith, D.H. Computer-assisted examination of compounds for common three-dimensional substructures. *Journal of Chemical Information and Computer Sciences* **23**, 186-197.

[13] Gabanyi, Z., Surjan, P. and Naray-Szabo, G. Application of topological molecular transforms to rational drug design. *European Journal of Medicinal Chemistry* **17**, 307-311, 1982.

[14] Jarvis, R.A. and Patrick, E.A. Clustering using a similarity measure based on shared nearest neighbours. *IEEE Transactions on Computers* **C-22**, 1025-1034, 1973.

[15] Johnson, M., Nain, M., Nicholson, V. and Tsai, C.C. Comparing the substructure metric to some fragment-based measures of inter-molecular structural similarity. In: Hadzi, D. and Jerman-Blazic, B. (editors) *QSAR in Drug Design and Toxicology* (in press).

[16] Lucarella, D. A document retrieval system based on nearest neighbour searching. *Journal of Information Science* **14**, 25-33, 1988.

[17] Randic, M. and Wilkins, C.L. Graph theoretical approach to recognition of structural similarity in molecules. *Journal of Chemical Information and Computer Sciences* **19**, 31-16, 1979.

[18] Rubin, V. and Willett, P. A comparison of some hierarchal monothetic divisive clustering algorithms for structure property correlation. *Analytica Chimica Acta* **151**, 161-166, 1983.

[19] Salton, G. and McGill, M.J. *Introduction To Modern Information Retrieval.* McGraw-Hill, New York, 1983.

[20] Sneath, P.H.A. and Sokal, R.R. *Numerical Taxonomy* Freeman, San Francisco, 1973.

[21] Topliss, J.G. and Edwards, R.P. Chance factors in studies of quantitative structure-activity relationships. *Journal of Medicinal Chemistry* **22**, 1238-1244, 1979.

[22] Varkony, T.H., Shiloach, Y. and Smith, D.H. Computer-assisted examination of chemical compounds for structural similarities. *Journal of Chemical Information and Computer Sciences* **19**, 104-111, 1979.

[23] Wilkins, C.L. and Randic, M. A graph theoretical approach to structure-property and structure-activity correlations. *Theoretica Chimica Acta* **58**, 45-68, 1980.

[24] Willett, P. A comparison of some hierarchal agglomerative clustering algorithms for structure-property correlation. *Analytica Chimica Acta* **136**, 29-37, 1982.

[25] Willett, P. Evaluation of relocation clustering algorithms for the automatic classification of chemical structures. *Journal of Chemical Information and Computer Sciences* **24**, 29-33, 1984.

[26] Willett, P. *Modern Approaches to Chemical Reaction Searching* Gower, Aldershot, 1986.

[27] Willett, P. *Similarity and Clustering in Chemical Information Systems* Research Studies Press, Letchworth, 1987.

[28] Willett, P. and Winterman, V. A comparison of some measures for the determination of inter-molecular structural similarity. *Quantitative Structure-Activity Relationships* **5**, 18-25, 1986.

[29] Willett, P., Winterman, V. and Bawden, D. Implementation of nearest neighbour searching in an online chemical structure search system. *Journal of Chemical Information and Computer Sciences*, **26**, 36-41, 1986.

[30] Willett, P. Recent trends in hierarchic document clustering: a critical review. *Information Processing and Management* (in press).

Prediction of Physicochemical Properties of Organic Compounds from Molecular Structure

P.C. Jurs, M.N. Hasan, P.J. Hansen, and R.H. Rohrbaugh

Department of Chemistry, Penn State University, University Park, PA 16802, USA

ABSTRACT. Relationships between molecular structure and biological activity or molecular structure and physical properties can be investigated for large sets of organic compounds using computer-assisted methods. Our research involves the design, implementation, testing, and application of computer software for the purpose of discovering structure-property and structure-activity relationships and thus developing the capability to predict properties for unknown compounds. The approach involves the graphical entry and storage of structures, three-dimensional molecular modeling, molecular structure descriptor generation, and analysis of the descriptors using pattern recognition methods or multivariate statistical methods. The computer-generated structural descriptors represent the molecules topologically (*e.g.*, path counts, molecular connectivity), geometrically (*e.g.*, molecular volume, surface area, principal moments), electronically (*e.g.*, partial charges, bond orders), and physicochemically (*e.g.*, log P, molar refractivity). A large, fully-integrated, interactive software system, called ADAPT for Automated Data Analysis and Pattern recognition Toolkit, has been developed to make such SAR and SPR research convenient. ADAPT is under continual development through the introduction of new molecular structure descriptors and new analysis methods. A number of successful studies have been reported in property prediction (prediction of boiling points of olefins, GC and HPLC retention indices, and simulation of ^{13}C NMR chemical shifts) and in the structure-activity area (pharmaceutical drugs, olfactory stimulants, mutagens, carcinogens, anti-tumor drugs). Examples of current studies include: SAR of anti-tumor retinoids, carcinogenicity of N-nitroso compounds, HPLC retention indices of PACs and the importance of molecular shape, GC retention indices of polychlorinated biphenyls, ^{13}C NMR simulation of substituted norbornanes.

Introduction

Values for the fundamental themodynamic and physicochemical properties of many chemicals are unavailable in the chemical literature, and their measurement can be a costly and time-consuming undertaking. However, values for such properties are needed for the design, control, and understanding of processes. A gap exists between the available information and what is needed. For this reason a need exists for estimation methods which are both reliable and accessible. Books (1,2) and numerous papers have been published devoted to methods for the estimation of thermophysical properties.

C. Jochum M.G. Hicks J. Sunkel (Eds.)
Physical Property Prediction
in Organic Chemistry

This paper deals with computer-assisted methods for the prediction of physicochemical properties from molecular structure. The attempt to rationalize the connections between the molecular structures of organic compounds and their physicochemical properties comprises the field of structure-property relations (SPR) studies. The approach presented here involves analyzing a set of compounds for which the property of interest is known by representing their molecular structures by a set of descriptors and then developing multiple regression model equations to relate the descriptors to the property. The overall approach is shown in **Figure 1**.

The main steps involved in performing a computer-assisted SPR study are as follows: (a) Input and store the molecular structures under investigation as well as the physicochemical property of interest for each compound. (b) Generate three-dimensional molecular models using a molecular mechanics approach. (c) Calculate computer-generated molecular structure descriptors for each molecule in the data set. The descriptors are typically derived directly from the stored topological representations of the structures. When appropriate, they are derived from the calculated three-dimensional molecular models. Test the descriptors for their significance and save for consideration only those descriptors which are not too highly correlated with each other. (d) Use multivariate statistical methods to develop quantitative mathematical models that relate the molecular structures as represented by the calculated descriptors to their properties. Use pattern recognition methods to search for the best sets of descriptors and to map and display the data for analysis. (e) Systematically focus on which of the molecular structure descriptors are the most useful in developing the quantitative predictive model. (f) Test the predictive ability of these quantitative models with new compounds not used during the development stage of the model.

The SPR studies described here have been done with the ADAPT (Automated Data Analysis and Pattern recognition Toolkit) computer software system (3). It currently consists of more than 90 independent FORTRAN modules that are executed interactively on a time-sharing

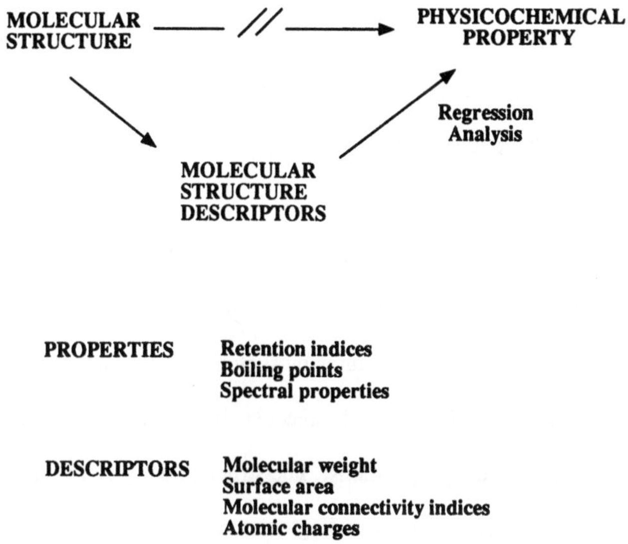

Figure 1. Schematic diagram of computer-assisted approach to structure-property relationship studies.

PRIME 750 computer system. The system has been designed and implemented to provide the user with the capabilities necessary to perform SPR studies on sets of up to several hundred compounds at a time. The ADAPT system can be viewed as a set of software tools with which to perform SPR experiments. The system has been implemented in modules. This provides the user with the freedom to choose sequences of operations in a flexible way suitable to the problem at hand. It affords the developers and users the ability to create new modules and plug them into the system with minimal or no disruption of other routines.

Entry of Molecular Structures

The ADAPT computer system has an interactive graphics facility that enables the user to enter, modify, retrieve, and display molecular structures of organic compounds. The routines allow the convenient, interactive entry of structures by sketching them on the screen of a graphics display terminal or by building them from stored molecular fragments. No special techniques or codes must be learned to perform these operations. The structure entry can be done in thirty seconds to several minutes per compound, depending on structural complexity, enabling large sets of structures to be entered into the ADAPT disk files in reasonable amounts of time. The structure files are stored permanently on disk files for further processing by the other modules of ADAPT. Information saved for each compound includes a compressed connection table, ring information, a list of associated numerical information, an identification number, the chemical name of the compound, and the two-dimensional coordinates of the atoms as entered (for possible redrawing later or to be used as starting coordinates for three-dimensional model building).

Molecular Mechanics Model Builder

A molecular mechanics model building routine has been developed to allow the generation of geometrical information about the compounds under investigation. The routine is capable of developing quite good conformations relatively quickly starting from two-dimensional models. These models are then used later by the geometric descriptor development routines during the descriptor development phase of a study. Interface programs (a preprocessor and a postprocessor) also have been developed to allow models to be passed to the MM2 program of Allinger (4,5) for more sophisticated conformational analysis if necessary.

Descriptor Generation

The heart of the ADAPT system is the set of molecular structure descriptor development modules. They fall into four classes: topological, geometrical, electronic, and physicochemical.

a. Topological. These descriptors are derived from the connection table representation of the structure and include such information as atom and bond counts, substructure counts, molecular connectivity indices, substructure environment, and path descriptors. Unique programs for representing molecular symmetry also exist. Many of the topological descriptors are based on a graph theory perspective of molecular structure.

ADAPT has a general substructure searching routine that is used to develop substructure-based descriptors. Each of the structures comprising a set of compounds under study is searched for the presence of the substructure of interest. The descriptor value for a given substructure in a particular molecule is the number of occurrences of that substructure within the molecule. The substructures to be used are problem dependent and must be chosen by the user through chemical reasoning. ADAPT also has routines that can be used to aid the chemist in searching for useful substructures by examining a large set of structures to locate common substructures.

While the presence or absence of substructures is indicated by the substructure count descriptors, their manner of connection into the molecular structure is missing. Environment descriptors are designed to supply information about such connections by coding the immediate surroundings of substructures. To generate an environment descriptor, the molecule being coded is searched for the presence of the substructural fragment that forms the heart of the environment being sought. If no match is found, then the descriptor is given the value of zero. If the substructure is found to be present, then the environment descriptor is computed by performing a calculation on a pseudomolecule comprised of the substructural atoms and the neighboring atoms. Several different calculations are implemented in ADAPT to code the electronic or steric surroundings of the substructure as imbedded within the structure being coded. The objective is to provide molecular structure descriptors that will code the environment in which the substructure is found. A number of the environment descriptors have been found to be important in previous SPR and structure-activity studies.

The molecular connectivity (6) of a molecule is a measure of the degree of branching of the structure. It is a graph theory based descriptor that depends only on the topology of the structure. Strong correlations between molecular connectivity indices and a large number of physicochemical properties have been presented in the literature over the past ten years. Recently, Kier (7-9) has developed a set of molecular connectivity indices, called kappa indices ($^{n}\kappa$), that are designed to encode steric aspects of molecular structure. Balaban has recently reported (10) the average distance-sum connectivity, and we have found it to be a useful topological descriptor in several SPR studies (*e.g.*, 11).

b. Geometrical. These descriptors are derived from the three-dimensional molecular models of the structures and include such information as the principal moments of inertia, molecular volume, and surface area. Recent developments in the area of geometrical descriptors designed to represent molecular shape (shadow areas) are discussed in conjunction with some shape-related studies below.

c. Electronic. These are quantities characterizing the structure with partial atomic charges, dipole moments, bond strengths, etc. derived either from Del Re sigma charge computations (12) or extended Huckel calculations. The types of electronic information that can be included in the descriptor sets are partial charges on atoms and a number of reactivity indices such as bond orders, free valence index, partial charges, energies of HOMO and LUMO molecular orbitals, etc.

d. Physicochemical. Parameters such as log P (the logarithm of the partition coefficient of a compound between water and 1-octanol), molar refraction, and molecular polarizability are often correlated with other physicochemical properties. Descriptors that are either estimations of such quantities or calculable measures that are simply related to the physicochemical descriptors can be calculated in ADAPT modules.

ADAPT includes a routine that allows the user to input any additional descriptors from outside the system. This allows the user to study laboratory data (*e.g.*, retention indices, spectroscopic data) along with the calculated descriptors. There is also a general purpose descriptor file management routine that allows the review or manipulation of any stored descriptor. Another routine allows mathematical manipulation of descriptors such as addition, multiplication, logarithmic transformation, exponentiation, etc.

The development of adequate sets of descriptors for the compounds forming a data set is the most critical, yet the most difficult, part of any SPR study. With an adequate set of descriptors available, the analysis portion of an SPR study is relatively straightforward. Lacking such a set of descriptors, one has no choice but to keep searching for better descriptors. Thus, descriptor development and testing is the most time-consuming and most experimental part of any SPR study.

Pattern Recognition and Statistical Analysis

The analysis portion of ADAPT has routines implementing various statistical and pattern recognition methods. The object of the analysis phase is to find discriminants that separate subsets of the data into the categories (if category data is being studied), or to build model equations (if quantitative data is available). This phase of an SPR study is guided by the user in a highly interactive manner in order to search through the available descriptors to find the best set of descriptors to be used in the equations or discriminants. Pattern recognition, a subfield of artificial intelligence, includes a collection of parametric and nonparametric techniques used to study data sets that may not conform to well-characterized probability density functions. A large and growing literature describes this field (*e.g.*, 13-18). All four of the major types of pattern recognition methods are available in ADAPT: mapping and display, discriminant development, clustering, and modeling.

The data to be analyzed -- quantitative descriptions of the compounds of interest -- are represented by points in a high-dimensional space. Each compound in the data set is represented as follows: $X = (x_1, x_2, x_3, ..., x_j, ..., x_d)$, where x_j is the value of the j-th descriptor for this compound. The expectation is that the points representing compounds of common property values (*e.g.*, highly volatile compounds) will cluster in one limited region of the space, while the points representing the compounds of another property class (*e.g.*, nonvolatile compounds) will cluster elsewhere. The clusters are regions of high local density which are relatively well separated in space. Pattern recognition consists of methods for investigating data represented in this manner to assess the degree of clustering and general structure of the data space.

Parametric methods of pattern recognition attempt to find classification surfaces or clustering definitions based on statistical properties of the members of one or both classes of points. For example, Bayesian classification surfaces are developed using the mean vectors for the member of the classes and the covariance matrices for the classes. If the statistical properties can not be calculated or estimated, then nonparametric methods are used. Nonparametric methods attempt to find clustering definitions or classification surfaces by using the data directly, without assuming any distribution. Nonparametric discriminant development methods available in ADAPT include the linear learning machine (13), a method based on simplex optimization (3), an iterative least squares procedure (3), and adaptive least squares (19). Once discriminants have been found that do separate the data set into the appropriate classes, then these discriminants can be used to predict the property for unknown compounds.

An alternative way to investigate the structure of the points in a data space is to use clustering methods (14). These methods search for intrinsic relationships among clusters of points using interpoint distances. ADAPT has several clustering routines, including hierarchical clustering (20), ISODATA (21), the K-means algorithm (22) and the maximum-minimum algorithm (23). Mapping and display methods can also be used to seek clusters visually, *e.g.*, principal components analysis and nonlinear mapping.

When quantitative property data are available, then quantitative mathematical models can be constructed using statistical methods. ADAPT contains several multiple linear regression analysis modules: stepwise, forward-selection and backward-deletion routines (24,25), a leaps-and-bounds routine (26), and ridge regression (27). Multiple linear regression analysis is used in ADAPT for two major purposes: to investigate multicollinear relationships among subsets of the molecular structural descriptors, and to build quantitative predictive models for SPR studies. A number of additional supportive multivariate statistical routines are present, such as provision to produce adjusted variable plots (28), and a method for determining optimum variable subsets for discriminant analysis (29). Nonlinear as well as linear regression analysis methods are available.

The final output of an ADAPT-based SPR study is the identity of the descriptors shown to be correlated with the property of interest and the model equations that were developed. Possession of these allows the prediction of the property for unknown compounds in the future.

STRUCTURE-PROPERTY RELATIONSHIP STUDIES

A number of studies of the application of these computational techniques to the problem of searching for relationships between molecular structure and physicochemical properties have been reported. Among such studies are retention index predictions in gas chromatography (30,31) and HPLC (32,33), prediction of boiling points of olefins (34), prediction of relative retention times of polychlorinated biphenyls (35), and prediction of impact sensitivity of explosives (36).

Shape Descriptors and HPLC Retention Indices (32,37)

Several parameters related to molecular shape, including six new parameters, were studied for the purpose of predicting retention in high performance liquid chromatography (HPLC) for polycyclic aromatic hydrocarbons (PAHs). The study was performed using data collected on both polymeric and monomeric HPLC stationary phases. The goal of this study was to test the hypothesis that molecular shape plays a more important role in retention on polymeric phases than on monomeric phases, as suggested by Sander and Wise (38).

The molecular shape parameters used in this study included: surface area and volume calculated using Pearlman's algorithm (39), length-to-breadth ratio as used by Sander and Wise (L/B) (38,40) and by Radecki (41), the κ shape indices proposed by Kier (7-9), and a new set of parameters called shadow areas (37). Other calculated molecular structure descriptors used in the study were moments of inertia, molecular connectivities, log P, molecular polarizability, and several sigma-charge based electronic descriptors.

Six new geometric indices, called shadow areas, were developed. These indices are based on a molecule's three-dimensional structure. The molecule being represented is put in a given orientaion and is flattened into a plane by disregarding the third dimension. The area of the molecule which is projected onto the remaining two dimensions defines the shadow area of interest. A simple analogy, from which the name is derived, is to take the area of the shadow produced when light is directed perpendicularly at the molecule. **Figure 2** shows a pictorial representation of the shadow descriptors for trichlorotoluene. The algorithm for calculating the area is a two-dimensional version of that presented by Stouch and Jurs (42). The six indices are defined as follows:

SHDW1 - The area of the shadow in the X-Y plane.
SHDW2 - The area of the shadow in the X-Z plane.
SHDW3 - The area of the shadow in the Y-Z plane.
SHDW4 - The standardized area of the shadow in the X-Y plane (area of shadow in X-Y plane divided by the area of a box defined by the maximum X and Y dimensions).

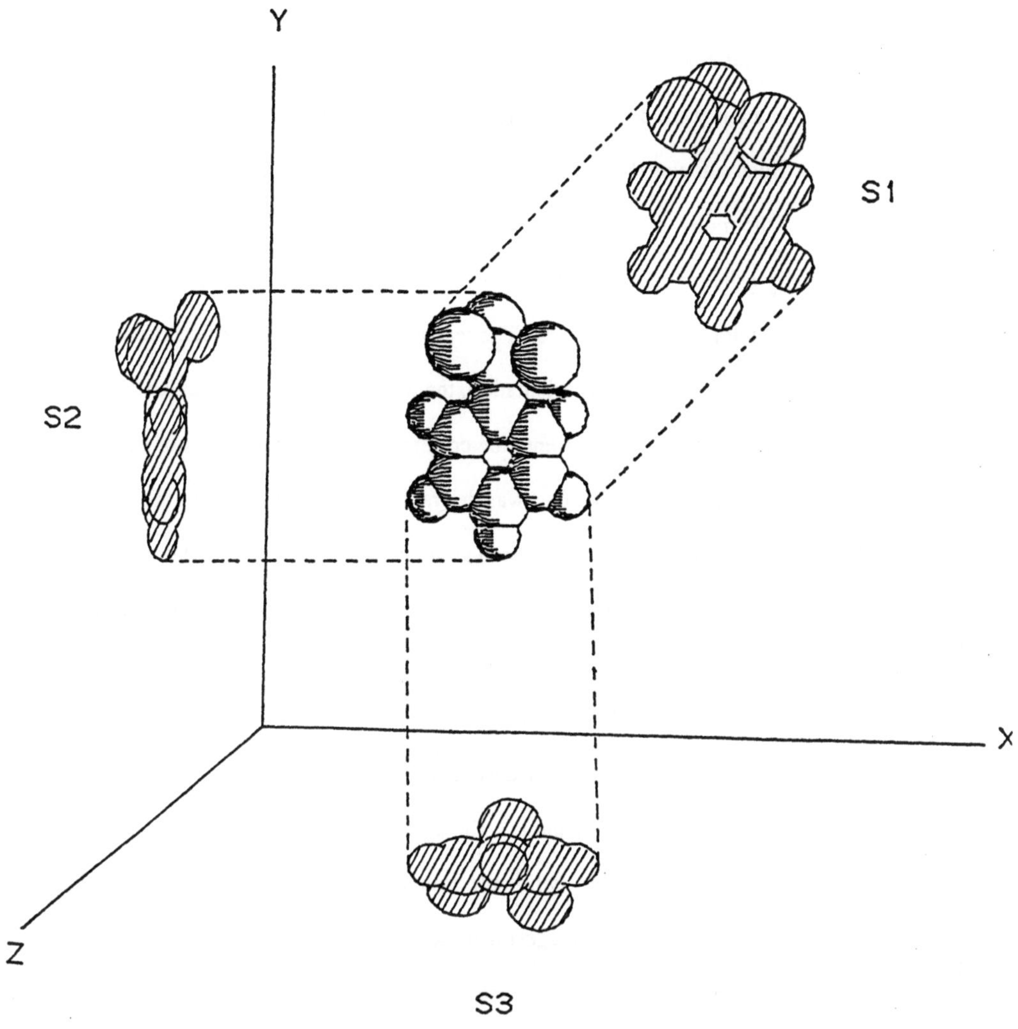

Figure 2. Pictorial representation of the three shadow descriptors for trichlorotoluene.

SHDW5 - The standardized area of the shadow in the X-Z plane.
SHDW6 - The standardized area of the shadow in the Y-Z plane.

The standarized areas were developed to give a measure of shape independent of molecular size. The values calculated for the shadow descriptors are dependent on the initial absolute orientation of the molecule, so a standard orientation must be defined. For this study, the molecules were initially oriented such that the first and second principal moments of the compound were aligned with the X and Y axes. The reason for choosing this orientation is related to the flatness of the PAHs. In this orientation, the main bulk of the molecule will lie in the first shadow area, or X-Y plane. This orientation corresponds to the alignment of a molecule in the slot model of Sander and Wise (40). This was desirable because we wanted to focus on shape/steric effects. Another alternative would be to align the first two charge vectors of the molecule with the X and Y axes. However, for the case at hand (*i.e.*, nonpolar aromatics) these vectors would be too small to be significant use.

Eight-variable equations were developed for both the polymeric and monomeric phase HPLC retention indices. Each equation contained molecular polarizability, L/B ratio, third moment of inertia, $^{3}\kappa$ index, and log P descriptors. The equation for the polymeric phase also included the shadow 3, 4, and 5 descriptors, while the equation for the monomeric phase included three electronic descriptors. These results suggest that molecular shape parameters are indeed more significant in the prediction of retention for the polymeric phase, while electronic parameters predominate in the prediction of retention for the monomeric phase.

Prediction of Boiling Points of Olefins (34)

Values for many fundamental properties of chemicals are unavailable in the chemical literature, and their measurement can be costly. Reliable, accurate, and easily accessible methods are needed to estimate such properties. We performed a study to estimate normal boiling points because the available estimation methods are poor and/or narrowly applicable. The normal boiling points for 123 olefins with two to ten carbons atoms were obtained from the literature (43). The molecular structure descriptors used were topological descriptors derived directly from molecular structure. The four descriptors that correlated most highly on an individual basis with olefin boiling point were number of carbon atoms, number of single bonds, molecular weight, and path-one molecular connectivity. The square roots of these four descriptors were also highly correlated with boiling point. Predictive equations having from one to eight independent variables were obtained by using stepwise multiple linear regression analysis. The best overall equation developed was as follows:

$$
\begin{aligned}
BP = & \ (54.62 \pm 2.19)[\text{sqrt of molecular weight}] \\
+ & \ (4.53 \pm 0.70)[\text{count of ring atoms}] \\
+ & \ (7.10 \pm 1.43)[\text{count of clusters of size 3}] \\
+ & \ (8.98 \pm 0.82)[\text{degree of alkene substitution}] \\
- & \ (12.33 \pm 1.77)[\text{count of paths of length 2}] \\
- & \ (5.71 \pm 1.09)[\text{count of methyl groups}] \\
+ & \ (7.84 \pm 1.69)[\text{path-3 molecular connectivity}] \\
- & \ (2.02 \pm 0.67)[\text{count of ethyl groups}] \\
- & \ (393.04 \pm 13.35)
\end{aligned}
$$

$$n = 123 \quad s = 1.83 \quad r = 0.999 \quad F(8,114) = 7120$$

The boiling points for the 123 member training set predicted by this model are plotted against the observed boiling points in **Figure 3**. The estimated adjusted standard deviation was 1.83 °C and the multiple correlation coefficient, r, was 0.999. The information content of the eight descriptors in this model was analyzed and discussed. The molecular weight, ring information, degree of branching, and the size of the molecule are all reasonable correlates for boiling point. Thus, it is possible to build simple model equations that predict boiling point for olefins with high accuracy.

Prediction of Gas Chromatographic Retention of PCBs (35)

The widespread occurence and importance of polychlorinated biphenyls (PCBs) in the environment has made their determination a very important analytical task. Capillary column gas

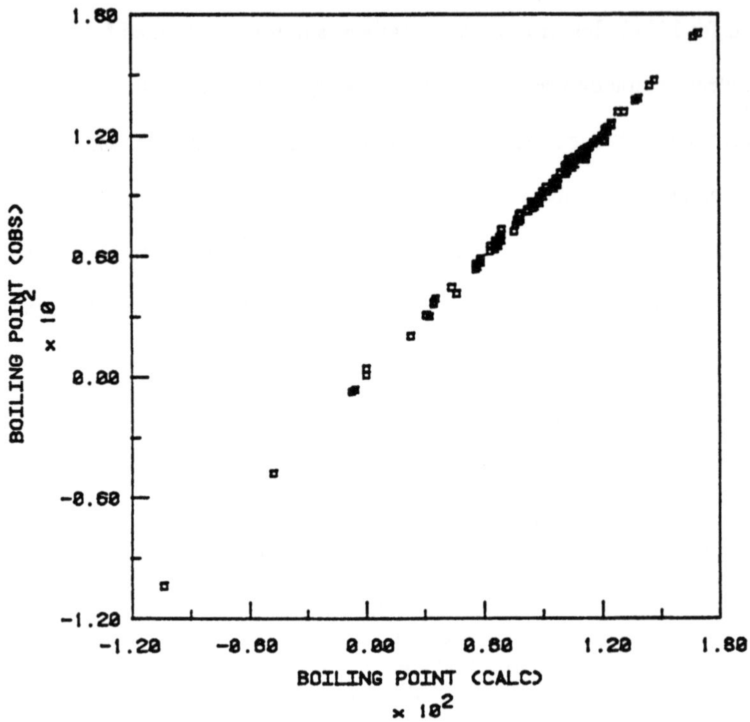

Figure 3. Plot of observed vs. calculated boiling points for 123 olefins.

chromatography is the most commonly used method of identification and quantification of PCBs in such samples. This study is an attempt to develop molecular structure-retention relationships for PCBs using computer-assisted techniques.

The data set used in this study consisted of retention data for all 209 PCB congeners taken from the literature (44). The separation was carried out on capillary column coated with SE-54 and the retentions were expressed relative to octachloronaphthalene (RT = 124.9 min) as a standard. Relative retention times (RRT) resulted.

With the ADAPT system we were able to enter and manipulate the molecular structures af all the compounds in the data set and then genarate three-dimensional models using force-field modeling programs. From these modeled structures were generated molecular structure descriptors, including topological, substructure, electronic, geometric, and molecular connectivity descriptors. Objective feature selection was used to prune the number of descriptors by eliminating those with high multicollinearities. The remaining descriptors were used as independent variables in several stepwise multiple linear regression analysis routines. Regression equations were formed from subsets of the molecular structure descriptors.

We have been able to generate a number of equations with r^2 values greater than 0.99 and relative standard deviations less than 4%. One of the best equations is as follows:

$$\begin{aligned} RRT = \; & (0.128 \pm 0.0047)[\text{no. of chlorines}] \\ - \; & (0.0327 \pm 0.0033)[\text{no. of ortho substituents}] \\ + \; & (0.0254 \pm 0.0021)[\text{no. of adjacent chlorines}] \\ - \; & (0.00482 \pm 0.00041)[(\text{no. of chlorines})^2] \\ + \; & (0.0000268 \pm 0.0000046)[\text{second principal moment}] \\ + \; & (0.0205 \pm 0.012) \end{aligned}$$

$$n = 209 \quad r = 0.997 \quad s = 0.0100 \quad F(5,203) = 13177$$

The variables are listed in the order they were chosen by the stepwise multiple linear regression anaysis procedure, that is, their order of statistical significance. The 95% confidence limits of the coefficients are also given. The small value of the standard deviation (s=0.0100) indicates the high precision of this equation. This is approximately 2% of the mean relative retention time. A plot of experimental against predicted relative retention times for the 209 PCB compounds is shown in **Figure 4**.

An important question to be addressed in the construction of models for prediction of physicochemical properties is how many observations (compounds) are necessary to generate reliable predictive models with the necessary accuracy? We have been able to address this question as part of this study of the capillary GC retention of PCBs (35). In our study, fifty training sets were generated by randomly selecting n compounds from the 209 PCBs in the full data set. For each training set, an equation was developed using the fivevariable equation shown above. The equations were then used to predict the relative retention times of those compounds not included in the training set. Comparisons were made against the observed values, and the residual mean square was calculated as a measure of precision. The entire procedure was repeated with n values between 10 and 100. **Figure 5** shows the prediction standard deviation, s, plotted against the size of the subset, n. The s value decreases rapidly as n increases and stabilizes when n is about 40. Thus, a subset of 40/209 = 19% of the entire data set was sufficiently large to allow the generation of a predictive equation of high quality. In this particular experiment, the mean value of s for n = 40 observations is approximately the same as the standard error of fitting the whole data set. This experiment was repeated using different training sets, and the same results were obtained. These results suggest that a subset of a closed set of data may be sufficient to generate high-quality predictive models for the entire set.

Impact Sensitivity of Explosives (36)

Impact sensitivity is an experimental measurement that is relatively easily done and that gives an indication of likely safety problems with an energetic material. Impact sensitivity is conveniently expressed as log H_{50}, the logarithm of the drop height at which the material is 50% likely to detonate. This variable has been correlated with several parameters derived from the

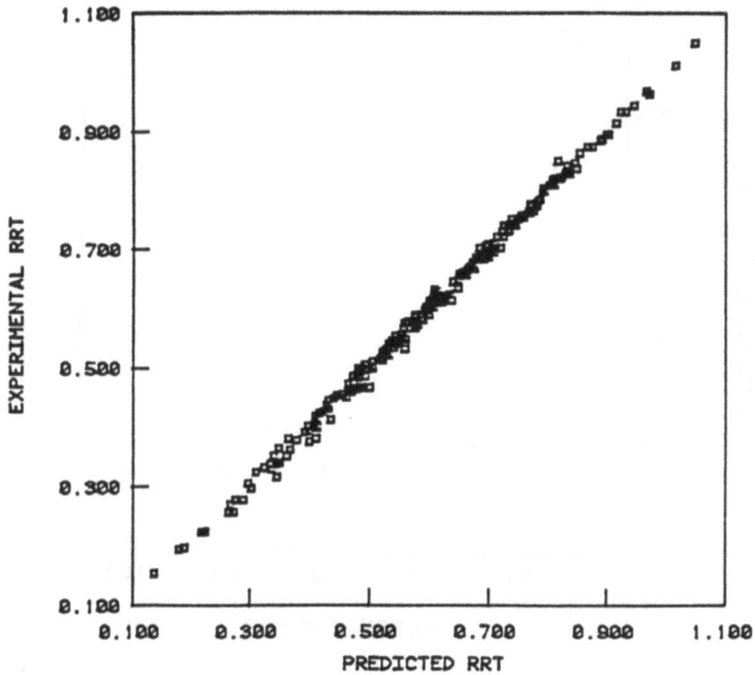

Figure 4. Plot of observed vs. calculated relative retention times for 209 PCBs.

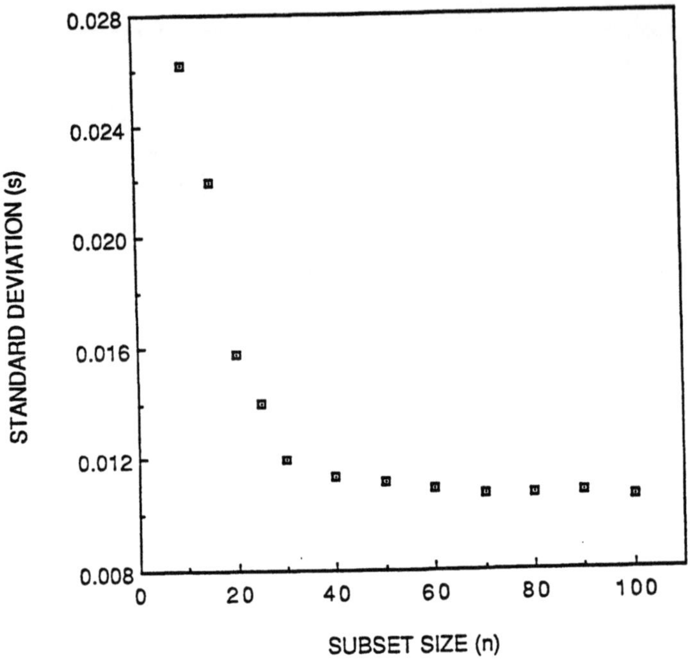

Figure 5. Plot of the prediction standard deviation vs. the subset size for model generation with PCBs.

molecular structures of materials (36,45-49), most notably oxygen balance. Oxygen balance is the number of equivalents of oxidant per hundred grams of explosive above the amount required to burn all the hydrogen to water and all carbon to carbon monoxide. For molecules containing C, H, N, and O oxygen balance has been defined as follows:

$$OB_{100} = 100\,(2n_O - 2n_H - 2n_C - 2n_{COO}) / (\text{Molecular Weight})$$

where n_O, n_H, n_C are the numbers of atoms of the respective elements O, H, C in the molecule, and n_{COO} is the number of carboxyl groups present. A number of values of log H_{50} have been reported for a variety of energetic materials (45,46).

Several investigators have published studies of relationships between molecular structure and impact sensitivity. For small sets of polynitroaliphatic and polynitroaromatic compounds Kamlet has found (45,46) excellent correlations between log H_{50} and OB_{100}, and has reported several equations with R values greater than 0.95.

For a set of 61 nitrated compounds which had their log H_{50} values measured at Los Alamos National Laboratory, we have found several regression equations that relate structural descriptors to impact sensitivity. The best equation found to date follows:

$$\begin{aligned} \log H_{50} = & - (0.242 \pm 0.025)[OB_{100}] \\ & - (0.411 \pm 0.045)[\text{ENVR } 1] \\ & + (0.077 \pm 0.020)[\text{ENVR } 2] \\ & - (4.53 \pm 1.43)[\text{ENVR } 3] \\ & + 6.71 \end{aligned}$$

$$n = 61 \quad r = 0.873 \quad s = 0.23$$

where the descriptors labelled ENVR are molecular connectivity environment descriptors for the following three substructures: 1: the azo linkage; 2: a secondary nitrogen linking an aromatic ring and a carbon atom having two single bonds and one double bond; 3: the nitro group. These three environmental descriptors generate information relevant to the surroundings of azo linkages, amine linkages, and nitro groups as they are bound in the molecular structures. For this set of compounds, oxygen balance is linearly correlated with log H_{50} only with an r value of 0.511, a much weaker association that with the smaller sets of nitrated compounds studied by Kamlet. Nonetheless, with the structurally-based descriptors included, the overall equation is quite significant.

Predicting Viscosity from Molecular Structure (50)

The viscosity at 210 ^{0}C for 48 compounds was predicted with a linear equation based on only two structural descriptors. The molecules considered were 26-carbon hydrocarbons with long saturated chains and some rings. Four example compounds are 11-n-butyldocosane, 3-phenyleicosane, 1,4-di-n-decylcyclohexane, and 2-n-dodecylperhydrophenanthrene. The viscosity at 210 ^{0}C ranges from 2.72 to 9.10 for these compounds; on the logarithmic scale this becomes 0.435 to 0.959. A number of topological and substructural descriptors were calculated and evaluated, and a good model was generated with only two descriptors for the set of compounds with three outliers deleted from consideration:

$$\log V_{210} = 0.017[\text{Number of ring atoms}] + 0.039[\text{Path-2 molecular connectivity}] + 0.106$$

$$n = 45 \quad r = 0.95 \quad s = 0.03$$

The standard deviation of 0.03 is a small fraction of the range of log V_{210} values, which is 0.50. Thus, it was possible to develop a simple equation based on readily calculated descriptors that could predict viscosity at 210 ^{0}C with good accuracy.

SUMMARY

Computer-assisted methods can be used to develop statistically strong equations that relate calculated molecular structure descriptors to physicochemical properties of the compounds. The methodology to perform such studies on sets of hundreds of organic compounds exists and is applicable to this class of problems. Examples of successful studies include gas chromatographic retention indexes, thermodynamic properties such as boiling points, engineering parameters such as impact sensitivity of energetic materials, and viscosity of liquids.

CREDIT

Financial support from the National Science Foundation through Grant No. CHE-8503542 is acknowledged.

REFERENCES CITED

1. R.C. Reid, J.M. Prausnitz, T.K. Sherwood, *Properties of Gases and Liquids*, 3rd Ed., McGraw-Hill, New York, 1977.

2. W.J. Lyman, W.F. Reehl, D.H. Rosenblatt, *Handbook of Chemical Property Estimation Methods*, McGraw-Hill, New York, 1982.

3. A.J. Stuper, W.E. Brugger, P.C. Jurs, *Computer Assisted Studies of Chemical Structure and Biological Function,* Wiley-Interscience, New York, 1979.

4. N.L. Allinger and Y.H. Yuh, Molecular Mechanics, Operating Instructions for MM2 and MMP2 Programs,1977 Force Field, Quantum Chemistry Program Exchange, QCPE Program No. 395, 1980.

5. U. Burkert and N.L. Allinger, *Molecular Mechanics,* American Chemical Society, Washington, DC, 1982.

6. L.B. Kier and L.H. Hall, *Molecular Connectivity in Structure-Activity Analysis,* John Wiley and Sons, Inc., New York, 1986.

7. L.B. Kier, A Shape Index from Molecular Graphs, *Quant. Struct.-Act. Relat.*, **4**:109 (1985).

8. L.B. Kier, Shape Indexes of Orders One and Three from Molecular Graphs, *Quant. Struct.-Act. Relat.*, **5**:1-7 (1986).

9. L.B. Kier, Distinguishing Atom Differences in a Molecular Graph Shape Index, *Quant. Struct.-Act. Relat.*, **5**:7-12 (1986).

10. A.T. Balaban, Highly Discriminating Distance-Based Topological Index, *Chem. Phys. Letters* **89**:399-404.

11. P.A. Edwards and P.C. Jurs, Correlation of Odor Intensities with Structural Properties of Odorants, *Chemical Senses,* in review.

12. G. Del Re, A Simple MO-LCAO Method for the Calculation of Charge Distributions in Saturated Organic Molecules, *Jour. Chem. Soc.*, 4031-4040 (1958).

13. N.J. Nilsson, *Learning Machines,* McGraw-Hill Book Co., New York, 1965.

14. J.T. Tou and R.C. Gonzalez, *Pattern Recognition Principles,* Addison-Wesley, Reading, Mass., 1974.

15. K. Varmuza, *Pattern Recognition in Chemistry,* Springer- Verlag, Berlin, 1980.

16. D.D. Wolff and M.L. Parsons, *Pattern Recognition Approach to Data Interpretation,* Plenum Press, New York, 1983.

17. M.A. Sharaf, D.L. Illman, B.R. Kowalski, *Chemometrics,* Wiley, New York, 1986.

18. D.L. Massart, B.G.M. Vandeginste, S.N. Deming, Y. Michotte, L. Kaufman, *Chemometrics: A Textbook,* Elsevier Scientific Publishers, Amsterdam, 1987.

19. I. Moriguchi, K. Komatsu, Y. Matushita, Adaptive Least Squares Method Applied to Structure-Activity Correlation of Hypotensive N-Alkyl-N"-cyano-N'-pyridylguanidines, *Jour. Med. Chem.*, **23**:20-26 (1980).

20. W.T. Williams and G.N. Lance, Hierarchical Classificatory Methods, in *Statistical Methods for Digital Computers,* K. Enslein, A. Ralston, H.S. Wilf (Eds.), Wiley-Interscience, New York, 1975.

21. G.H. Ball and D.J. Hall, Isodata, an Iterative Method of Multivariate Analysis and Pattern Classification, Proceedings of the IFIPS Congress, 1965.

22. J. MacQueen, Some Methods for Classification and Analysis of Multivariate Data, *Proc. of the 5th Berkeley Symposium on Probability and Statistics,* University of California Press, Berkeley, CA, 1967.

23. B.G. Batchelor and B.R. Wilkins, Method for Location of Clusters of Patterns to Initialize a Learning Machine, *Electronics Letters,* **5**, 481-483 (1969).

24. N.R. Draper and H. Smith, *Applied Regression Analysis,* 2nd. Ed., Wiley, 1981.

25. D.A. Belsley, E. Kuh, R.E. Welsch, *Regression Diagnostics: Identifying Influential Data and Sources of Collinearity,* Wiley, 1980.

26. G.M. Furnival and R.W. Wilson, Jr., Regressions by Leaps and Bounds, *Technometics,* **16**:499 (1974).

27. D.W. Marquardt and R.D. Snee, Ridge Regression in Practice, *Amer. Stat.*, **29**:3-20 (1975).

28. J.M. Chambers, W.S. Cleveland, B. Kliener, P.A. Tukey, *Graphical Methods for Data Analysis,* Duxbury Press, Boston, 1983.

29. G.P. McCabe, Jr., Computations for Variable Selection in Discriminant Analysis, *Technometrics,* **17**:103-109 (1975).

30. R.H. Rohrbaugh and P.C. Jurs, Prediction of Gas Chromatographic Retention Indexes of Polycyclic Aromatic Compounds and Nitrated Polycyclic Aromatic Compounds, *Anal. Chem.* **58**:1210-1212 (1986).

31. R.H. Rohrbaugh and P.C. Jurs, Prediction of Gas Chromatographic Retention Indexes of Selected Olefins, *Anal. Chem.* **57**:2770-2773 (1985).

32. R.H. Rohrbaugh and P.C. Jurs, Molecular Shape and the Prediction of High-Performance Liquid Chromatographic Retention Indexes of Polycyclic Aromatic Hydrocarbons, *Anal. Chem.* **59**:1048-1054 (1987).

33. M.N. Hasan and P.C. Jurs, Computer Assisted Prediction of Liquid Chromatographic Retention Indices of Polycyclic Aromatic Hydrocarbons, *Anal. Chem.* **55**:263-269 (1983).

34. P.J. Hansen and P.C. Jurs, Prediction of Olefin Boiling Points from Molecular Structure, *Anal. Chem.* **59**:2322-2327 (1987).

35. Mohamed Noor Hasan and P.C. Jurs, Computer-Assisted Prediction of Gas Chromatographic Retention Indexes of Polychlorinated Biphenyls, *Anal. Chem.*, in press.

36. N.R. Greiner, L.E. Wangen, P.C. Jurs, R. Heaton, G. Peterson, C.B. Storm, M. Phillips, A Chemometrics Approach to Impact Sensitivity, Working Group Meeting on Sensitivity of Explosives, CETR, Socorro, NM, Mar 1987.

37. R.H. Rohrbaugh and P.C. Jurs, Descriptions of Molecular Shape Applied to StructureActivity and Structure-Property Relationship Studies, *Anal. Chim. Acta* **199**:99-109 (1987).

38. L.C. Sander and S.A. Wise, Investigations of Selectivity in RPLC of Polycyclic Aromatic Hydrocarbons, in *Advances in Chromatography,* J.C. Giddings (Ed.), Vol. 25, 1986, pp. 139-218.

39. R.S. Pearlman, Molecular Surface Areas and Volumes and Their Use in Structure/Activity Relationships, in *Physical Chemical Properties of Drugs*, S.H. Yalkowsky, A.A. Sinkula, S.C. Valvani (Eds.), Marcel Dekker, New York 1980.

40. L.C. Sander and S.A. Wise, Synthesis and Characterization of Polymeric C-18 Stationary Phases for Liquid Chromatography, *Anal. Chem.* **56**:504-510 (1984).

41. A. Radecki, H. Lamparczyk, R. Kaliszan, A Relationship between the Retention Indices on Nematic and Isotropic Phases and the Shape of Polycyclic Aromatic Hydrocarbons, *Chromatographia*, **12**:595-599 (1979).

42. T.R. Stouch and P.C. Jurs, A Simple Method for the Representation, Quantification, and Comparison of Volumes and Shapes of Chemical Compounds, *Jour. Chem. Inf. Comp. Sci.* **26**:4-12 (1986).

43. TRC Thermodynamic Tables - Hydrocarbons, Thermodynamics Research Center, Texas A&M University, College Station, TX, 1986; Vol. I, Part a.

44. M.D. Mullin, *et al.*, High-Resolution PCB Analysis: Synthesis and Chromatographic Properties of All 209 PCB Congeners, *Environ. Sci. Tech.* **18**:468-476 (1984).

45. M.J. Kamlet, The Relationship of Impact Sensitivity with Structure of Organic High Explosives. I. Polynitroaliphatic Explosives, in *Proc. 6th Symposium (International) on Detonation,* San Diego, CA Aug 1976; ONR Report ACR 221, p. 312.

46. M.J. Kamlet and H.G. Adolph, The Relationship of Impact Sensitivity with Structure of Organic High Explosives. II. Polynitroaromatic Explosives, *Propellants and Explosives,* **4**:30-24 (1979).

47. M.J. Kamlet and H.G. Adolph, Some Comments Regarding the Sensitivities, Thermal Stabilities, and Explosive Performance Characteristics of Fluorodinitromethyl Compounds, in *Proc. 7th Symposium (International) on Detonation*, Anapolis, MD, Jun 1981; NSWC MP 82-334, p. 84.

48. H.G. Adolph, J.R. Holden, D.A. Cichra, Relationships Between the Impact Sensitivity of High Energy Compounds and Some Molecular Properties which Determine their Performance: N, M, and rho(0), NSWC TR 80-495, Apr 1981.

49. I. Fukuyama, T. Ogawa, A. Miyake, Sensitivity and Evaluation of Explosive Substances, *Propellants, Explosives, Pyrotechnics* **11**:140-143 (1986).

50. Predicting Viscosity from Molecular Structure, Applications Note, Molecular Design, Ltd., 1985.

Current Problems in Quantitative Structure Activity Relationships

Hugo Kubinyi

BASF AG, 6700 Ludwigshafen,
Federal Republic of Germany

1. INTRODUCTION

Quantitative structure activity relationships (QSAR) are to be understood as a consequence of the fact that the interactions of drugs with their biological counterparts are determined by intermolecular forces. Thus the structural dependence of biological activities can be described either by physicochemical parameters (Hansch analysis) or by indicator variables encoding different structural features (Free Wilson analysis, pattern recognition).

2. INTERMOLECULAR BINDING FORCES

While covalent bonds are exceptional cases in the binding of drugs to their receptor (e.g. alkylating agents, penicillin, suicide inhibitors), all different types of electrostatic, polar and hydrophobic interactions play an important role.

DRUG RECEPTOR INTERACTIONS

Type	Energy, $kJ.mol^{-1}$
Covalent bonds	170 - 460
Ionic bonds	20 - 40
Ion-dipole interactions	4 - 17
Dipole-dipole interactions	4 - 17
Hydrogen bonds	4 - 17
Charge transfer complexes	4 - 17
Hydrophobic interactions	4
van der Waals interactions	2 - 4

While the electrostatic interactions are responsible for a first contact of the ligand to its binding site and for the orientation of the ligand, the weak hydrophobic and polar interactions sum up to high binding energies when both structures are complementary not only in their topology but also in their surface properties.

C. Jochum M. G. Hicks J. Sunkel (Eds.)
Physical Property Prediction
in Organic Chemistry

3. FREE WILSON ANALYSIS AND HANSCH ANALYSIS

The concept of Free Wilson analysis [1] is simple and easy to apply. The biological activity of a compound is described as the sum of the (theoretical) biological activity value μ of a lead compound and of the activity contributions a_i of all substituents differing from the corresponding substituents in this reference compound. Only the structures and the biological activity values are needed for a Free Wilson analysis. On the other hand, only predictions for new substituent combinations are possible.

FREE WILSON ANALYSIS (1964)

$$\log 1/C = \Sigma\, a_i + \mu$$

Hansch analysis [2] goes a step further: since the differences in the biological activity values must be caused by the differences in the physicochemical and steric properties of the substituents, the biological activities are described by linear combinations of physicochemical parameters, like log P (P = octanol/water partition coefficient), π (lipophilicity parameter), MR (molar refractivity), σ (Hammett parameter), E_s (Taft steric parameter), etc. In addition, nonlinear lipophilicity activity relationships can be described by a parabolic model, including $(\log P)^2$ or π^2.

HANSCH ANALYSIS (1964)

$$\pi_X = \log P_{R-X} - \log P_{R-H}$$

$$\log 1/C = a\pi + b\sigma + cE_s + \text{const.}$$

$$\log 1/C = a(\log P)^2 + b \log P + c$$

To illustrate the practical application of both models, the results of a Free Wilson analysis and of two different Hansch analyses on the antiadrenergic activities of a series of 22 N,N-dimethyl-α-bromo-phenethylamines are given below. The correlation coefficients of all analyses indicate the good fit of the data. While this is not surprising in the case of the Free Wilson analysis, because the activity values of 22 compounds are described by 10 variables, Hansch analysis gives an excellent quantitative description of the data with only two or three variables. Predictions outside the substituent matrix can be made from the Hansch analyses and, in addition, some conclusions can be drawn on their mode of action: the bromo-phenethylamines are unstable compounds, presumably reacting irreversibly with the adrenergic receptor via aziridinium ions. The better description of the data by σ^+ (rather than by σ) provides a strong support for this hypothesis.

SUBST. N,N-DIMETHYL-α-BROMO-PHENETHYLAMINES

Adrenolytic Activity in the Rat [3]

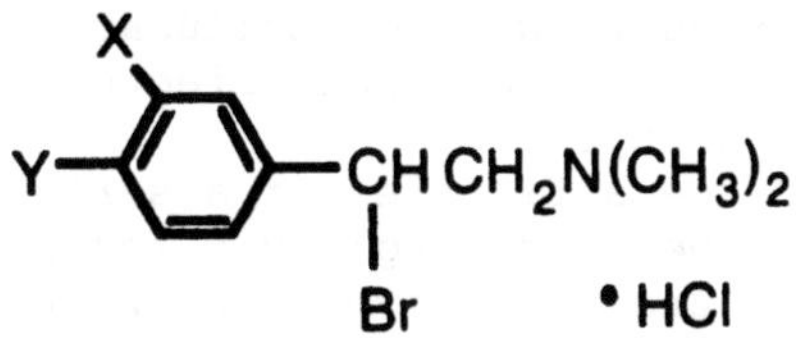

meta	para	log 1/C	meta	para	log 1/C
H	H	7.46	Cl	F	8.19
H	F	8.16	Br	F	8.57
H	Cl	8.68	Me	F	8.82
H	Br	8.89	Cl	Cl	8.89
H	J	9.25	Br	Cl	8.92
H	Me	9.30	Me	Cl	8.96
F	H	7.52	Cl	Br	9.00
Cl	H	8.16	Br	Br	9.35
Br	H	8.30	Me	Br	9.22
J	H	8.40	Me	Me	9.30
Me	H	8.46	Br	Me	9.52

Results of Free Wilson and Hansch analyses [4]

a) Free Wilson analysis $\mu = 7.82$

a_i values	H	F	Cl	Br	J	Me
meta	0.00	-0.30	0.21	0.43	0.58	0.45
para	0.00	0.34	0.77	1.02	1.43	1.26

$(n = 22;\ r = 0.97;\ s = 0.19)$

b) Hansch analysis

$$\log 1/C = 1.15\ \pi - 1.47\ \sigma^+ + 7.82$$
$$(n = 22;\ r = 0.94;\ s = 0.20)$$

$$\log 1/C = 1.26\ \pi - 1.46\ \sigma^+ + 0.21\ E_s^{meta} + 7.62$$
$$(n = 22;\ r = 0.96;\ s = 0.17)$$

4. ADDITIVITY AND NONADDITIVITY OF ACTIVITY CONTRIBUTIONS

Many successful applications of Hansch and Free Wilson analysis confirm the validity of the basic concept of the additivity of group contributions to biological activity values. However, in addition to the exceptions resulting from nonlinear lipophilicity activity relationships there are some other effects causing deviations from the additivity concept.
Abramson et al. [5] studied the affinities of a series of quaternary ammonium compounds to the postganglionic acetylcholine receptor. While a Free Wilson analysis gives a perfect description of the data [6], as long as different groups X are considered as individual substituents, the additivity concept fails when these groups are dissected into smaller segments: the differences in the affinity now depend on the nature of the substituents already present in the group X.

FREE WILSON ANALYSIS OF THE AFFINITIES OF QUATERNARY AMMONIUM COMPOUNDS TO THE POSTGANGLIONIC ACETYLCHOLINE RECEPTOR [5,6]

$X\text{-}CH_2CH_2\text{-}N^+(R_1R_2R_3)$	a_i	$X\text{-}CH_2CH_2\text{-}N^+(R_1R_2R_3)$	a_i
$CH_3CH_2O\text{-}$	-2.479	$(C_6H_5)_2CHCOO\text{-}$	0.872
$CH_3CH_2CH_2\text{-}$	-2.175	$(C_6H_5)_2CHCH_2O\text{-}$	-0.070
$C_6H_5CH_2COO\text{-}$	-1.228	$(C_6H_5)_2CHCH_2CH_2\text{-}$	0.374
$C_6H_5CH_2CH_2O\text{-}$	-1.177	$C_6H_5(C_6H_{11})CHCOO\text{-}$	2.035
$C_6H_5CH_2CH_2CH_2\text{-}$	-0.909	$(C_6H_5)_2C(OH)COO\text{-}$	2.047
$C_6H_{11}CH_2COO\text{-}$	-1.035	$(C_6H_{11})_2CHCOO\text{-}$	1.467
$C_6H_{11}CH_2CH_2O\text{-}$	-0.819	$(C_6H_{11})_2CHCH_2O\text{-}$	0.806
$C_6H_{11}CH_2CH_2CH_2\text{-}$	-0.683	$C_6H_5(C_6H_{11})C(OH)COO\text{-}$	2.975
$\mu = 6.499$		$(n = 128;\ r = 0.991;\ s = 0.231)$	

Differences in receptor affinity (logarithmic scale), by changing R in group X from H to C_6H_5 or H to C_6H_{11} or C_6H_5 to C_6H_{11}

	H to C_6H_5	H to C_6H_{11}	C_6H_5 to C_6H_{11}
$R\text{-}CH_2CH_2CH_2\text{-}$	1.266	1.492	0.226
$R\text{-}CH_2CH_2O\text{-}$	1.302	1.660	0.358
$C_6H_5CH(R)COO\text{-}$	2.100	3.263	1.163
$C_6H_{11}CH(R)COO\text{-}$	3.070	2.502	-0.568

5. NONLINEAR LIPOPHILICITY ACTIVITY RELATIONSHIPS

A nonlinear dependence of biological activity on lipophilicity is observed in all cases where the lipophilicity of the compounds under investigation covers a sufficiently wide range. Many different reasons were discussed for such nonlinear relationships [7]. However, the kinetics of drug transport through biological tissues seems to be the most general explanation in the case of complex biological systems; for simple in vitro systems, like enzyme inhibition or receptor binding, the equilibrium distribution of a ligand between polar and hydrophobic sites and the actual binding site and/or a limited space at the binding site may cause a nonlinear dependence on lipophilicity (or molar refractivity).

POSSIBLE REASONS FOR NONLINEAR LIPOPHILICITY ACTIVITY RELATIONSHIPS

- kinetic control of drug transport
- equilibrium control of drug distribution
- limited binding space / steric hindrance
- allosteric effects
- different metabolism
- lower solubility of lipophilic analogs
- micelle formation
- end product inhibition
- minimal receptor occupation

Besides the parabolic model [7] several other models have been derived for the quantitative description of such nonlinear dependencies; while the probability model [8] and the equilibrium model [9] are only suited for special cases, the bilinear model [10] is applicable to all types of nonlinear relationships.

MATHEMATICAL MODELS FOR NONLINEAR LIPOPHILICITY ACTIVITY RELATIONSHIPS

Parabolic Model (C. Hansch, 1964)

$$\log 1/C = a(\log P)^2 + b \log P + c$$

Probability Model (J. McFarland, 1970)

$$\log 1/C = a \log P - 2a \log(P + 1) + c$$

Equilibrium Model (R. Hyde, 1975)

$$\log 1/C = \log P - \log(aP + 1) + \text{const.}$$

Bilinear Model (H. Kubinyi, 1976)

$$\log 1/C = a \log P - b \log(\beta P + 1) + c$$

The example given below illustrates the use of the bilinear model for the quantitative description of the relationships between different biological models. Timmermans et al. [11,12] determined partition coefficients, central hypotensive and peripheral hypertensive activities and binding affinities to α_1- and α_2-receptors of a series of clonidine type α-adrenoceptor agonists; the data of the whole animal studies could be successfully correlated to the binding affinities measured in the much simpler in vitro models.

QSAR OF CLONIDINE TYPE α-ADRENOCEPTOR AGONISTS

$\log P_{app}$ in octanol/buffer pH 7.4

pC_{25} = central hypotensive activity (anesthetized normotensive rats, i.v.)

pC_{60} = peripheral hypertensive activity (pithed rats, i.v.)

$IC_{50}\alpha_1$ = binding affinity to α_1-adrenoceptors (displacement of prazosin)

$IC_{50}\alpha_2$ = binding affinity to α_2-adrenoceptors (displacement of clonidine)

$$pC_{60} = 1.16(\pm 0.21)\log(1/IC_{50}\alpha_2) - 0.96$$

$$(n = 21;\ r = 0.936;\ s = 0.317)$$

$$pC_{25} = 0.81(\pm 0.22)\log P - 3.37(\pm 1.02)\log(\beta P + 1) + 1.07(\pm 0.20)\log(1/IC_{50}\alpha_2) - 1.16$$

$$\log \beta = -1.99 \qquad \log P_{opt} = 1.48$$

$$(n = 21;\ r = 0.971;\ s = 0.284)$$

$$pC_{25} = 0.78(\pm 0.26)\log P - 3.69(\pm 1.39)\log(\beta P + 1) + 0.83(\pm 0.20)pC_{60} - 0.19$$

$$\log \beta = -2.08 \qquad \log P_{opt} = 1.51$$

$$(n = 21;\ r = 0.954;\ s = 0.354)$$

While the peripheral activities are a linear function of the binding affinities, for the quantitative description of the central activity a nonlinear lipophilicity relationship has to be included to account for the fact that the compounds have to pass the blood brain barrier to reach their site of action. As a consequence, the central hypotensive activities and the peripheral hypertensive activities can be correlated by using a corresponding nonlinear lipophilicity relationship.

6. DISSOCIATION AND IONIZATION OF DRUGS

Many drugs are either weak acids or weak bases, being ionized in biological fluids to a more or less extent, depending on their pK_a values. Additional complications result from this phenomenon for quantitative structure activity relationships.

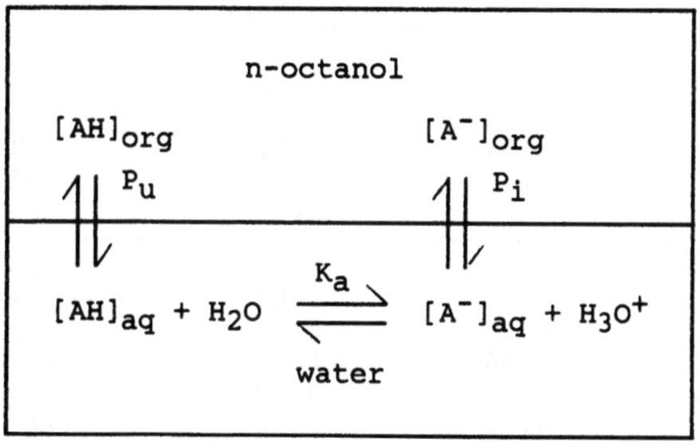

The apparent partition coefficients can be described as functions of the partition coefficient of the unionized form, the partition coefficient of the ionized form, and the pK_a and pH values. On the other hand, these values can be estimated from the pH dependence of the partition coefficient.

PARTITIONING OF ACIDS AND BASES

$$P_{app} = P_u f_u + P_i f_i \approx P_u f_u$$

ACIDS: $AH + H_2O = A^- + H_3O^+$

$$P_{app} = \frac{P_u \cdot 10^{pK_a} + P_i \cdot 10^{pH}}{10^{pK_a} + 10^{pH}} \approx \frac{P_u}{1 + 10^{pH-pK_a}}$$

BASES: $B + H_3O^+ = BH^+ + H_2O$

$$P_{app} = \frac{P_u \cdot 10^{pH} + P_i \cdot 10^{pK_a}}{10^{pK_a} + 10^{pH}} \approx \frac{P_u}{1 + 10^{pK_a-pH}}$$

Scherrer and Howard [13] were the first who proposed the use of apparent partition coefficients in QSAR. Much better correlations could be obtained as in earlier investigations, especially when the bilinear model is used as the nonlinear function [14].

BUCCAL ABSORPTION OF ACIDS AND BASES [14,15]

Compound	pH	log P_{app}	log k_{abs}
propranolol,	5.08	-1.04	-2.19
	6.02	-0.10	-1.71
log P = 3.33,	7.00	0.88	-1.22
pK_a = 9.45	7.93	1.81	-0.79
	8.94	2.70	-0.53
	9.93	3.21	-0.35
p-hexylphenyl-acetic acid	4.00	4.20	-0.46
	5.00	3.63	-0.54
	6.00	2.72	-0.72
log P = 4.25,	7.00	1.72	-1.07
pK_a = 4.36	8.00	0.72	-1.44
	9.00	-0.28	-1.78

$$\log k_{abs} = 0.45\ (\pm 0.05)\ \log P_{app} - 0.45\ (\pm 0.05)\ \log\ (0.0016\ P_{app} + 1) - 1.69$$

$$(n = 12;\ r = 0.988;\ s = 0.102)$$

Much more complex models have been used by Schaper [16] for the quantitative description of the absorption of ionizable drugs.

7. THE TOPOLOGY OF DRUGS AND THEIR BINDING SITES

The three-dimensional structure of a drug is of utmost importance for its biological activity. However, in classical QSAR there is no way to quantify the geometry of a compound in a manner comparable to the description of hydrophobic, polar and electronic properties. Only steric properties of substituents or of certain substructures can be adequately described.

While there are many different approaches to explore the conformational space of a drug molecule, the three-dimensional structure of its binding site can only be deduced from X-ray analysis. This has been done for many enzymes and some enzyme inhibitor complexes, but not yet for membrane bound receptors or ion channels. However, the rapid progress in the genetic expression of receptor and ion channel proteins and the successful X-ray structure analysis of the membrane bound photosynthetic reaction centre of Rhodopseudomonas viridis [17] gives some hope for future developments in that field.

CONFORMATIONS OF DRUGS

in vacuo
- molecular mechanics calculations
- quantum mechanical methods

in the crystal
- X-ray analysis

in aqueous solution
- 2D-NMR analysis
- molecular dynamics calculations

at the binding site
- X-ray analysis of enzyme inhibitor complexes
- molecular dynamics simulations
- design and testing of rigid analogs

A large increase in our knowledge about the interactions between drugs and their binding sites came from the investigation of the inhibition of dihydrofolate reductases from different species [18]. Only some aspects will be discussed here. The Hansch equations for the inhibition of DHFR from Escherichia coli and from Lactobacillus casei show significant differences: while 5-substituents increase the affinity to E. coli DHFR, this is not the case for the L. casei DHFR. A comparison of the X-ray structure analyses gives a reasonable explanation for that phenomenon: at the binding site of the 5-substituent there is a flexible methionine side chain in the E. coli DHFR, while in the L. casei DHFR there is a relatively rigid leucine side chain, giving unfavourable steric interactions with the 5-substituent [19].

INHIBITION OF BACTERIAL DHFR BY SUBST. BENZYLPYRIMIDINES

E. coli dihydrofolate reductase

$$\log 1/K_i = 0.75\ \pi_{3,4,5} - 1.07 \log(\beta.10^{\pi_{3,4,5}} + 1)$$

$$+ 1.36\ MR'_{3,5} + 0.88\ MR'_4 + 6.20$$

$$\log \beta = -0.12$$

$$(n = 43;\ r = 0.903;\ s = 0.290)$$

L. casei dihydrofolate reductase

$$\log 1/K_i = 0.31\ \pi_{3,4} - 0.88 \log(\beta.10^{\pi_{3,4}} + 1) +$$

$$+ 0.95\ MR'_{3,4} + 5.32$$

$$\log \beta = -1.33$$

$$(n = 42;\ r = 0.876;\ s = 0.222)$$

From the structural similarity between two different DHFR inhibitors, methotrexate and trimethoprim, Kuyper et al. [20] concluded that the exchange of one methoxy group of trimethoprim against an acidic side chain could lead to new inhibitors with higher affinity to the enzyme.

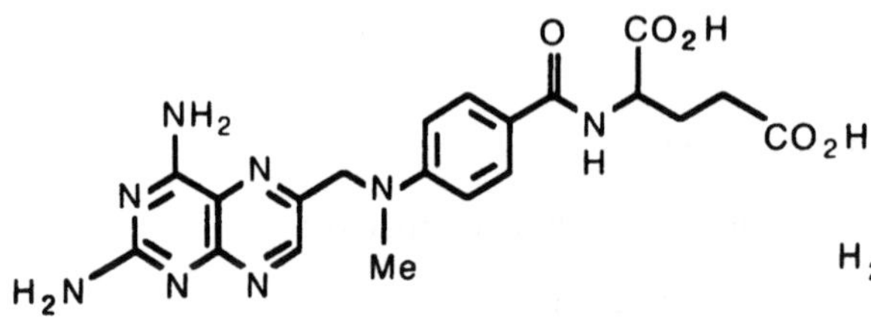

Methotrexate

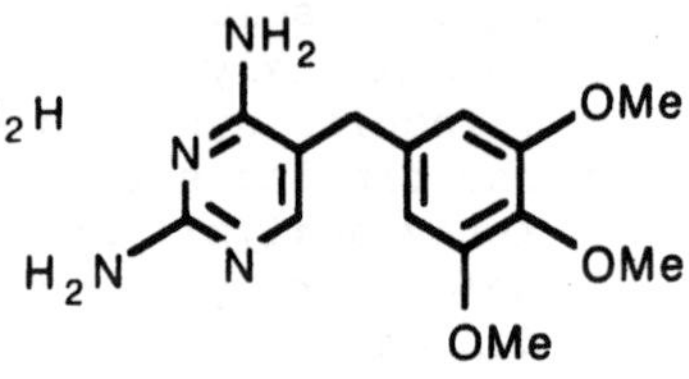

Trimethoprim

This is indeed the case; modelling studies demonstrated that a new interaction takes place between this acidic side chain and the arginine in position 57 of the E. coli DHFR. However, due to the fact that an arginine is present also in the human enzyme in a corresponding position, the advantage of the high selectivity against bacterial enzymes is lost. In addition, the new analogs are active only in vitro, because they cannot pass the bacterial cell wall. Seydel [21] followed this approach to design a hybrid structure between trimethoprim and the antileprosy drug diaminodiphenylsulfone, which is active against Mycobacterium lufu DHFR in vitro and is also active against whole cells, probably due to an increase in lipophilicity of this compound; the growth inhibition of E. coli whole cells is low (although the inhibition of E. coli DHFR is extremely high), indicating that both bacteria differ largely in the permeability of their cell walls.

In the last years increasing evidence came from modelling studies, from X-ray analyses and even from classical QSAR analyses, that closely related compounds may bind to the same site in different geometries. In addition, the flexibility of the binding site and the insertion of water molecules between a ligand and its binding site may lead to further complications in the understanding of drug receptor interactions, especially when the three-dimensional structure of the binding site is not known.

Starting from the crystal structure of cytochrome P-450_{cam} from Pseudomonas putida, which converts camphor to 5-hydroxy-camphor, Poulos and Howard [22] elucidated the three-dimensional structures of the inhibitor complexes with 1-phenyl-, 2-phenyl- and 4-phenylimidazole. While for 1-phenyl- and 4-phenylimidazole one nitrogen atom of the imidazole ring is the sixth ligand at the central iron atom of the porphyrin ring (instead of molecular oxygen in the camphor complex), 2-phenylimidazole is bound in a totally different manner: the imidazole ring is now located between a tyrosine

(which normally binds to the keto group of camphor) and via an intermediate water molecule to the side chain of an aspartic acid. This aspartic acid is removed from its normal position toward the active site, to bind to the new water molecule which is further stabilized in its position by a threonine, whose side chain was rotated by 180° around the C_{α}-C_{β} bond. Movements of up to 2.0 Å are observed in the backbone of the protein.
While we can hope to proceed in our understanding of the binding of ligands, we are far from a deeper knowledge of the dynamic processes that make a ligand to be a receptor agonist. In the case of the calcium channel modulating agents we have the situation that for certain optically active analogs, e.g. Bay K 8644 [23], one enantiomer is a calcium channel antagonist, while the other enantiomer is an agonist (from the classical definition of agonists and antagonists both enantiomers are agonists, the so-called antagonist stabilizing the ion channel in its closed form, while the agonist stabilizes the channel in its open form). Obviously both enantiomers induce different allosteric changes in the channel proteins; however, without information on the three-dimensional structure all hypotheses about the mode of binding of these drugs are mere speculation.

8. SUMMARY AND PERSPECTIVES

Classical QSAR methods and molecular modelling differ significantly in their practical applicability. The linear free energy related methods are suited to analyze transport and distribution and the surface properties of drugs; molecular modelling and X-ray analysis contribute to our knowlegde of the topology and the conformational flexibility of the substances.

CLASSICAL QSAR vs. MOLECULAR MODELLING and X-RAY ANALYSIS

Molecular Property	QSAR	Molecular Modelling	X-Ray Analysis
Transport and Distribution			
Nonlinear Relationships	+	-	-
Dissociation and ionization	+	-	-
Intermolecular forces			
Hydrophobic interactions	+	±	±
Electrostatic interactions	+	±	+
Polar interactions	+	±	±
Topology			
Steric interactions	±	+	+
Ligand conformations	-	+	+
Flexibility of binding site	-	±	±
Different binding modes	-	±	+

While it is obvious that the concepts, ideas and strategies of QSAR aim to a large extent to the rational design of drugs, many problems complicate the practical use as a routine method, i.e.

- nonlinear dependence of the biological activity on lipophilicity and/or polarizability due to a limited area of the binding site,
- different modes of binding of closely related analogs,
- conformational flexibility (neither the drug molecule nor the binding site are rigid systems; a flexible fit occurs when a drug approaches its binding site, only determined by the energy of the stabilizing interactions between both partners),
- replacement of water molecules in the active site and/or insertion of water molecules between the ligand and the binding site,
- entropy effects,
- cooperativity and other allosteric effects,
- lacking relations between the binding affinity and the intrinsic activity of drugs,
- nonlinear dependence of biological activity on lipophilicity due to drug transport and drug distribution in the biological system,
- dissociation and ionization of acidic and basic drugs, and
- differences in bioavailability, metabolism and elimination.

Thus drug design is often nothing more than a ligand design [24]. Nevertheless, the concepts of QSAR have aimed in the understanding of drug receptor interactions. Together with molecular modelling and X-ray crystallography enzyme inhibitors can be designed on a more rational basis than years before [24-29]. In addition, the methods of multivariate statistics help in the evaluation of in vitro tests, like enzyme inhibition, binding assays or cell culture models, which substitute the classical animal models in the screening of drugs to an ever increasing extent.

Dedicated to Professor Dr. Corwin Hansch on the occasion of his 70th birthday.

REFERENCES

[1] S. M. Free, Jr. and J. W. Wilson, J. Med. Chem. 7, 395-9 (1964)

[2] C. Hansch and T. Fujita, J. Amer. Chem. Soc. 86, 1616-26 (1964)

[3] J. D. P. Graham and M. A. Karrar, J. Med. Chem. 6, 103-7 (1963)

[4] H. Kubinyi and O.-H. Kehrhahn, J. Med. Chem. 19, 578-86 (1976)

[5] F. B. Abramson, R. B. Barlow, M. G. Mustafa and R. P. Stephenson, Br. J. Pharmac. 37, 207-33 (1969)
[6] H. Kubinyi, in: Comprehensive Medicinal Chemistry, Edited by C. Hansch, Vol. 4, Pergamon Press, Oxford, England, in press
[7] C. Hansch and J. M. Clayton, J. Pharm. Sci. 62, 1-21 (1973)
[8] J. W. McFarland, J. Med. Chem. 13, 1192-6 (1970)
[9] R. M. Hyde, J. Med. Chem. 18, 231-3 (1975)
[10] H. Kubinyi, J. Med. Chem. 20, 625-9 (1977)
[11] P. B. M. W. M. Timmermans, A. de Jonge, J. C. A. van Meel, F. P. Slothorst-Grisdijk, E. Lam and P. A. van Zwieten, J. Med. Chem. 24, 502-7 (1981)
[12] P. B. M. W. M. Timmermans, A. de Jonge, M. J. M. C. Thoolen, B. Wilffert, H. Batink and P. A. van Zwieten, J. Med. Chem. 27, 495-503 (1984)
[13] R. A. Scherrer and S. M. Howard, J. Med. Chem. 20, 53-8 (1977)
[14] H. Kubinyi, QSAR Des. Bioact. Compd., Edited by M. Kuchar, Prous, Barcelona 1984, Spain, p. 321-46
[15] R. A. Scherrer and S. M. Howard, in: Computer-Assisted Drug Design, ACS Symposium Series 112, 507-26 (1979)
[16] K.-J. Schaper, Quant. Struct. Act. Relat. 1, 13-27 (1982)
[17] J. Deisenhofer, O. Epp, K. Miki, R. Huber and H. Michel, Nature 318, 618-24 (1985)
[18] J. M. Blaney, C. Hansch, C. Silipo and A. Vittoria, Chem. Rev. 84, 333-407 (1984)
[19] R.-L. Li, C. Hansch, D. Matthews, J. M. Blaney, R. Langridge, T. J. Delcamp, S. S. Susten and J. H. Freisheim, Quant. Struct. Act. Relat. 1, 1-7 (1982)
[20] L. F. Kuyper, B. Roth, D. P. Baccanari, R. Ferone, C. R. Beddell, J. N. Champness, D. K. Stammers, J. G. Dann, F. E. A. Norrington, D. J. Baker and P. J. Goodford, J. Med. Chem. 25, 1120-2 (1982)
[21] K.-H. Czaplinsky, M. Kansy and J. K. Seydel, Quant. Struct. Act. Relat. 6, 70-72 (1987)
[22] T. L. Poulos and A. J. Howard, Biochemistry 26, 8165-74 (1987)
[23] G. Franckowiak, M. Bechem, M. Schramm and G. Thomas, Eur. J. Pharmacol. 114, 223-6 (1985)
[24] P. M. Dean, Molecular foundations of drug-receptor interaction, Cambridge University Press, Cambridge 1987, England
[25] P. J. Goodford, J. Med. Chem. 27, 557-64 (1984)
[26] P. J. Goodford, J. Med. Chem. 28, 849-57 (1985)
[27] C. Hansch and T. E. Klein, Acc. Chem. Res. 19, 392-400 (1986)
[28] G. R. Marshall, Ann. Rev. Pharmacol. Toxicol. 27, 193-213 (1987)
[29] R. L. DesJarlais, R. P. Sheridan, G. L. Seibel, J. S. Dixon, I. D. Kuntz and R. Venkataraghavan, J. Med. Chem. 31, 722-9 (1988)

Computation of Volumes and Surface Areas of Organic Compounds

Mario Marsili

Istituto di Chimica, Universitá de l'Aquila, L'Aquila, Italy

Molecules, due to their electron density distribution, are conveniently described as entities with a finite extension in physical space, i.e. as solid bodies. This is reflected in the popular CPK plastic models that almost every chemist has used for educational or visualization purposes.

A computerized binary representation of a solid molecular body was developed [1,2] allowing the computation of molecular volumes, molecular surface areas and the calculation of an empirical degree of molecular shape similarity. We want to discuss this particular approach because it uses directly the intrinsic capabilities of a computer as a binary machine, highlighting the applicability and the relevance of Boolean operators in chemical problem solving.

A portion of physical space is described by a tensor Γ (t,t,t) of rank 3 having t components per axis. The space included by Γ can be represented as a virtual cube, subdivided into t^3 subspaces, each of them defined by an index triple (i,k,l). A properly scaled v.d.Waals molecular model, looking like a usual CPK model, can be placed inside the cube spanned by Γ . We can now obtain a purely logical encoding of the space-filling properties of the molecular model in the following simple manner: All the subspaces located inside the v.d.Waals model are labelled with "1", those outside are labelled with "0". This is therefore a boolean description of the pattern of distributed solid matter in 3D space. In a computer representation the 1's and the 0's are equivalent to TRUE and FALSE, which are logical variables. This method does not merely encode the outer shell of

C. Jochum M.G. Hicks J. Sunkel (Eds.)
Physical Property Prediction
in Organic Chemistry

the molecular model, but also its interior as matter filled solid body.

Our experience showed that to provide for a good resolution of the molecular shape a minimum of at least about one million subspaces in the virtual cube seems adequate.

To optimize memory requirements the virtual cube must be represented by the tensor $\Gamma(t,t,t,)$, in which the t components per axis can conveniently be chosen coherent with the CPU architecture. For example, to save core memory, a compact bitwise encoding can be preferred to the use of simple logical variables or even integer numbers. Exploiting a 32-bit machine, an immediate way to obtain about one million subspaces (= bits inside the cube) is to generate them from an equivalent number of bits in a program-internal tensor $\Gamma_P = \Gamma_P\ (96,96,3)$, in which we have 96 words each for the x- and the y- axis tensor components, and three words, each of 32 bits, spanning the 96 components along the z- axis. Compared to an integer number codification this technique saves core memory by a factor of 32. Special bit-addressing routines are then responsible for switching on and off a specific bit inside a certain word in Γ_P .

The program segment here below illustrates the bit-addressing technique that we have chosed to create a bit mask, that is a specific word that contains one bit only with value 1 in a given required position, all other bits having value 0. The bit mask is used iteratively to fill up the extension of the v.d.Waals model inside of Γ with 1's.

```
      SUBROUTINE SETMASK (MASC, IDIG, WIZ)

C     INDEX IS THE POSITION NUMBER OF A SPECIFIC BIT INSIDE A 32-
C     BIT WORD; WIZ IS THE NUMBER OF THE ACTUAL WORD ALONG THE Z-
C     AXIS, ITS VALUE CAN BE  1,2 OR 3. IDIG IS THE ABSOLUTE
C     POSITION OF A CERTAIN BIT ALONG THE  Z-AXIS,  ITS VALUE
C     RANGING FROM 1 TO 96;  HAMMER FORCES A 1 (HEXADECIMAL FORM)
C     AT BIT POSITION 32, NORMALLY UNADDRESSABLE BECAUSE USED AS
C     PARITY BIT IN THE CPU
```

```
      INTEGER MASC, IDIG, WIZ, INDEX, HAMMER/Z80000000/
      WIZ = 1 + (IDIG - 1)/32
      INDEX = IDIG - (WIZ - 1) *32
      IF (INDEX .EQ. 32) MASC = HAMMER
      IF (INDEX .NE. 32) MASC = 2**(INDEX - 1)
      RETURN
      END
```

The following routine shows the simple mechanism of how TRUE bits inside Γ can be summed up to provide a) a measure for the molecular volume of b) a measure of the molecular surface area

```
      SUBROUTINE SUMBIT  (LSPACE, CUNTER)
C     IN AN IBM 370 VERSION AN EQUIVALENCE MUST BE DEFINED
C     BETWEEN THE INTEGER AND LOGICAL VARIABLES INVOLVED IN
C     BOOLEAN OPERATIONS TO ALLOW BITWISE PROCESSING OF
C     WHOLE WORDS. THIS APPROACH IS NOT NECESSARY ON DEC MACHINES

      INTEGER MASC, CUNTER, HAMMER/Z80000000/
      LOGICAL LSPACE (96,96,3), LMASC
      EQUIVALENCE  (LMASC, MASC)
      CUNTER  = 0
      DO 300 IJ = 1,32
      IF (IJ .EQ. 32) MASC = HAMMER
      IF (IJ .NE. 32) MASC = 2 **(IJ - 1)
      DO 300 I = 1,96
      DO 300 J = 1,96
      DO 300 K = 1,3
      IF (LSPACE(I,K,L) .AND. LMASC) CUNTER = CUNTER + 1
 300  CONTINUE
       .
       .
       .
```

Boolean Tensor Operations

For each molecular model a specific bit tensor can be generated. These tensors can be easily handled by boolean operators to yield some quantities useful in molecular modelling. The molecular volume V is obtained by summing up all TRUE bits $(ikl)_T$ inside the virtual cube

eq. 1 $$V = \Sigma\, (ikl)_T$$

The molecular area is calculated by the sum of all TRUE bits which are placed next to any FALSE bit, which are evidently defining the outer bit shell at the v.d.Waals border of the molecular model.
The boolean processing of two bit tensors Γ_1 and Γ_2 results in another bit tensor Γ_3. The resulting bit pattern describes quantitatively special space regions, the so-called common and residual regions following to the action of molecular superposition. We can define following molecular body regions derived from the spatial overlap of two virtual v.d.Waals molecular models: The union of a first bit tensor Γ_1, encoding molecule A, with a second tensor Γ_2, encoding molecule B, results in a new tensor Γ_U encoding the union body, or superbody, of the two molecules in the chosen specific superposition (eq. 2).

eq. 2 $$\Gamma_U = \Gamma_1 \text{ OR } \Gamma_2$$

The space region belonging to both superposed molecules is the common body, represented by Γ_C (eq.3).

eq. 3 $$\Gamma_C = \Gamma_1 \text{ AND } \Gamma_2$$

Subtracting the common body from the union body, the residual body, expressed by Γ_R, is obtained (eq.4).

eq. 4 $$\Gamma_R = \Gamma_U \text{ AND NOT } \Gamma_C$$

this equation can be rewritten into eq. 5 using Γ_1 and Γ_2

eq. 5:

$$\Gamma_R = (\Gamma_1 \text{ AND NOT } (\Gamma_1 \text{ AND } \Gamma_2)) \text{ OR } (\Gamma_2 \text{ AND NOT } (\Gamma_1 \text{ AND } \Gamma_2))$$

The first term in eq. 5 encodes the residual body of the first molecule, the second term the residual body of the second molecule. Using the above bit-counting method the volumes V_u, V_c, and V_R of these secondary bodies are quantitatively evaluated. In a case of an isomorphic superposition of two identical molecules (such that V_R is zero) one obtains

$$V_u = V_c$$

and therefore

$$V_c/V_u = 1$$

For every other case

$$V_c/V_u < 1.$$

An empirical shape similarity index S can now be proposed as

eq. 6 $$S = V_c/V_u$$

The advantage of the boolean approach for encoding molecular models can be seen in the simplicity of handling the boolean tensor. Other methods, geared on difficult dissections and geometric parametrizations of the molecular model, have been developed to compute surface areas and molecular volumes [3,4], but they do not provide a code for the molecular body. The bit code can be canonized by orienting the molecular moments of inertia along the tensor axes, and placing the molecular center of gravity in the middle of the cube.

The computed volumes have been successfully correlated with some volume-dependent physico-chemical parameters like the vaporization enthalpy [1] and the boiling point [5] of homologue series of organic compounds.

References:

1. M.Marsili, P.Floersheim and A.S.Dreiding, Comp.Chem., 7, 175, 1983

2. M.Marsili and P.Floersheim, Analyt.Chem.Symp.Ser., 15, 332, 1983, Elsevier, Amsterdam

3. M.L.Connolly, J.Appl.Cryst., 16, 548, 1983

4. A.Gavezzotti, J.Am.Chem.Soc., 105, 5220, 1983

5. M.Marsili, C.del Buono, unpublished results

Total System of Molecular Design

Shin-ichi Sasaki, Yoshimasa Takahashi, and Kimito Funatsu
Toyohashi University of Technology, Tempaku, Toyohashi 440, Japan

(1) Generation of Structure with Potential Effective Action (TUTORS)

1. Introduction

Research and development of molecular design support systems has been carried out actively in Japan as well as in the U.A. and Europe. All systems proposed so far, however, are designed only for management of data on chemical compounds or for statistic or various molecular calculations. The authors have been engaged in development of a comprehensive molecular design support system, called TUTORS[1], that, in addition to serving for the above-mentioned purposes, can generate and propose candidate structures of new promising compounds that are expected to have specific functions. This is the ultimate goal of studies in the field of molecular design. Some unique techniques are also developed and incorporated. The present report describes the present status of the development of the TUTORS system, centering on the basic concept of computer-assisted molecular design and major functions of the system.

2. Basic Concept of System

The system consists of two basic subsystems, called TUTORS-DB and TUTORS-SG. TUTORS-DB serves for management and analysis of data on chemical compounds and many other related purposes.

C. Jochum M.G. Hicks J. Sunkel (Eds.)
Physical Property Prediction
in Organic Chemistry

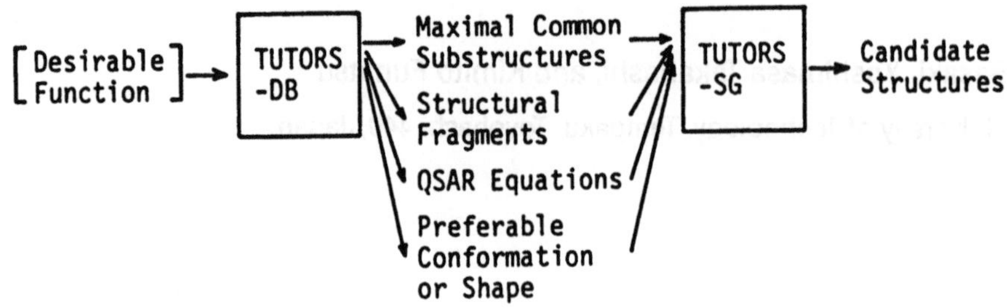

Fig. 1-1 A strategy of computer aided molecular design in the TUTORS system.

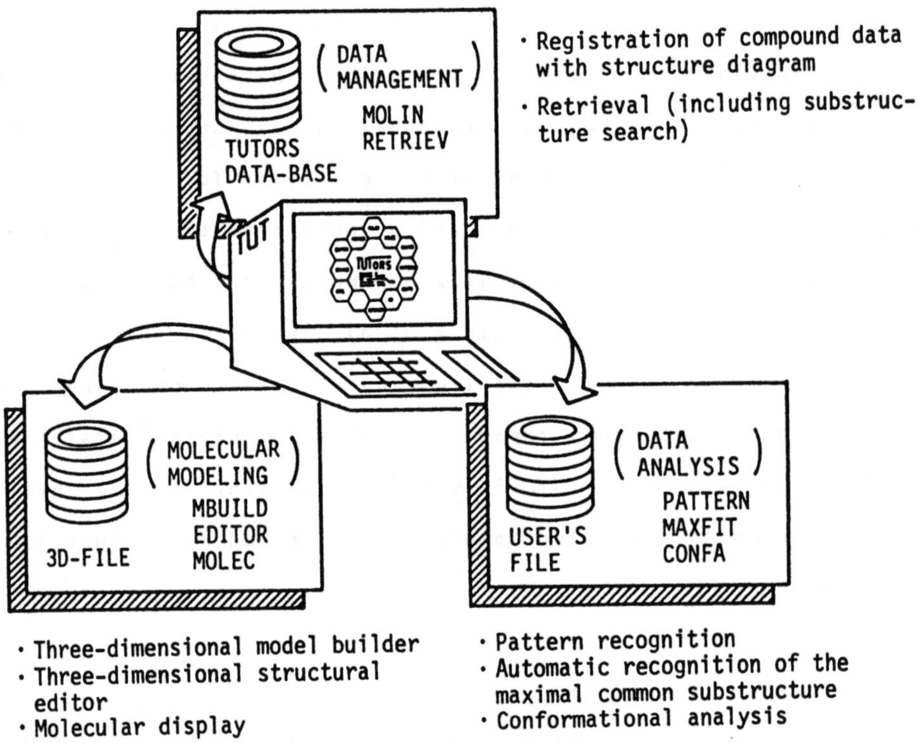

Fig. 1-2 A system overview of TUTORS-DB.

The information on various structure-activity correlations obtained from detailed data analysis by this subsystem is then processed by TUTORS-SG to generate and propose specific candidate structures of new promising compounds that may have desired functional properties (Fig. 1-1). From the viewpoint of molecular design support, the former is aimed at supporting the acquisition of required knowledge data while the latter performs the real design of chemical structures using such knowledge obtained from the former part.

3. TUTORS-DB Subsystem

Fig. 1-2 outlines the modules making up the TUTORS-DB subsystem. The subsystem can be roughly divided into the following three parts according to the functions of the modules incorporated: (1) data management part that mainly performs registration of data on compounds in data base and retrieval of them from the data base; (2) molecule modeling part that can perform display of various molecular structures as well; and (3) data analysis part that consists of many program packages designed to analyze structure-activity correlations. Major modules contained in these parts are described below focusing on their functions.

3.1 Data Management

Management of chemical compound data including those on chemical structures is one of the most important issue not only in the field of molecular design but also in various other chemical areas where computer-assisted investigations are required.

In view of such a situation, the present system has been designed to have the capability to construct chemical compound data base consisting of data on the structures and biological activity of chemical compounds and to perform retrieval and management (up-dating, etc.) of these data.

What is important here is how to express data related to chemical structural formulae, which are serving as the "common language" in the field of chemistry, and how to treat them on the computer. With the structure input module (MOLIN), a chemist can use template structures or the free drawing mode to allow a structural formula on the graphic terminal by a procedure like drawing it on paper. For processing by the computer, structural data drawn and entered are represented by a matrix composed of diagonal elements for expressing data on the atomic species and electric charges and non-diagonal elements expressing data on the bonds among the atoms. This module serves also for entering initial data for molecule modeling, in addition to registering structural formulae and related data into the data base.

With the retrieval module (RETRIEV), on the other hand, a user can have access to the chemical compound data base by using such various retrieval keys as follows: registration number, name of chemical compound, molecular formula, molecular weight (designation of range) and other key words arbitrary provided at the time of registration. In addition, the module can perform retrieval in terms of structures, including full structure and substructure search. It is also possible to execute logical operations for separate retrieval results given by these different keys so as to combine them using such operators as AND, OR and NOT.

3.2 Molecule Modeling

One of the most important things for chemists is to understand the three-dimensional structure of the molecule of each chemical compound. It is often difficult, however, to determine the three-dimensional structures of a compound only from experimental data. Thus, it is required to employ other methods such as calculation to allow the three-dimensional structure of a particular compound to be inferred from data on other compounds with known structures. Furthermore, visualization of the structures of such molecules can serve for easy understanding of the three-dimensional arrangement of functional groups in the molecules and their three-dimensional shapes. To provide a tool to this end, we developed separate modules to automatically estimate three-dimensional coordinates, edit three-dimensional structures and display molecular structures. Their major functions are described below.

The three-dimensional structure modeling module (MBUILD) automatically estimate three-dimensional coordinates of a molecule on the basis of the two-dimensional structural formula given by the above-mentioned MOLIN. The module can carry out various other operations. For example, the three-dimensional arrangement can be inverted about an symmetric carbon atom and the dihedral angles can be varied. All these operations are executed interactively. With the three-dimensional structure editor (EDITOR), calculation and modification of data on molecular geometry can be conducted interactively. It performs display and correction of the inter-atomic distances, bond angles, bond lengths and twist angles of a molecule designated by the user. It also can delete part of a structure or construct a new three-dimensional molecular structure by adding another three-dimensional (partial) structure. To display the structures of these molecules, we have developed a multifunctional graphic tool (MOLEC) for molecular structure display. This module incorporates various molecular model display modes, including skeleton display, ball-and-stick representation, space-filling model and dot surface model. In each model display mode, views from three different angles can be drawn and superimposed structures of different molecules can be displayed. Other functions of the module include stereopair display as well as rotation, expansion and contraction of a molecule.

To carry out these operations, a variety of data are required concerning the molecule in question and its constituent atoms (e.g., valence angles between atoms, normal inter-atomic bond lengths, stereochemical rules). Processing operations different from those for usual two-dimensional structural formulae are needed for the system to represent and extract required data, though detailed discussions are not made here.

The development of the above module has made it extremely easy to perform a series of operations for three-dimensional molecular structure modeling, ranging from estimation of three-dimensional coordinates to correction and modification of the structure and display of the molecular structure, by using as initial data a two-dimensional chemical structural formula drawn and entered through MOLIN described above. Another module, called CONFA, is also involved in the present system, that performs conformation analysis required to determine a conformation of a molecule in the most stable energy state from three-dimensional coordinates produced.

3.3 Data Analysis

Needless to say, an important clue to efficient molecular design is information on the chemical structures and actions of a set of chemical compounds, which constitutes the basis for it. To collect such information, detailed analysis of related data accumulated so far should be carried out positively to provide knowledge that supplements existing empirical findings. To this end, the TUTORS system incorporated two modules, called PATTERN and MAXFIT, that we have developed for the analysis of structure-activity correlations.

PATTERN is a program package that consists of various pattern recognition programs useful to analyze structure-activity correlations. The purposes of major programs incorporated are listed in Table 1-1. The approach to knowledge acquisition consists of two procedures: parameterization of data on the structures of compounds using various characteristic values, and the use of them in combination of a mathematical model to formulate the correlations of the structure with activities of the compound such as biological activity. We have applied each of these programs to studies on structure-activity correlations to determine their performances[2).

Table 1-1 Programs in TUTORS/PATTERN

SPLX ···	Binary classification by the use of simplex technique. (active/inactive)
MCSPLX···	Multicategorical classification by pair-wise application of the binary classification. (type of activity)
ORMUCS···	Ordered multicategorical classification using simplex technique. (potency of activity; 1+, 2+, 3+, ···)
CLUST ···	Hierarchical cluster analysis. Q-mode and R-mode analysis Centroid, Nearest-neighbour Euclidian distance, Correlation coef. (including minimal spanning tree, MST)
KLPLOT···	Linear mapping using Karhunen-Loeve method.
HANSCH···	Linear regression analysis for QSAR.

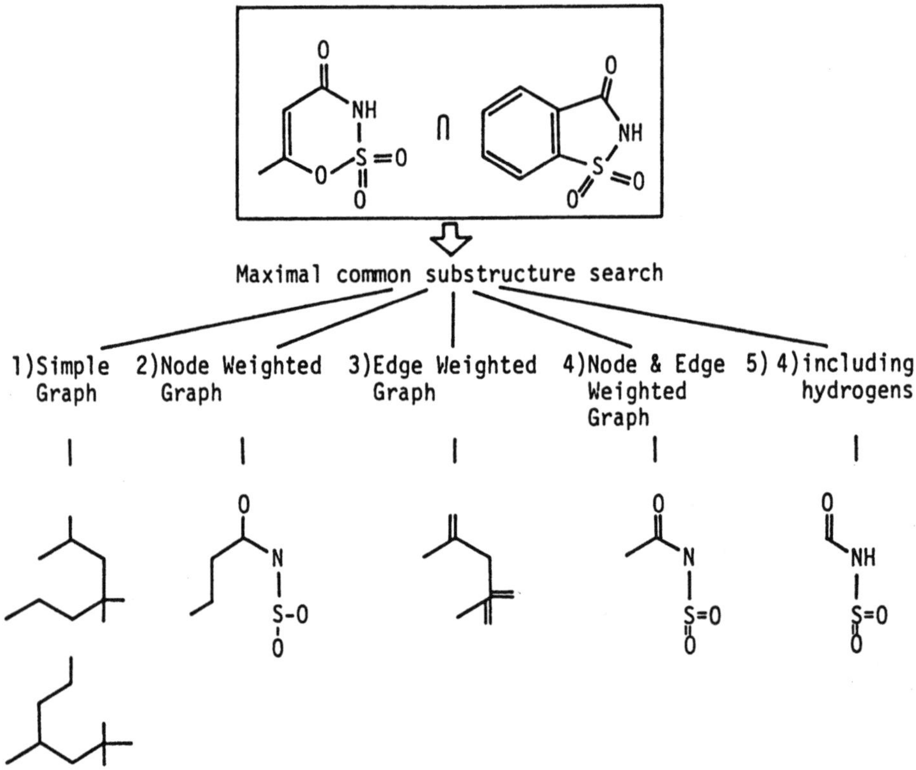

Fig. 1-3 Maximal common substructures between acesulfame and saccharin found by the MAXFIT program.

Also important is the classic approach to structure-activity correlations, which resorts to investigations on the differences and similarities between structures in terms of chemical structural formulae as has been coventionally conducted by organic chemists. Thus, effective methods are required which can directly treat chemical structural formulae familiar to chemists to compare and analyze common structural features among compounds with similar actions and to automatically detect and display such common structural features. For this kind of data analysis, we have developed another module, MAXFIT, which is designed for automatic recognition of the largest common substructure[3]. MAXFIT uses graph theoretical representations of the structures of chemical compounds to permit search operations at the following five different structural representation levels: 1) simple graphic representation of simple backbone (hydrogen atoms are omitted in 1) to 4), 2) node-weighted graph with only atomic species being distinguished, 3) edge-weighted graph with only bond types being distinguished, 4) node- and edge-weighted graph with both atomic species and bond types being distinguished, and 5) graph similar to that in 4) but with hydrogen atoms being incorporated. An example is given in Fig. 1-3, which illustrates the results of search for the largest common partial structure between two artificial sweeteners, acesulfame and saccharin, carried out at each representation level.

MAXFIT automatically recognizes the largest common substructure at a level designated by the user researcher. The user, therefore, is not required to take these structural representations into consideration when entering data. In studying structure-activity correlations, the largest common substructure is not always the only important feature and therefore, the module is designed to generate all possible common substructures, including smaller ones, at each representation level. MAXFIT with the capacity for these operations is one of the extremely important elements as a tool to ensure that specific building blocks required to construct some candidate structure of interest are supplied to the structure generation process in the TUTORS-SG subsystem described below.

4. TUTORS-SG Subsystem

The development of TUTORS-SG has been conducted with the idea that structural design by this subsystem will be performed in the following three processes: 1) inference and extraction of a set of structural fragments to serve as building blocks required to construct structures that meet requirements for the manifestation of an action in question, 2) construction of logically possible chemical structural formulae from a given set of fragments (structure construction), and 3) selection of candidate structures based on simulation, etc. The procedure for the processing is as follows. First, the processes 1) and 2) are carried out to generate primary candidate structures that consist of relatively large partial structures which are meaningful in connection with structure-activity correlations. These candidate structures should also meet the requirements for the target structure. Then, such operations as activity prediction by model equations, three-dimensional modeling and conformation analysis are carried out for a set of structures generated above. Results obtained are used to carry out simulation to compare them with known active substances and select out more promising candidate structures. Thus, a set of final candidate structures are proposed. A conceptual diagram of the processing by TUTORS-SG is illustrated in Fig. 1-4.

In the first step, relatively large elemental substructures and requirements (constraints) are assumed. The structures may include the largest common substructure and other meaningful substructures while the constraints may include requirements for the range of molecular weight and other properties such as the optimum hydrophobicity. In most cases, however, it is almost impossible to construct a perfect candidate structure that meets the requirements specified for the given set of structural fragments (fragment set). Therefore, some additional fragments should be selected and added to allow the fragment set to meet the requirements. The fragment set thus prepared is then sent to the next structure construction step. In this structure construction step, all possible structures (candidate structures) that are consistent with chemical assumptions (rules and knowledge) are generated from the given fragment set. In this way, a set of primary candidate structures is obtained in this step. It is expected, however, that the set produced here contains a rather large number of candidate structures, because the structure construction process inevitably involves combinatorial problems even if a variety of constraints are imposed. This means that another process is required for further selection from among the primary set of candidate structures. In other words, such a

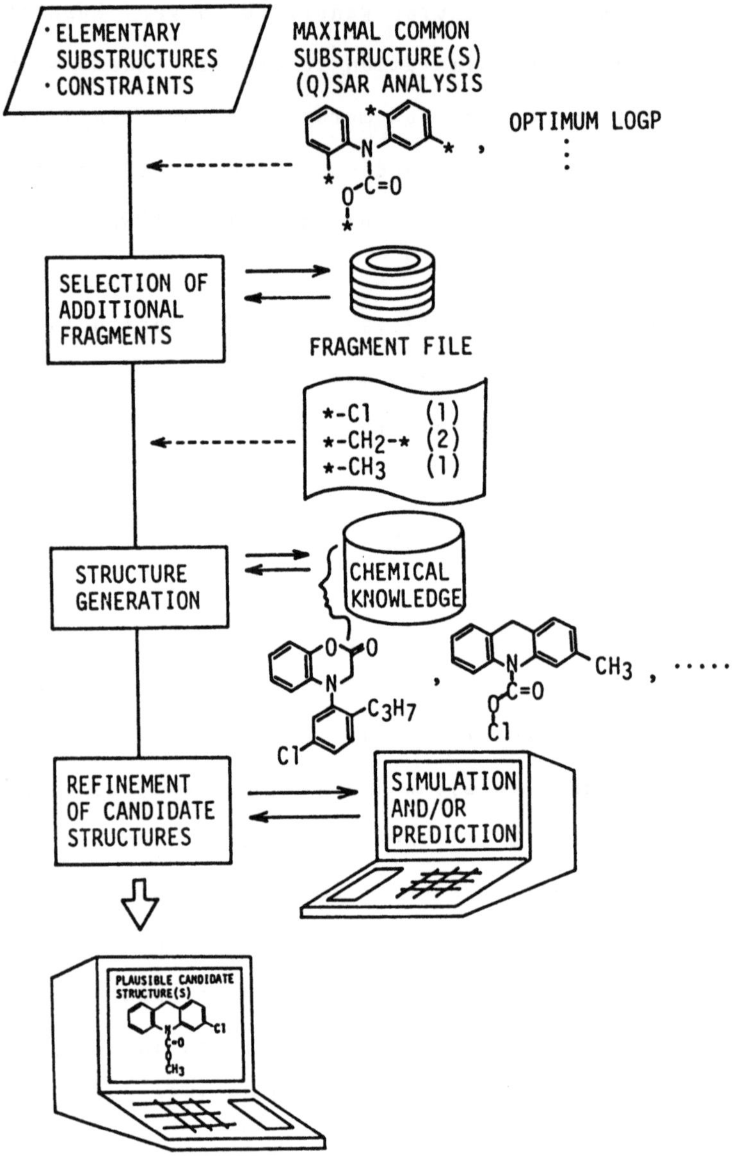

Fig. 1-4 A schematic flow diagram for structural design in TUTORS-SG.

process should be designed to pick up more promising candidates. There are a variety of possible procedures to be used for this process, ranging from simple screening in terms of partial structures to evaluation based on various simulation results.

To construct a structural designing system consistent with the above concept, it is indispensable to develop a module for structure generation (structure generator) which constitutes the core of such a system.

4.1 Candidate Structure Generation Program (CAST)

Following the above idea, we have developed a candidate structure generation program, CAST (CAndidate STructure generator), specially performed in such a way that meet the following requirements.

1. All given partial structure (structural fragments) should be reflected in the candidate structures.
2. Each fragment should be independent of the others.
3. The following operations are prohibited.
   Formation of a bond between any two dummy atoms that are connected to the same atom (formation of a loop).
   Formation of a direct bond between any two dummy atoms that are connected to the same aromatic ring.
   Formation of a bond between any two dummy atoms that are different in the multiplicity of bonding.
4. Stereochemistry is not taken into consideration.

<Outline of Algorithm>

The process of structure construction by CAST is roughly divided into the following four steps.

(I) Modification of fragments into abstracted forms.

(II) Generation of abstracted tree graphs.

(III) Formation of list of dummy atom pairs corresponding to the abstracted tree graphs.

(IV) Addition of remaining dummy atom pairs.

In Step I, each structural fragment in a given fragment set is modified into an abstracted form using the superatom representation so as to simplify the graph production operation. All possible unique tree graphs (abstracted tree graphs) are produced using these superatoms. This operation is designed to prevent the formation of separated structures in the initial step in structure construction. Only the node number corresponding to each superatom, the number of dummy atoms and their bond multiplicity are used for the processing to produce the abstract tree graphs. A set of tree graphs more rational for the final structure construction processing can be obtained if checking for isomorphic structures within each fragment and weighting of the superatoms are carried out previously. Then, operation is performed to produce a list of all corresponding dummy atom pairs for each tree graph. In general, several lists are produced from one tree. Thus, several labeled tree graphs are generated from one tree graph. Based on the remaining dummy atoms in the edge-labeled graphs, an additional atom pair list is added in the next step to complete a structure. Each individual processing operation is not discussed in detail here. An example of operations carried out by CAST prepared using these algorithms is given in

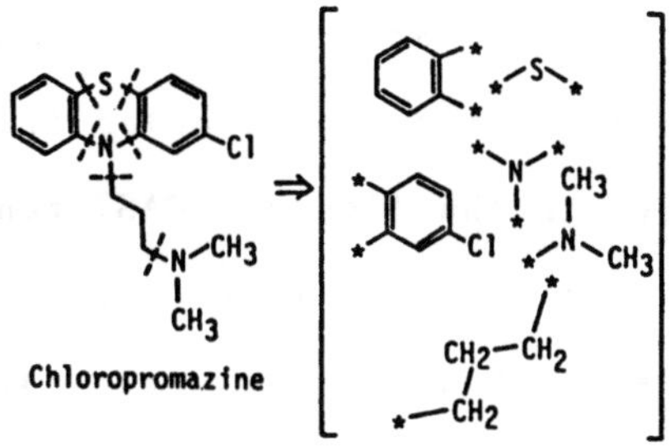

Fragments : 6
Dummy atoms : 12
Generation mode : [COMPLETE]
Structures : 114
CPU time(sec.) : 146.33

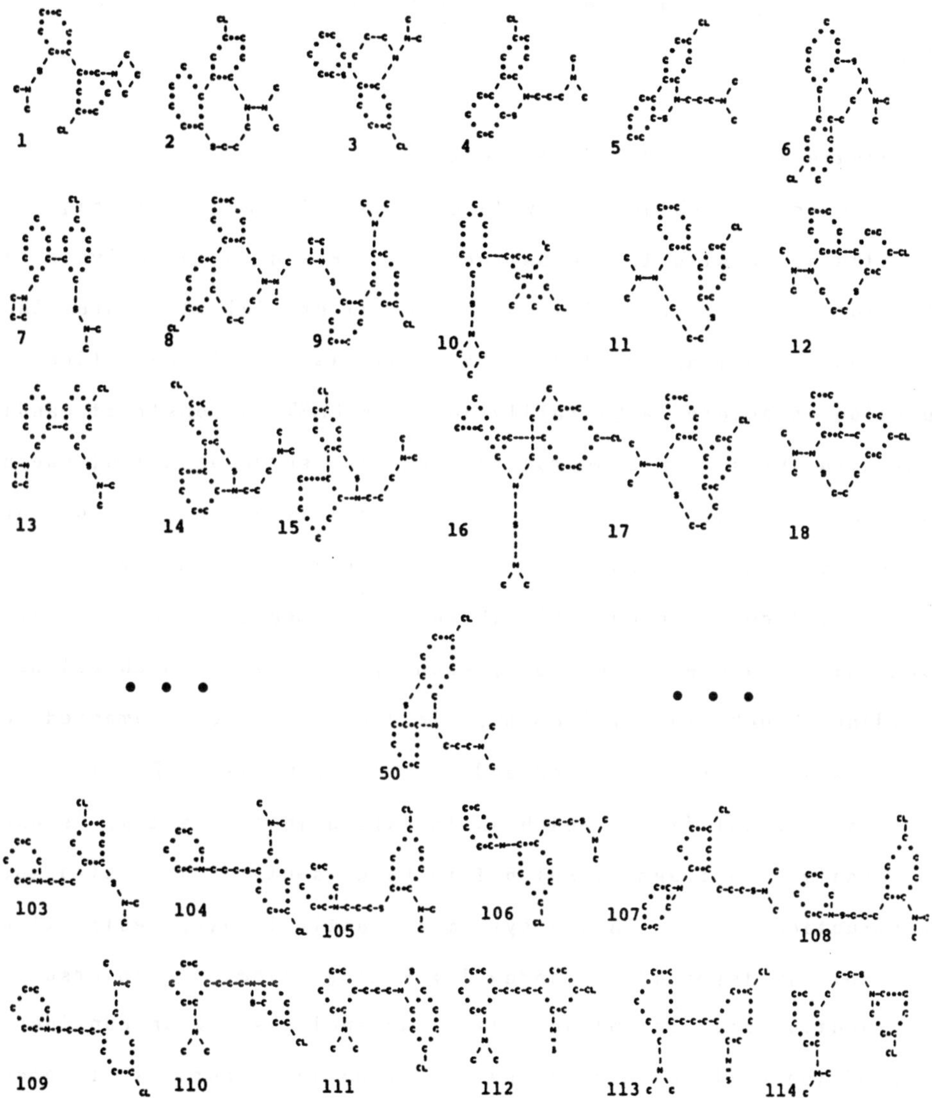

Fig. 1-5 An example for construction of chemical structures with a set of structual fragments based on chloropromazine by the CAST program.

Fig. 1-5, in which the molecule of chloropromazine is divided into six fragments and the structure is re-constructed from the resultant fragment set. As shown in the figure, CAST generated 114 structures which were different in each other. The prediction of synthetic route for one of those structures (actually chloropromazine itself) will be discussed in the next chapter of AIPHOS.

4.2 Fragment Set as Building Block

The next problem is how to develop a fragment set required for the above structure construction. Based on an analysis of structure-activity correlations, a set of partial structures that are considered meaningful for the manifestation of the action in question is entered arbitrarily to allow CAST to construct candidate structures. This operation alone can serve as a tool useful for many purposes. However, since a variety of restrictions are to be imposed simultaneously on each candidate structure, it is often difficult to meet all these requirements using a given fragment set alone. Therefore, each fragment set, which act as a building block for TUTORS-SG, is designed to be composed of elementary substructures and additional fragments. For the latter, a fragment file, which is installed in the system, is used to modify the fragment set before being sent to the structure construction part. In the system currently in use, restrictions are set in terms of the index log P (logarithm of inverse of distribution coefficient in water/1-octanol system) in consideration of the well-known fact that hydrophobic properties of molecules have close relations with their various biological activities. We have developed an additional fragment search program (FRSEL) to be applied to modifying a fragment set under the restrictions. Basically, the Rekker's hydrophobic fragment constant, f, is used to determine the sum of f's for the entered partial structures, and appropriate additional fragments are selected out from the fragment file to adjust the difference between actual log P and required log P (Δlog P). The flow of these operations are illustrated in Fig. 1-6. Needless to say, there are many other restrictions that should or can be taken into considerations with respect to structural features, physicochemical characteristic values, molecular size, etc. Thus, there remain a number of problems as to, for example, how to analyze and represent them in the form of knowledge and how and in which stage in the system they should be utilized. Further studies are required to solve these problems.

5. Conclusions

A study is now under way to develop a process for the screening of candidate structures. Though there have been many studies on molecular designing, almost no efforts have been made in pursuit of techniques for practical structure designing operations or generation of candidate structures. We hope to find technical possibilities and new research directions toward the fulfillment of the goal while carrying out the research and development of the TUTORS system.

Preposition: $\log P = \sum_i f_i$

f_i: hydrophobic fragmental constant of fragment i.

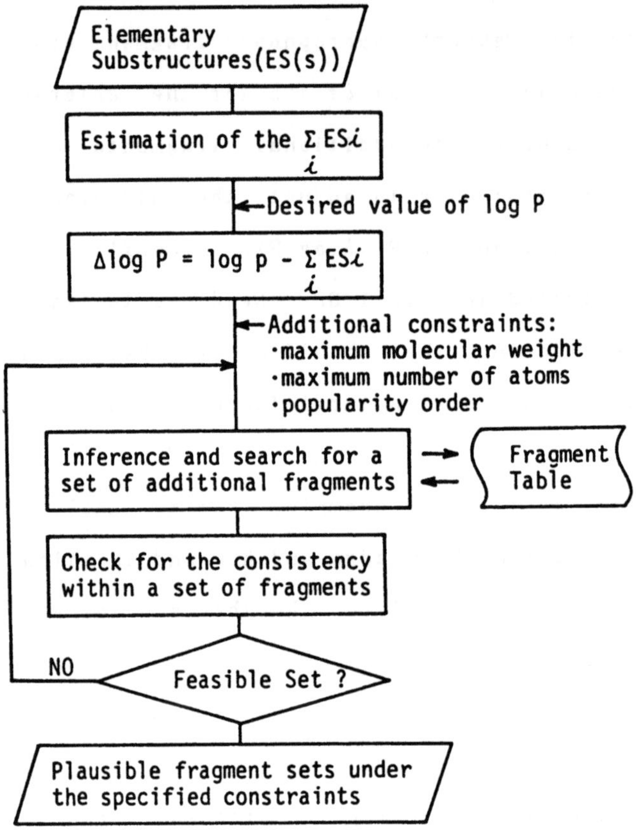

Fig. 1-6 Preparation of fragment set using some constraints with a required value of log P.

References

1) Y. Takahashi and S. Sasaki, "Biorational Approach for Pesticide Design", Agricultural and Biological Chemistry ser. 9, Ed. The Agricultural Chemical Society of Japan (Soft Science Co.), 1986, p.63.; Y. Takahashi, K. Hosokawa, F. Yoshida, and S. Sasaki, Anal.Chim. Acta (Special Issue for Proceedings of 8th ICCCRE) in press.

2) S. Sasaki, Y. Takahashi, Hazard Assessment of Chemicals, (Ed., J. Saxena) 5, 61(1986).

3) Y. Takahashi, Y. Satoh, H. Suzuki, and S. Sasaki, Anal.Sci., 3, 23(1987).

(2) Production of Compound with Proposed Structure (AIPHOS)

1. Introduction

A variety of systems have emerged during the past 20 years or so, as a result of efforts to make use of computers as a tool for organic synthesis design. Most of these systems are either of a data-base oriented (experience oriented) type or theory and logic oriented type. With a system of the former type, simulation can be performed only in conformity with empirical facts accumulated so far. Thus, once data base is prepared using, for example, appropriately processed data on potentially useful reactions, the system will successfully carry out designing operations within the scope of the data base. Under the current situation, however, which is characterized by diversified research activities in the field of organic synthetic chemistry, development of new synthetic processes is important as well as simply manipulating known reactions. In this respect, logical approach will be necessary, though this alone cannot be practical since past experience or accumulated data will have to be ignored. In applying a logical procedure to derivation of precursors for synthesis, it is often found impossible to set reasonable reaction conditions. In actual cases, reaction conditions have rather large effects on the synthetic processes conducted. For the setting of reaction conditions, therefore, the system is required to have a capability to propose suitable conditions after automatically searching for and enumerating possible cases picked up from past experiences and data. The authors have been carrying on the development of such a system, called AIPHOS, for organic synthesis design and reaction prediction, following the idea that the application of computers to organic synthesis design is particularly useful in that they help develop a system incorporating accumulated data, knowledge and logical procedures to propose all possible cases through calculations beyond human capabilities.[1)]

2. System Configuration of AIPHOS

AIPHOS is roughly divided into three subsystems as shown in Fig. 2-1. Their functions are as follows:

(1) Logical data processing mainly by the reaction generator. Major operations include the recognition of significant features of the target structure in relation to organic synthesis, and disconnection and re-connection of the struc-

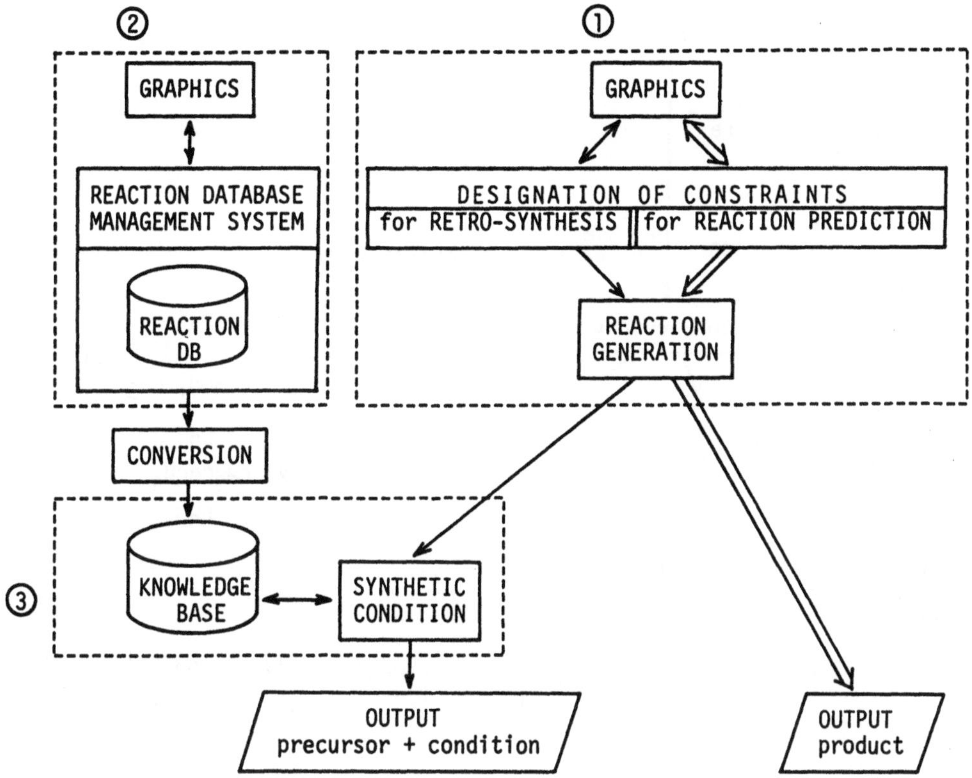

Fig. 2-1 Block diagram of AIPHOS.

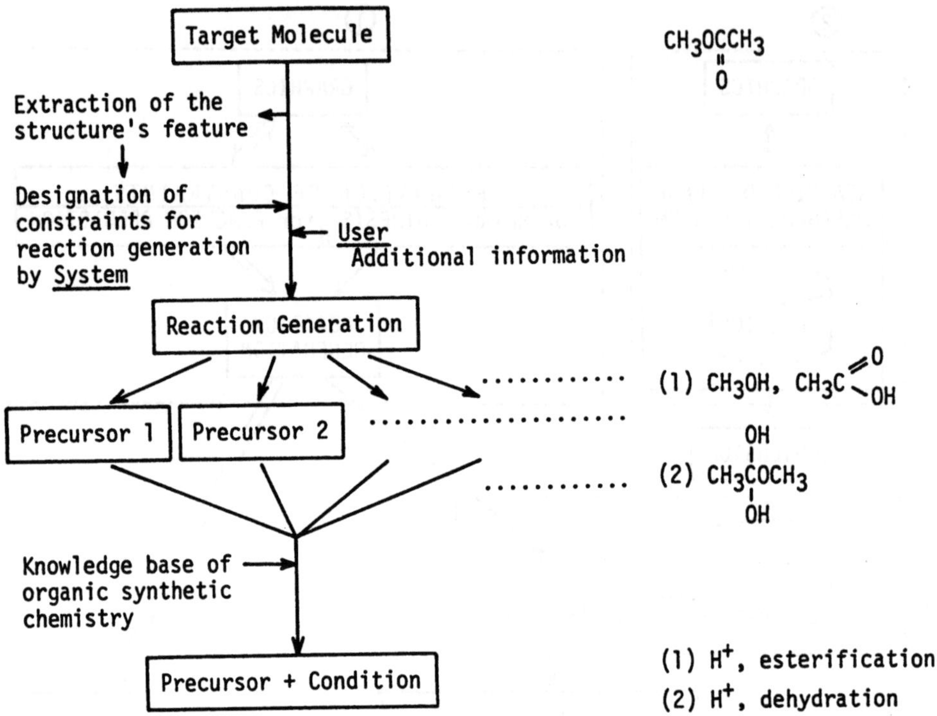

Fig. 2-2 General flow of logical part in AIPHOS.

ture at the disconnection points proposed after considering the above features.

(2) Utilization and management of data base consisting of individual data on specific reactions.

(3) Knowledge base and its operation programs are incorporated. The former is developed by selecting out required data from the specific reaction data base and generalizing and formulating organic chemical reactions. Reaction conditions are proposed for the reaction processes generated through the operations (1).

The first major problems encountered in making a logical approach to organic synthetic designing is how to define reactions and represent them for the computer. In AIPHOS, a reaction is treated as a "inter-molecular isomerization" process from several molecules to different several molecules. First, a case is considered below where all possible synthetic precursors that may constitute the target synthetic compound is derived through operations by subsystem (1). As shown in Fig. 2-2, the target structure is entered from the graphic terminal and various topological features (type of atomic species, ring, aromaticity, bond, etc.) of the target structure are recognized in the next step. Then, these features are subjected to logical processing operations to infer partial structure units important for the synthesis, and disconnection points in the target compound are proposed which will be required in considering the reversed synthesis process. Concerning a portion to be modified, the system places limits on the modification for each unit that constitutes the portion. All these are then used as constraints in generating precursor. The intention is to utilize these constraints in deriving precursors required to construct the target compound in the next step. For rather complex molecules, it will be beyond human capabilities to pick up all of possible precursors satisfying the constraint conditions. As a tool to carry out this, AIPHOS incorporates a reaction generator. It was developed by modifying the structure generator for CHEMICS, an automatic structure elucidation system, so as to treat organic reactions. Some of the most important modifications of the structure generator consist in the fact that program can now generate separated structures and the corresponding relation between the number and sort of atoms with regard to the precursors and the targets is maintained. Especially, because of the latter fact, it is possible to control the variation in the reaction sites by means of various information.

Fig. 2-3 illustrates precursors generated by the reaction generator which are required in synthesizing methyl acetate under constraints assumed from structural features recognized by the system. In the first stage, based on the recognized structural features of methyl acetate, the system indicates that both of the two methyl groups can be maintained all through the reaction with other portions modifiable within limits associated with their functionality. It is also indicated that the bonds represented by thick lines must not be disconnected while the others may be disconnected. Thus, under the above constraints, the reaction generator generates the first possible case where the bond (1) is

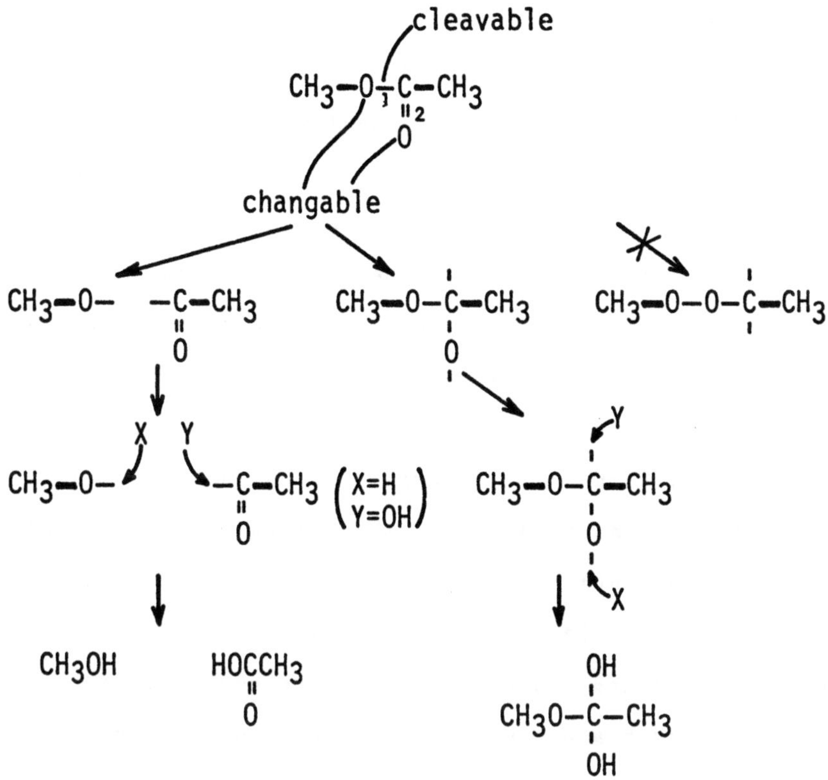

Fig. 2-3 Generation of precursors.

disconnected while the bond (2) is maintained. Then, the dummy atoms, i.e. cationic dummy X and anionic dummy Y, connect the disconnected sites in a way consistent with the ionic properties of the sites. If X and Y are assumed to be H and OH, respectively, the precursors generated here become a pair consisting of methanol and acetic acid. In another possible case, the bond (1) is maintained while the bond (2) is a single bond instead of a double bond. This also leads to the generation of gem-diols as precursors following the connection of the disconnected sites by the dummy atoms X and Y. Though it is known that this cannot exist as unstable intermediates, the current system outputs this because it does not have a function to evaluate this aspect. Still another possible case is that both the bonds (1) and (2) are disconnected and two oxygen atoms are connected each other. However, since such a precursor as the one possessing a substructure -O-O- is not to be generated, which is registered beforehand in the forbidden substructure list created by a user, the peroxide-type precursor will not be generated. The user may register some items (triple bond in three-membered ring, cumulene, etc.) in the forbidden list. From the items registered by the user, the system selects out only such ones that can be utilized for the relevant operation. The user also can modify the list appropriately. In this test run, two structures were generated as possible precursors that meet all the requirements. In this way, the reaction generator, which constitutes the logical core in this system, generates all possible precursors based on data acquired by the system and those entered by the user.

It is expected that the use of a logical approach can help to find interesting or new reactions. The major problem here is to find reasonable reaction conditions for the reaction process to produce the target compound using all of the synthetic precursors generated. Though the proposal of interesting reactions is helpful itself, it cannot serve practicality if required reaction conditions remain unidentified. A module designed for AIPHOS to support this is currently under development, which consists of knowledge base automatically produced from a variety of specific reactions in data base, combined with a program to manage it. In the knowledge base, reactions are described at a more general level than in the specific reaction data base, making it possible to cover a larger number of reactions than that of reaction examples registered in the specific reaction data base from which the knowledge base is derived. Thus, it becomes possible to check a reaction process generated by the reaction generator, through the procedure of identifying required reaction conditions by means of the knowledge base. For methyl acetate, for instance, the system is expected to show the required reaction conditions as "H^+ esterification" and "H^+ dehydration" for the two reaction processes output.

Important functions of this knowledge base are described below, citing another example. For the reaction process from methyl p-nitrobenzoate to p-nitrobenzylalcohol as given in Fig. 2-4, the reaction condition may be the use of $LiAlH_4$ as a reaction reagent if attention is paid only to the reaction sites,

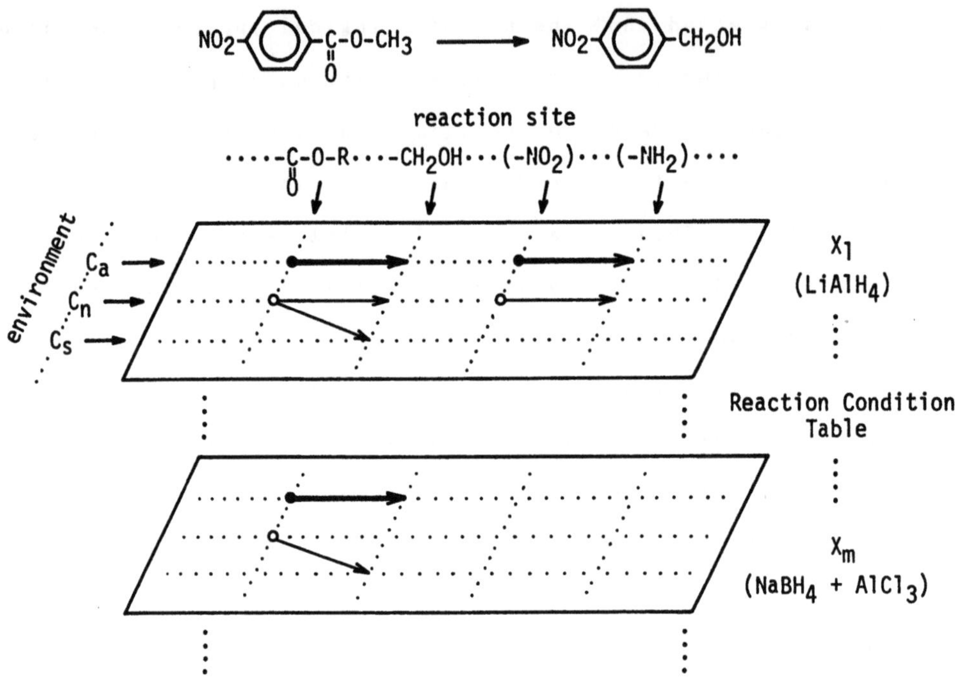

Fig. 2-4 Evalustion of synthetic condition by the knowledge base.

namely, the conversion of the ester group into the alcohol group. This reagent, however, also acts to convert the nitro group into the amino group in a minor part of the reaction. Thus, the system makes the decision that the target reaction is not feasible, and rejects the use of $LiAlH_4$ as a reaction condition for this reaction. Then, another condition is examined. It may be, for example, the use of $NaBH_4 + AlCl_3$. In this case, the use of this reagent will be proposed to the user as a candidate reaction condition to allow the ester to be converted into the alcohol without damaging the nitro group. In other words, the system judges that this reaction process is feasible. The most important function of this module is to recognize the reaction site and the structural environment around it, examine structural features other than the reaction site, and verify the overall consistency of the reaction process in question.

Thus, the logical module incorporated in AIPHOS executes a series of sequential operations as follows: taking in appropriate information selected out of data acquired by the system or entered by the user; generating all reaction processes that cover all of them while receiving additional information from the system to improve incomplete data;examining the feasibility of the reaction processes through the procedure of giving reaction conditions by means of knowledge base that consists of generalized data taken from past experiences and findings; and finally proposing the appropriate reaction process together with required reaction conditions. The present report has been focused on the logical subsystem, especially the reaction generator incorporated, and knowledge base that serves, as one of its functions, for the identification of reaction conditions. A module to support synthesis design based on reaction prediction is also under development at present, though it is not presented here.

3. Prediction of Synthetic Precursor with Structure Proposed by TUTORS

The procedure for predicting synthetic precursors having the promising candidate structure (chloropromazine already known as tranquilizer being taken as example here) proposed by TUTORS is outlined below in connection with the synthetic precursor prediction operations by AIPHOS described above. For the structural formula of chloropromazine entered from the graphic terminal, the structural feature recognition component of the logical information processing subsystem extracts various data from the standpoint of organic reaction chemistry and, based on its results, proposes disconnection points (strategy bonds and atoms) to be taken into account in considering the reverse synthesis. The structure given in the graphic display is shown in Fig. 2-5. The wave lines represent the strategy area proposed by the above subsystem. The user may modify the results. Receiving this plan, the reaction generator will propose all possible synthetic precursors conformable to the requirements, while employing dummy atoms in some cases, according to the procedures described above. In Fig 2-6 are listed some of the synthetic precursors of chloropromazine predicted by AIPHOS. Although the examination of reaction conditions for these synthetic processes proposed by

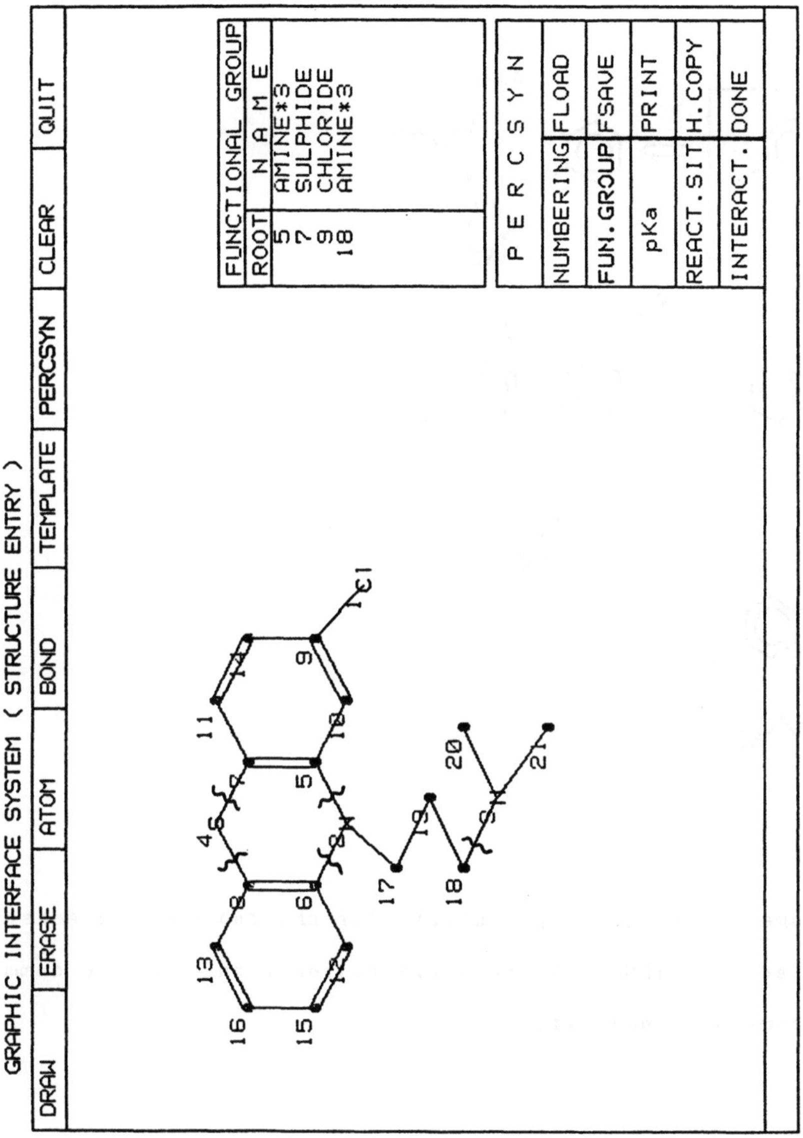

Fig. 2-5 Strategy area proposed by AIPHOS for the running sample, chloropromazine (wave lines represent it)

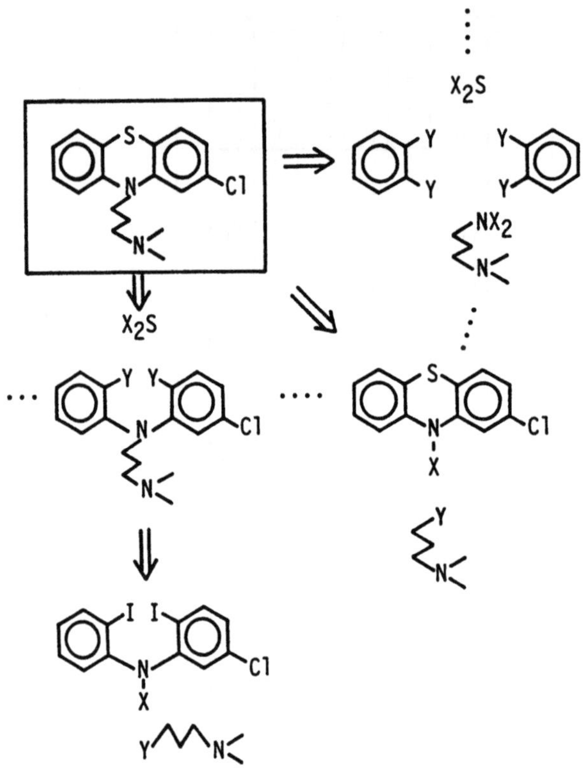

Fig. 2-6 Some of synthetic precursors for chloropromazine generated by AIPHOS (X and Y are cationic and anionic dummy atoms or atomic group).

AIPHOS has not been done because of imperfection of knowledge base capability, each result seems feasible from practical point of view, dummy atoms X and Y being regarded as cationic species Na or H and anionic species Br or Cl, respectively.

References

1) K. Funatsu and S. Sasaki, "Computer-Assisted Organic Synthesis Design and Reaction Prediction System, AIPHOS. -Overview-", Tetrahedron Computer Methodology (submitted).

2) K. Funatsu, C.A. Del Carpio, and S. Sasaki, "automatic Perception of Reactivity Characteristics of Molecular Structures Directed to The Planning of Organic Synthesis", ibid. (submitted).

3) K. Funatsu, T. Endo, N. Kotera, and S. Sasaki, "Automatic Recognition of Reaction Site in Organic Chemical Reactions", ibid. (submitted).

(3) Structure Elucidation of the Synthesized Compound (CHEMICS)

1. Outline of Current CHEMICS

The CHEMICS system, developed by the authors, is a computer-assisted structure elucidation system for organic compounds, which depends on the way of structure generation method; that is, the most probable structure is generated by the automated analysis of data (also for instance, chemical spectra) of an unknown using empirical and theoretical rules.[1] The principle of the system is that all possible structures, which are known to exist or which might exist on chemical grounds, are listed in a computer. The number of the structures in a particular case is then narrowed down by successively entering information from spectroscopic measurements. CHEMICS is designed to store all the substructures (called 'components') necessary for building any likely structures. At present, CHEMICS contains 630 components for the structure elucidation of organic compounds consisting of C, H, O, N, S and halogen atoms (Table 3-1). The set of components has been devised so that it is possible to construct any structures by selecting appropriate components from the complete set. To store such a set of components in a computer is synonymous with storing all the complete structures which could be present.

The current CHEMICS system is composed of the following four functional modules, as shown in Fig. 3-1: a) Data analysis, b) Structure generator, c) Stereo-generator, d) Input of macrocomponent (partial structure).

a) Data analysis: Among the components which have survived because they are consistent with the molecular formula, some can be subsequently discharged because they are inconsistent with H-1 and C-13 NMR chemical shift values, or IR data. In the selection of components by CHEMICS these spectral data measured on the sample are compared by computer with those in component/chemical shift or component/wave number correlation tables, so that only components consistent with these data are left. Part of the correlation table showing ppm ranges for H-1 and C-13 shifts is shown in Table 3-2. The next step is to make component sets by use of the components which have been selected as being not

Table 3-1 Component set for structure elucidation of organic compounds containing C,H,O,N, S, and halogens.

No.	Component		No.	Component	
1	tert-Bu–	(S)	372	$>C-$	(I)
2		(ND)	373		(Br)
⋮			374		(Cl)
51	CH_3CH_2-	(CD)	⋮		
52		(CT)	403	$>N\rightarrow O$	(Y)
53		(CS)	404	$>S\rightarrow O$	(Y)
⋮			⋮		
185	$CH_3-S(\rightarrow O)(\rightarrow O)-$	(O)	547	–OH	(CD)
186		(Y)	548		(CT)
⋮			549		(CS)
351	$>C=NH$	(F)	⋮		
352		(S)	626	–F	
353		(ND)	627	–Cl	
⋮			628	–Br	
			629	–I	
			630	–D–	

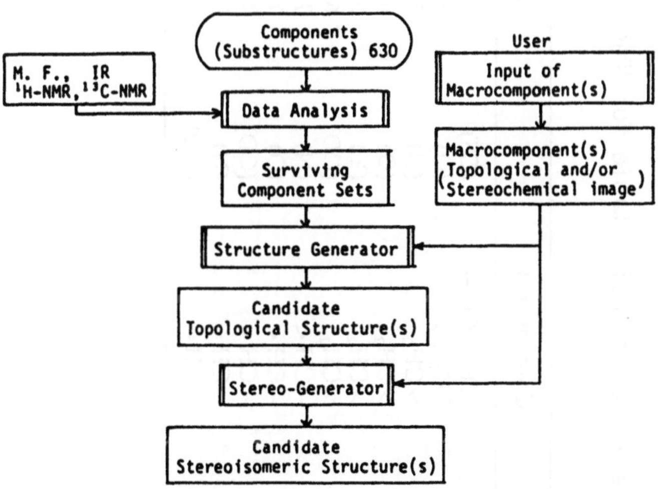

Fig. 3-1 Block diagram of the current CHEMICS.

Table 3-2 Correlation table for NMR analyses.

No.	Components		^{1}H-NMR (ppm)				^{13}C-NMR (ppm)	
16	$(CH_3)_2CH-$	(CS)	1.20	0.50	28.9	13.4	40.7	18.2
132	CH_3CO-	(O)	2.50	1.80	24.2	17.7	174.0	165.8
196	$>CH_2$	(N)	5.60	1.10	75.5	25.7		
197	$>CH_2$	(O)	6.10	2.30	88.6	43.2		
198	$>CH_2$	(Y)	5.40	1.70	60.4	6.6		
199	$>CH_2$	(CD)	6.20	0.50	58.0	11.9		
208	$-CH<$	(N)	7.30	1.10	96.9	27.1		
209	$-CH<$	(O)	7.70	1.80	111.1	41.3		
211	$-CH<$	(CD)	4.40	0.80	75.8	15.4		
274	$-CH=$	(N)	9.60	6.60	184.5	91.6		
277	$-CH=$	(CD)	9.00	4.50	165.0	90.1		

contradictory to the molecular formula and spectral data.

b) Structure generator: This step is to generate structures from the individual component combinations. The generation is carried out taking all possibilities into account, in due consideration of the principle that most of the components can only be linked to a limited number of species. On the basis of specially designed logic, connectivity stack, when the system functions properly it does not reproduce the same structure nor does it fail to generate any structure which can justifiably be built.

c) Stereo-generator: The major role of the above module is generation of constitutional isomers. On the other hand, this module has a function for generating all possible stereo-isomeric structures due to asymmetric carbon, double bond and so on using topological information of the respective constitutional isomers generated by 'structure generator'.

d) Input of macrocomponent (partial structure): The chemist often has some information about the structure of a sample. This may be obtained from the past record of the sample or the experience in its laboratory handling. When the partial structure is entered by the user, the constitutional information is degraded into its components (described above), which are then compared with the components that the system has selected. The system will adopt the information entered only when all the components derived from the partial structure inserted have already been selected by CHEMICS. This means that the components which the system has selected with a full safety factor will take precedence over the additional information which has been entered manually. The stereochemical information of the macrocomponent is reflected on the final results according to other logic.

As obvious from the above explanation, the fragments for making up structures and carriers of spectral information are just components. The unit of examining the reasonable allocation of each component to NMR signals, is also component centering around the correlation table. Moreover, the examination of the input macrocomponent by spectral data is also based on the component unit. It is obvious that essentially the analytical ability of 'data analysis' in CHEMICS never exceeds what is provided by component units. Thus, correspondence of candidate structures with input data is said to become ambiguous in some cases. The number of candidates increases in proportion to the ambiguity. According to the principle of never missing correct solution, this result is said to be unavoidable. As one of the ideas for coping with this situation, CHEMICS-F, which has a file retrieval function, has been developed, and the modules for prediction of the number of C-13 NMR signals and judgement of probability on the basis of strain energy calculation, have been provided. These functions play an effective role after generation of whole structures.

On the other hand, the introduction of partial structures selected by the user, has enhanced the correctness and practicality of structure elucidation by our system. However, if possible, it seems to be one of the ideal features that partial structures entered should be determined by agreement with both deduction by the computer and judgement of it by the user. In order to realize about this situation an analytical way different from that in 'data analysis' of the current CHEMICS is required. In this sense, as a new approach of automated partial structure elucidation, an interdependent analytical way based on the relationships between H-1 and C-13 NMR chemical shifts for each atomic group with specified neighboring groups, has been developed. According to this method, rather big-sized partial structure can be generated automatedly which is most helpful for narrowing the number of candidates.[2)] Furthermore, the introduction of carbon-carbon signal connectivity information provided by 2-D NMR into CHEMICS has been accomplished so that the connectivity of carbons in skeletal structure of an unknown is automatically analyzed and elucidated at the both steps of 'data analysis' and 'structure generator' in generating whole candidate structures. This also results in the fewer number of candidates.[3)]

2. Examples of Structure Elucidation

The first example of the elucidation process of an unknown compound (actually, nicotine) is serially illustrated in Fig. 3-2. Chemical data of the unknown, molecular formula and spectroscopic data of IR, H-1 and C-13 NMR, are compared with 630 components at the step of 'data analysis' to give rise 65 components. The number of candidates will amount to more than ten thousand if all the possible structures are constructed on the basis of the 65 components. In this example, the interdependent analytical method based on the relationships between H-1 and C-13 NMR chemical shifts to suggest automatically the presence of '3-substituted pyridyl group' as a possible substructure in the

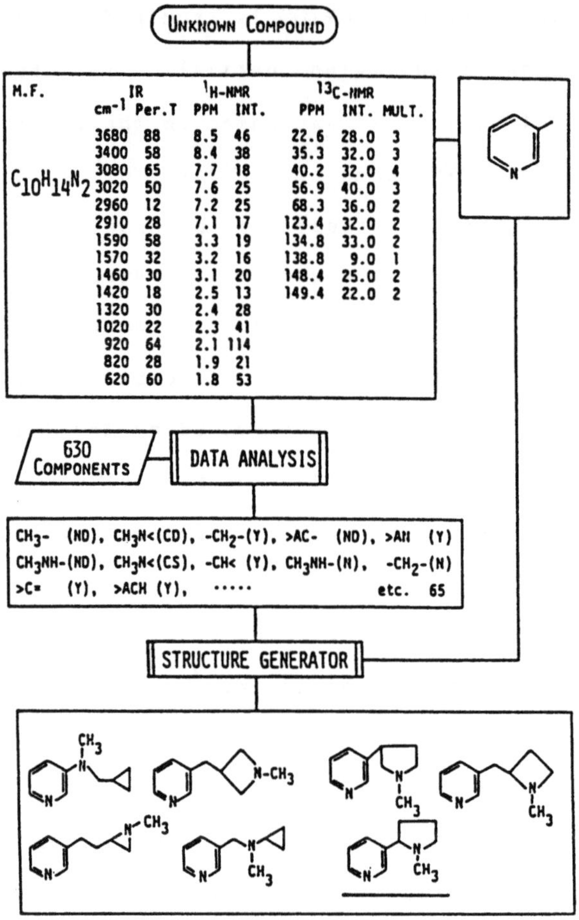

Fig. 3-2 The analytical processes using 3-substituted pyridyl group provided by the H-1 and C-13 NMR chemical shift interdependent analysis.

unknown structure so that the candidates' number drastically diminishes into only seven structures in which the underlined correct solution is involved.

The second example deals with an unknown compound a part of which is already identified from chemical or instrumental analysis made by chemist. The remaining part is generated automatically and connected to the known part to form the whole structure. Let's assume an unknown compound which has the molecular formula $C_{17}H_{17}O_6Cl$ and comprises a known substructure, denoted as A, as the host(Fig. 3-3). If the composition of A, i.e. $C_9H_7O_4Cl$, and the number of its single bonds, i.e. two, are entered in the computer, calculation will be performed on the assumption that the remaining part (counterpart) has the composition of $C_8H_{10}O_2$ and two single bonds. Analysis of this compound by CHEMICS in terms of its molecular formula, H-1 NMR and C-13 NMR allows 65 components to remain as the components consistent with these data. However, of the 65, 16 components are not consistent with input data in consideration of forming the counterpart, with the remaining 49 components being left for construction of the counterpart that is expressed as $C_8H_{10}O_2$. Thus, the operation, which is performed with reference to NMR data, generates four substructures (B, C, D, E) as candidates for the counterpart, as shown in Fig. 3-3. Finally, the four are connected with the host (A) to generate six structures, F, G, H, J, K and L, of which G correctly represents the structure of griseofulvin used as the unknown compound, as listed in Fig. 3-4.

The third example shows a procedure for applying 2-D NMR data. CHEMICS analysis of H-1 and C-13 NMR spectra of a certain

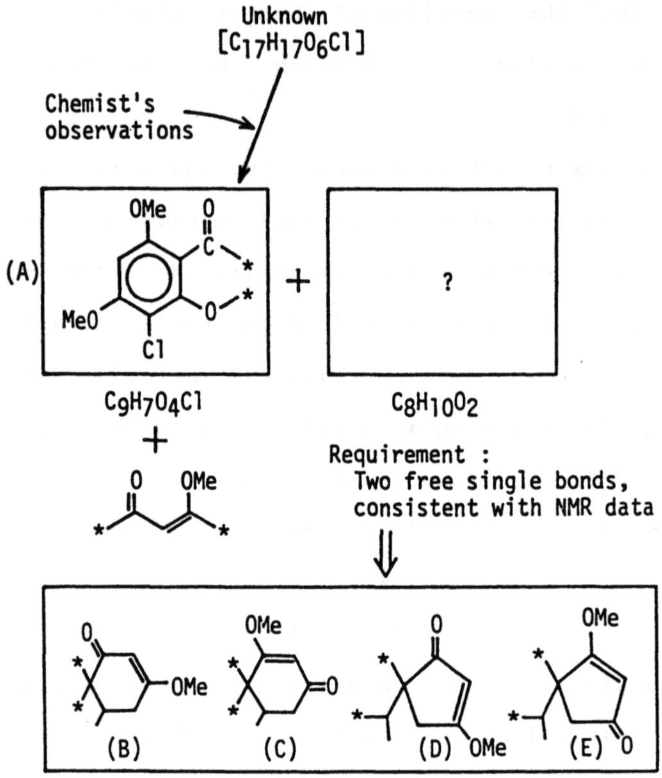

Fig. 3-3 The conditions and results for elucidating the counterpart structure.

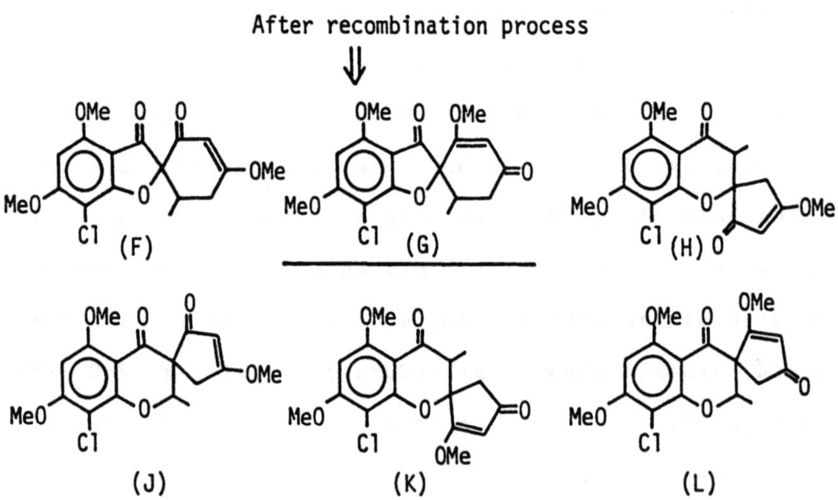

Fig. 3-4 The candidates provided by connection of counterpart structures with host (A).

compound ($C_{13}H_{20}O$) gives 39 components, from which 1417 candidate structures are generated. 2-D NMR data can serve to reduce this number drastically. Three types of carbon-carbon signal connectivity information (on C-13 NMR) to be entered in the computer are obtained from the 2-D NMR of this specimen as shown below.

Type 1: 1-10, 4-11, 7-9 (long range C-H cosy : path 2)

Type 2: 2-5, 6-12, 8-12 (combination of C-H and H-H cosy)

Type 3: 3-4, 3-7, 4-8, 6-7 (long range C-H cosy : path 3)

Receiving these data the computer carries out the following operations. The data of Type 1 are used mainly to decide on the possibility of existence of each component with two carbons contained in the group of 39 components. As a result, from the remaining 36 components 203 structures are generated without using Type 2 and Type 3 data. Next, if the Type 2 and Type 3 data are used together with Type 1 data, the number of generated structures is decreased to 36 by examination through the Type 2 data and then to only one, the target compound beta-ionone(I), through Type 3 data. The above results are summarized in Table 3-3. In Table 3-4 is summarized the number of candidate structures given by CHEMICS for menthol (II), linalool(III) and an isomer of thujopsene(IV), reported recently, with and without the aid of 2D-INADEQUATE information. In these samples, the number of candidates diminishes into one and only correct structure through four check stages.

3. Structural Examination of the Compound Synthesized According to the Proposal by AIPHOS

Table 3-3 The results of 2-D NMR analyses using H-H cosy and C-H cosy information.

Information used in CHEMICS	The number of candidates
Ordinary Analyses	1,417
1 - 10 4 - 11 7 - 9	203
2 - 5 6 - 12 8 - 12	36
3 - 4 4 - 8 3 - 7 6 - 7	1

STRUCTURE NO. = 1

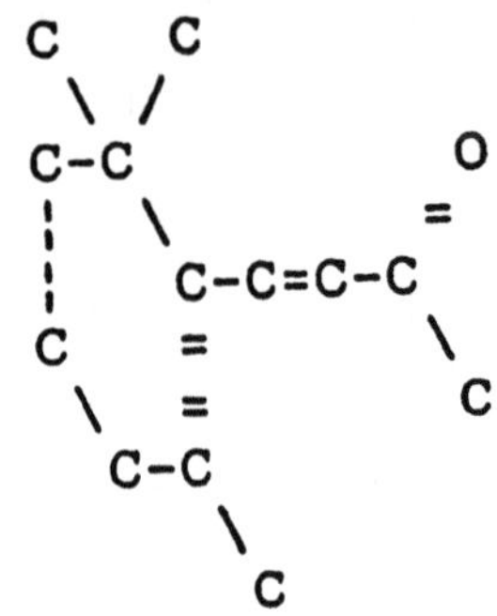

SAMPLE(I) : β-IONONE

Table 3-4 The results of 2-D NMR analyses using 2D-INADEQUATE information.

	The number of candidates		
	II	III	IV
Ordinary Analyses	219	48	4,450
by Check 1	46	3	920
by Check 1 Check 2	5	3	17
by Check 1 Check 2 Check 3	3	2	9
by Check 1 Check 2 Check 3 Check 4	1	1	1

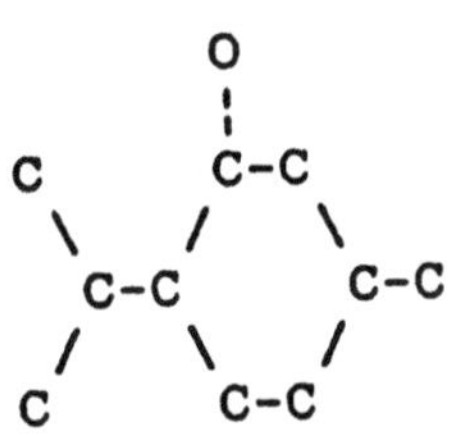

SAMPLE(II)
L-Menthol

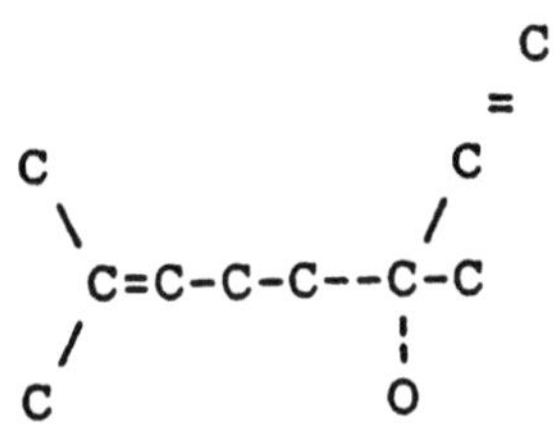

SAMPLE(III)
Linalool

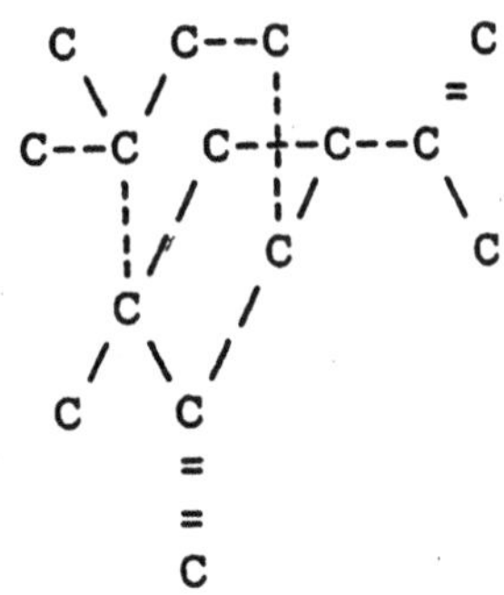

SAMPLE(IV)
an isomer of Thujopsene reported recently

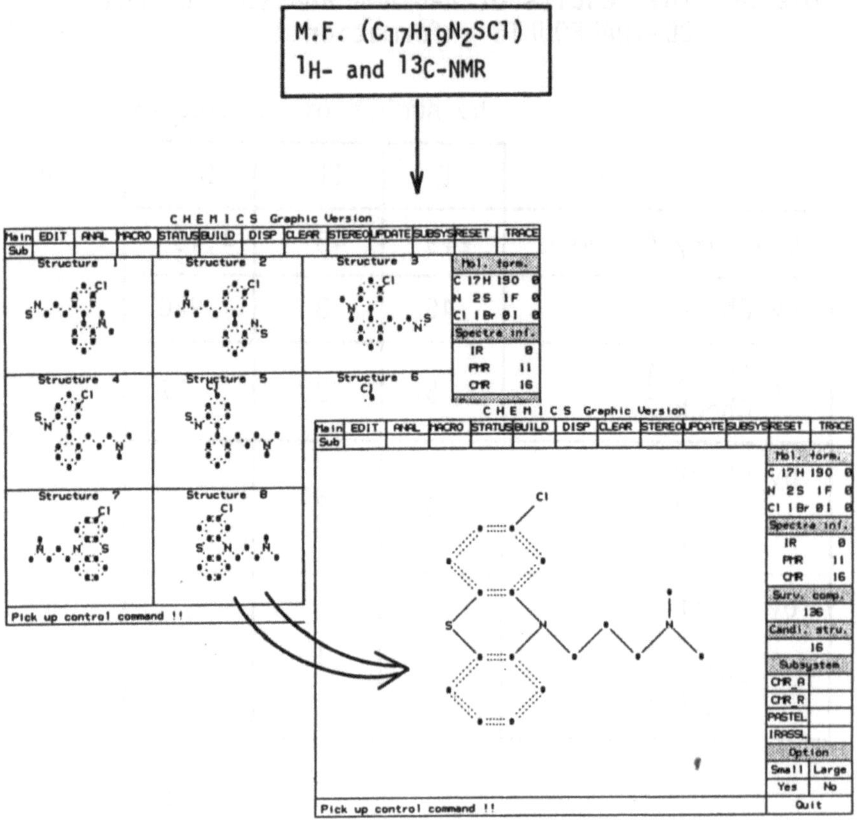

Fig. 3-5 Part of candidate structures for the running sample (chloropromazine) elucidated by CHEMICS.

In order to examined whether the synthesized compound has the target structure or not, we will utilize CHEMICS. Suppose the running sample (chloropromazine) for this purpose.

The molecular formula ($C_{17}H_{19}N_2SCl$), H-1 and C-13 NMR spectroscopic data are input into CHEMICS as the structural information of the unknown. Additionally, some substructural information are also input in consideration of the synthetic processes of the sample compound. Using these information, CHEMICS generates and proposes the candidates of the running sample on the basis of the logic mentioned above. The candidates generated are displayed on graphic terminal as shown in Fig 3-5, and no. 8 structure is just correct solution, chloropromazine, which can be enlarged if necessary. It may be concluded that generation of the corresponding structure to chloropromazine by CHEMICS as one of the candidates of the target supports the high possibility that the synthesized compound has the target structure.

Reference

1) K. Funatsu, N. Miyabayashi, and S. Sasaki, J.Chem.Inf.Comput. Sci., **28**, 19(1988).

2) K. Funatsu, C.A. Del Carpio, and S. Sasaki, Computer Enhanced Spectroscopy, **3**, 119(1986); K. Funatsu, C.A. Del Carpio, and S. Sasaki, ibid., **3**, 133(1986).

3) K. Funatsu, Y. Susuta, and S. Sasaki, J.Chem.Inf.Comput.Sci. (submitted).

XI

In order to examined whether the synthesized compound has the correct structure of [illegible]

[illegible]

Physico-chemical Data Estimation for Environmental Chemicals

R. Brüggemann and B. Münzer

Gesellschaft für Strahlen- und Umweltforschung mbH München, Projektgruppe Umweltgefährdungspotentiale von Chemikalien, Ingolstädter Landstrasse 1, 8042 Neuherberg, Federal Republic of Germany

ABSTRACT

The environmental fate and mobility assessment is an essential part of the process of selecting and identifying environmental chemicals. However, for most chemicals only fragmentary knowledge exists about those properties which determine their fate in the environment. Therefore, it is pertinent to estimate these physical-chemical data on chemicals. Due to the huge number of organics not only an appropriate package of formula but also a high degree of automatism for this task is needed.

There are two types of approaches to estimate properties. The first one is to use a databasis for fragments of the molecule. The second one is to use property-property- relationships which in general do not need the knowledge of (sub)structures.

Therefore, DTEST which is part of the EDP-code E4CHEM (acronym for "Exposure and Ecotoxicity Estimation for Environmental Chemicals") mainly property-property- relationships are used which allow a high degree of automatism. The relationship to calculate vapor pressure, partition coefficients, solubility and other environmental relevant properties are shown and discussed in DTEST. A brief explanation of the leading ideas in the program structure is given.

Keywords: property-property-estimation methods

C. Jochum M. G. Hicks J. Sunkel (Eds.)
Physical Property Prediction
in Organic Chemistry

Physico - chemical Data Estimation for Environmental Chemicals

1. Environmentally relevant chemical properties

In this talk I am going to present the computerized estimation program DTEST. This program was developed primarily as a part of a software package estimating the exposure and ecotoxicity of environmental chemicals. The design of DTEST will be best understood if I say some words on exposition models first.

The code E4CHEM (Exposure and Ecotoxicity Estimation for Environmental Chemicals) [1,2,3] was developed to find out which substances out of a large list of compounds were most likely to present an environmental hazard. Such a priority setting process needs data on the behaviour of the compound. To demonstrate the problem at hand, I will show you a list of data suggested for a fairly good hazard estimation:

Please keep in mind that for most substances involved in such an exercise, very few data are known, and that the set of known data is hardly the same for any two substances.

Obviously, we cannot just postpone the priority setting exercise until all these data are measured. We use a combination of data estimation and modelling. First we augment the data set, then we use models to estimate the expected effects on the environment. After this, we employ an further analysis to find out which effects could not be estimated satisfactorily and to identify the input data responsible for this. For these data, a closer literature research or even additional measurements may be necessary.

Identification of chemical
Name Empirical and structural formula CAS - number
Economic data
Production volume Import / export Use pattern
Occurrence in the environment
Concentrations in soil, air, water - " - in biota
Physical - chemical properties
Molecular weight Melting and boiling points Vapour pressure Density Surface tension Water solubility Solubility in organic solvents Partition coefficients (K_{ow}, K_{oc} etc.) Dissociation constant Flash point
Degradation and accumulation
Biodegradation Photolysis and hydrolysis Bioconcentration
Effects on biota
Ecotoxicity (daphnia, fish, algae etc.) Acute, chronic toxicity on mammals Genotoxicity

2. Using exposition models for priority setting

Mathematical models can give good results even if they are quite simple and use only few input data. For surface waters (for example), the simplest approach would be a dilution model. This kind of model does not need any substance data. It will be conservative because it will not take volatilisation and degradation processes into account, and thus the concentrations of the substance in the river will be overestimated; the user of this model will always be "on the safe side".

The next picture shows how the modelling results are much improved if substance parameters are known to calculate the fate of substances. In this example the Henry coefficient is used to estimate volatilisation. The computation was made with the steady - state model EXWAT [4,5,6] for Trichloroethene in the river Main.

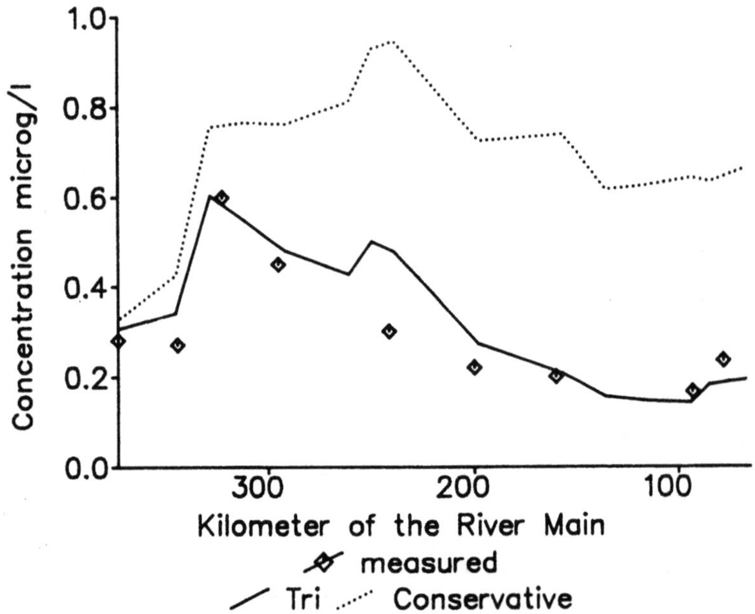

Figure 1: Estimated concentrations in the river Main

When we developed the program package E4CHEM, we combined two approaches to deal with the data problem :

I All models can work with a very small set of data. The more data are known, the more precise the results will be.

II To deal with the problem of gaps and lack of structure in the available data sets an information system for environmental chemicals [7] on one hand, and a system of chemical property estimation methods on the other hand have been developed.

3. The estimation program DTEST

3.1 A short description of DTEST

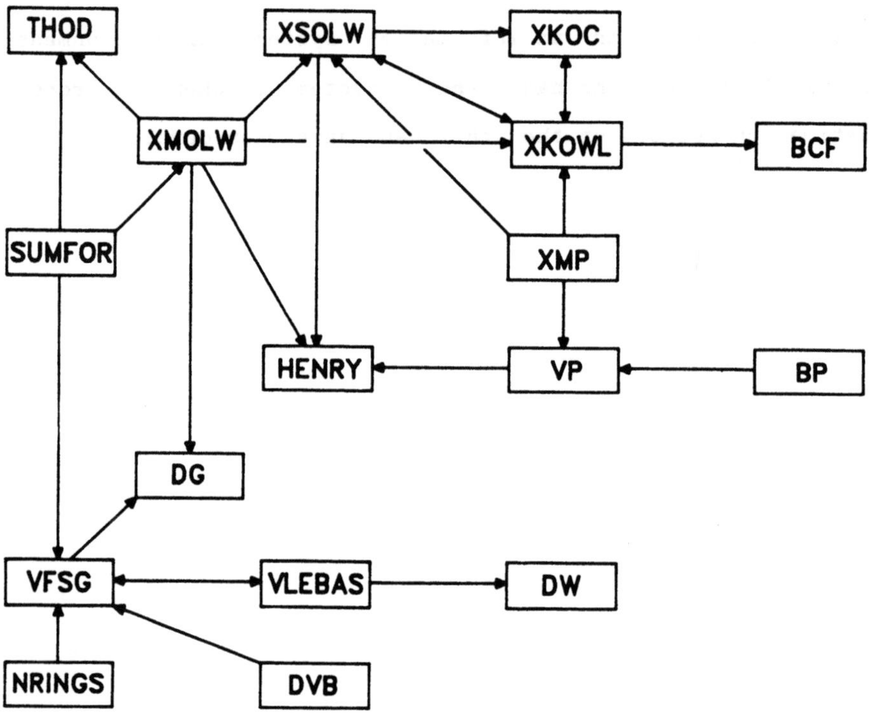

Figure 2: Property-property relations within DTEST

This diagram shows the most important substance data used in DTEST. An arrow leading from one parameter to another means that the former parameter is used to estimate the latter. Usually each estimation formula needs several data, i.e. the formula is represented by several "arrows" pointing to the estimated parameter. The program checks if all data needed for an estimation are known. If any are missing, an attempt is made to estimate the missing data, or an estimation formula using less input data is selected. After the execution of DTEST, all missing data that could possibly be estimated given the initial data set are calculated; and for each parameter the formula granting the most reliable result for the given data set has been selected.

DTEST runs automatically without any dialog once it has been started. The data can be entered during an E4CHEM session either manually or from a file or (as yet only in some mainframe implementations of E4CHEM) from a database. DTEST can be called from within any model of E4CHEM; running the program in batch mode is also possible.

3.2 Criteria for selecting estimation methods

DTEST has been developed to process large lists of substances. One consequence of this goal has been that recognizing known and missing data and selecting the best formulas for the situation at hand is done automatically without any dialog during the estimation. The other consequence was a strong restriction on the selection of estimation methods. I will return to this later.

We did not develop any new estimation formulas for DTEST; this was not within the scope of the project. Our sources were the already established methods described mainly in the Handbook of Chemical Property Estimation Methods [8] and the manual for the CHEMEST program [9]. DTEST contains a selection of estimation methods also used in CHEMEST :

	CHEMEST	E4CHEM
Water solubility	+	++
K_{oc}	+	++
Activity coefficient	+	-
Boiling point	+	++
Vapour pressure	+	++
Volatility		
- from water	+	++
- from soil	-	+
Henry coefficient	+	+
Melting point	+	-
Acid dissociation constant	+	-
Diffusion constants		
- in gas	-	++
- in water	-	++
Molecular volume	-	++
Theoretical oxygen demand	-	++
Relative molecular mass	+	++
K_{ow}	-	++
Root concentration factor	-	+
Plant concentration factor	-	+
Partition aerosol - gas	-	+
Correction acid/base	-	+
Dry deposition	-	+
Wet deposition	-	+
Element analysis	-	++
Environmental entrance media	-	++
Release rates	-	++

++) Computed in DTEST
+) Computed in submodels of E4CHEM
-) not computed

Structure - property relations have generally not been selected. The reasons are :

- It is not always clear which structural element is characterizing the compound. This is important if class - specific formulas are used.
- An automatic structure recognition needs a highly complex algorithm. The development of such an algorithm was out of the scope of this project.

Often there are several estimation methods suggested in the literature; especially in connection with the system BCF / K_{ow} / K_{oc} / water solubility (see e.g. Kenaga Goring [10]). The methods were selected according to the following criteria :

- A good correlation (r^2 large) for a large number of different substances.

  This was not easy to achieve, since most formulas (especially regression formulas) have been developed for a special class of substances.
- The priority was given to relations with a theoretical foundation.
- Some property - property relations need additional information on the molecular structure. Formulas requesting more information than the number of aromatic/heterocyclic rings were not selected.

3.3 Examples for estimation formulas in DTEST

3.3.1 Remark

There is no time to discuss every relation individually. Instead, I have selected some formulas to give you the flavour of how the design of such an algorithm is developed.

3.3.2 Water solubility

The most important substance data used in E4CHEM are the partition coefficients K_{ow}, K_{oc} and the dimensionless Henry coefficient. Several other data are only used to estimate these composite values. The water solubility plays the role of an important link, since for all three partition coefficients there are estimation formulas using the water solubility, and the water solubility itself can be estimated from either K_{ow} or K_{oc}.

Lyman [8] gives an overview of estimation methods for water solubility. An approach using activity coefficients seems to be promising, since these can be used to estimate other properties as well. However, the present state of knowledge requires either a classification of the substances according to their functional groups (Pierrotti method [11]) or a separation into constituing groups (UNIFAC, [12,13,14,15]). For the reasons mentioned above, these methods do not meet the simplicity requirements of DTEST.

The most attractive of the remaining methods is the computation of the water solubility from K_{ow} or from the boiling point. The methods using K_{ow} have a better theoretic foundation (Miller [16]), they are more reliable [8,9,17] and there are voluminous tables on K_{ow}. Therefore, we use only estimation methods with Kow as input parameter.

The following equation holds for not easily soluble substances [16] :

$$\log K_{ow} = m \log XSOLW + c \qquad (3\text{-}1)$$

K_{ow} Octanol - water partition coefficient

XSOLW water solubility

m a constant between 0.8 and 1.0

c depends weakly on molecular descriptors like mol volume and molecular surface (TSA)

We selected a relation developed by Hansch et al. [18] using a correction by Yalkowski et al. [19]. In the form used by DTEST it reads :

$$XSOLW = XMOLW * \frac{9.506}{K_{ow}^{\ 1.399}} * F \qquad (3\text{-}2)$$

$$F = \begin{cases} \exp(-0.0224 * (XMP - 293)) & ; XMP > 293 \\ 1 & ; \text{otherwise} \end{cases}$$

XSOLW water solubility in g/l

XMOLW molecular mass in g/mol

XMP melting point in K

This relation holds for many different substance classes and has a high correlation coefficient (r^2 = 87.4 %). It has been generated using 156 substances and the value of 0.75 for m is sufficiently close to the range demanded by Mackay. It is valid for

$0.3 < \log K_{ow} < 4.7$

$-5.2 < \log C_s < 0.7$ (C_s : solubility in mol/l)

For compounds containing nitrogen or sulphur this estimation fails; the estimated value can be orders of magnitude above the exact value. Generally, 77 % of the substances examined in [8] were within a factor of 10 and 93 % within a factor of 100 from the measured values.

Substances for which the previous equation does not apply are treated with the relation established by Mackay et al. [20] :

$$XSOLW = XMOLW * \frac{1.8}{K_{ow}} * F \qquad (3-3)$$

$$F = \begin{cases} \exp(-0.0224 * (XMP - 293)) & ; XMP > 293 \\ 1 & ; \text{otherwise} \end{cases}$$

F reflects the melting entropy assumed to be 13 cal/Mol K for stiff organic compounds (see Yalkowski).

The last formula was derived from 45 substances with XMOLW < 290. The set included no organic acids.

3.3.3 Computation of K_{OC}

The partition coefficient K_D describing the sorption to solids containing natural organic carbon (soil matrix, river sediments, suspended matter) can be estimated to a remarkable accuracy given the K_{OC} :

$$K_{OC} = \frac{\left(\dfrac{\text{g sorbed}}{\text{g organic C}}\right)}{\left(\dfrac{\text{g solved}}{\text{cm}^3\ \text{Water}}\right)}$$

K_{OC} is a constant depending on the substance only, while K_D also depends on the environmental medium. K_D is computed by multiplying K_{OC} with the organic carbon content of the medium.
To estimate K_{OC} from K_{OW}, the theoretically founded regression developed by Karickhoff [21] is used. (If K_{OW} is to be estimated from K_{OC}, the equation is simply reversed.) :

$$K_{OC} = 0.411 * K_{OW}$$

The original number of substances used to derive this formula was very small (N = 5) and only polycyclic aromatics had been selected, but it has since been validated for quite a heterogeneous set of substances (see table III in [21]) with acceptable results. For phenylated ureas however, the deviations are particularly high (one order of magnitude).
The validity of this equation is reported to range up to half of the saturated concentration. For environmental chemicals the typical concentrations are far below this limit.

If the water solubility is known instead of K_{ow}, a relation stated by Kenaga Goring [10] is used :

$$\log K_{oc} = 3.64 - 0.55 \quad \log XSOLW$$

XSOLW is the water solubility in ppm. The correlation coefficient using a comparatively large number of substances (N = 106) is moderate (r^2 = 70 %). This follows a recommendation in [8], regarding the generality in the application (no restriction regarding chemical classes or structures).

3.3.4 Vapour pressure estimation

For the vapour pressure, the following formula by Mackay [22] does not need any information on the chemical structure :

$$\ln p = - \frac{\Delta H_B}{R * BP} * (1+K) * (\frac{BP}{T} - 1) - K * \ln \frac{BP}{T} \qquad (3\text{-}4)$$

ΔH_B : evaporation enthalpy at boiling temperature

For ΔH_B, the following equation is used :

$\Delta H_B = 36.6 * BP + R * BP * \ln BP$

R : general gas constant (8.134 J / (K * Mol))

K : fitting parameter. Mackay uses K = 0.803

BP : boiling point at 1 atm

Using the suggestions made by Mackay, we get the following algebraic equivalent relation :

$$V_1 = 0.803 * \ln \frac{BP}{293} - 1.803 * (\frac{BP}{293} - 1)$$

$$VP = \exp((4.4 * \ln BP) * V_1) * 10^5 * F \qquad (3\text{-}5)$$

$$F = \begin{cases} \exp(-0.0224 * (XMP - 293)) & ; XMP > 293 \\ 1 & ; \text{otherwise} \end{cases}$$

Mackay remarks that this formula should only be used for hydrocarbons and halogenated hydrocarbons with a boiling point above 393 K. For compounds containing oxygen, nitrogen or sulphur this estimation seems to be less accurate. Table II in [22] at least shows that the order of magnitude comes out right. For the first step in priority setting, this is often good enough.

The Fromherz equation [23]

$$VP = 10^{-(5.4 * BP / 293 - 10.4)} \tag{3-6}$$

was suggested as an alternative to the previous formula. We had to find out under which circumstances any of the two equations was more reliable before including them in the DTEST algorithm.

3.3.5 Validation for the vapour pressure estimation

We compiled a testing set of 51 substances from several data sources [8,24,26,27,28]. This set contained the following classes of compounds :

Substance class	number of substances
Phenyls	14
polycondensated aromatics	6
Amine functions	6
Derivates of alkanes, alkenes, alkynes	18
Aromatics or arom. heterocyclic compounds	5
Halogenated compounds	8
Carbonyl functions	9
Ether	7
Alcohols	2
non - aromatic cyclic compounds	8
Nitro groups	3

The numbers do not add up to 51 because some classes apply to more than one substance. There were 40 liquids and 11 solids in the test set.

We evaluated the equations by Mackay (MAC) and Fromherz (FROM) with this reference set. To check on the restrictions suggested by Mackay for his method, we also included another method (MAC') using MAC within and FROM outside these bounds. As an additional plausibility check we compared these three to the regression derived from the 51 reference substances :

$$\ln VP = -0.0629 * BP + 32.2 \qquad \text{(TEST)}$$

To compare these four methods, we computed the regression equations for the measured versus the calculated vapour pressure :

$$\ln \widehat{VP} = a + b * \ln VP_{calc}$$

The coefficents in this equation and r^2_{DF} are a measure for the quality of the approximation :

Method	a	b	r_{DF}^2	Std.dev (b)
MAC	-0.412	1.05	98.7	0.02
FROM	-3.27	1.48	95.4	0.05
MAC'	-0.636	1.08	96.1	0.03
TEST	0.0	1.00	95.4	0.03

The statistic requirements (a -> 0, b -> 1, r^2 -> 100%) are best met by Mackay's formula with no constraints upon the substances used. The most distinct deviations in the four methods are caused by malathione, disulfotone and dieldrine. The first two substances are organophospate insecticides, and their boiling points had to be extrapolated. This lead to additional uncertainties. The deviation for dieldrine cannot be traced as easily. In any case this is a recommendation to check again upon the validity of substance data.

3.3.6 Henry coefficient

To estimate the equilibrium distribution between water and air, the equation

$$C_{air} = H * C_{water} \quad (3-7)$$

is used. The dimensionless Henry coefficient H is a partition coefficient like K_{ow} and K_{oc} and one of the most important data in E4CHEM. DTEST uses the following estimation formula [9,29,30] :

$$H = XMOLW * \frac{VP}{8314 * XSOLW * T} \quad (3-8)$$

XMOLW : relative molar mass (g/mol)

VP : vapour pressure (Pa)

XSOLW : water solubility (g/l)

T : temperature (K)

According to Prausnitz [31] this holds for partial pressures below 10 bar and a concentration in water of up to 1.7 mol/l. CHEMEST has 1.0 mol/l as the critical concentration. These values are not exceeded for most organic compounds in the environment (excluding accidents). For substances with high water solubility (e.g. methanol) activity coefficients should be used to compute the distribution between air and water instead of the Henry coefficient.

There are methods to estimate the Henry coefficient using molecular fragments (see [9]). For the reasons explained before, DTEST does not use fragments. However, if they are used anyway, an approach using activity coefficients at a given concentration is to be preferred which may estimated by UNIFAC [12,15].

Here is a comparison of estimated and measured Henry coefficients. Some substances did not have unique input data. For these, a minimum and a maximum value for the estimated Henry coefficient is shown in the table.

Nr.	Substance	(min)	$^{10}\log H$	(max)	(meas.)
1	Lindane	-4.69		-3.83	-3.68
2	HCB		-1.44		-2.02
3	DDT	-3.42		-1.40	-3.47
4	α-Endosulfane		-3.29		-3.54
5	β-Endosulfane		-4.65		-4.80
6	Aldrine	-3.19		-1.54	-2.74
7	Dieldrine	-4.40		-2.95	-3.39

The picture below visualizes these results :

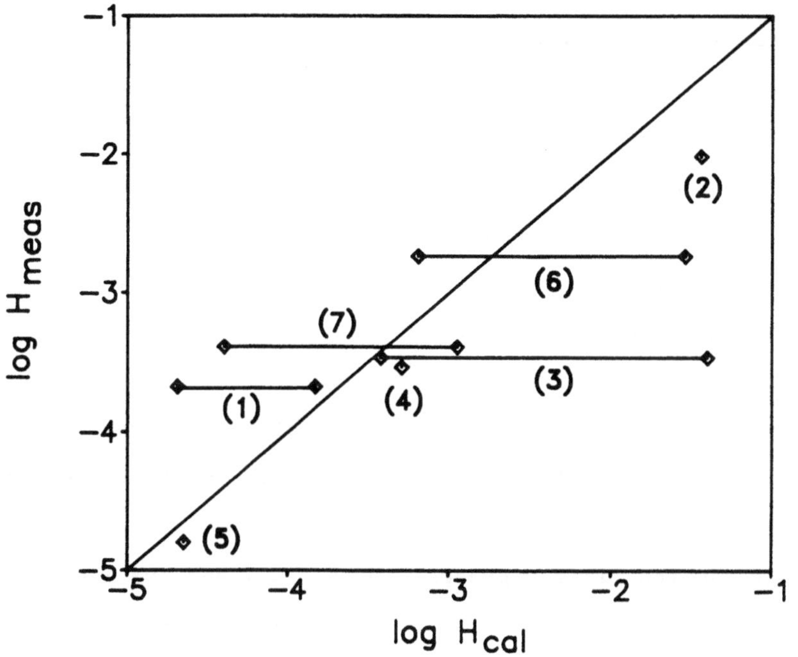

Figure 3 : Measured versus calculated Henry coefficients

The reason for the wide range of some estimates is the lack of data in the available sources. In such a case, the estimation has to be carried over several steps, maybe including estimation of vapour pressure and water solubility, until the Henry coefficient can finally be computed.

If a large list of substance is to be processed, it may not be possible to look for better data for each individual compound. However, if the data insecurity strongly affects the predicted exposition of a compound, more research is necessary to get useful results. A simple tool to find the substances with crucial data insecurity is the exposition map developed for E4CHEM.

4. Sensitivity analysis using the exposition map

At the beginning of this talk I showed an application of EXWAT as an example for an environmental model. Sensitivity analysis with a model like this has been done, but it is a matter too complex to be demonstrated here. Instead, I have selected EXTND, another submodel of E4CHEM that simulates a thermodynamic equilibrium between air, water and sorbens in a closed system. Only few environmental parameters are needed for this model (volumes, densities, organic carbon content).

One possibility to display the results is an "equilibrium distribution map" showing directly the effect of changes in the substance data (Henry coefficient and K_{oc}) on the distribution of the substance between the three media. In a log Henry/log K_{oc} - coordinate system the lines represent 50 %, 90 % and 99 % of the substance in each compartment. A substance is represented by a point in the coordinate system or - if the substance data are not known exactly - by a line or an area. The intersections of the percentage lines with the area covered by the data range shows how the data insecurity affects the quality of the results.

To demonstrate this, I have collected some K_{oc} values for the substances introduced before :

Nr.	Substance	(min)	$^{10}\log K_{oc}$	(max)
1	Lindane	3.0		3.3
2	HCB	4.3		4.7
3	DDT	5.4		6.8
4	α-Endosulfane		4.3	
5	β-Endosulfane		4.4	
6	Aldrine	5.1		7.0
7	Dieldrine		5.8	

This data collection has no claim to be complete or thoroughly validated; but it is sufficient to demonstrate our method of assessing data insecurities :

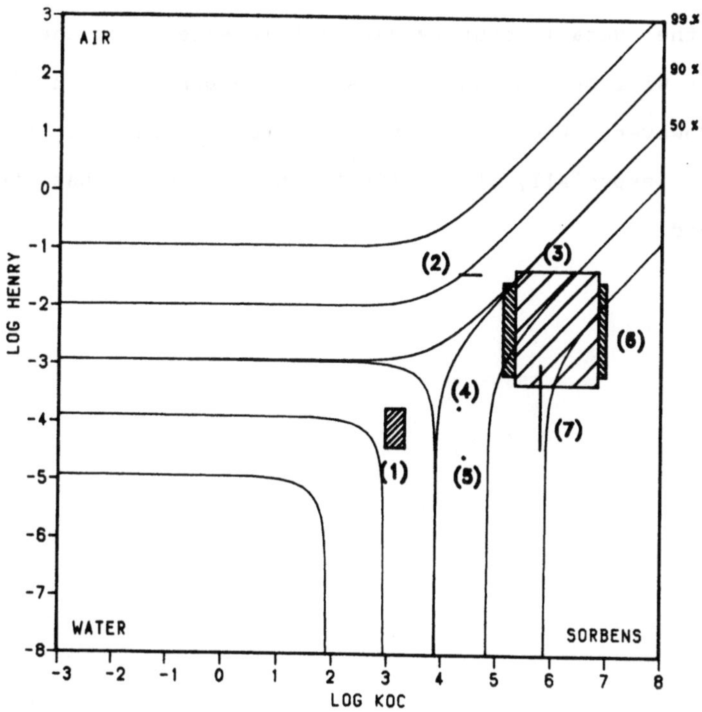

Figure 4: Equilibrium distribution map for seven compounds

The most striking feature is the wide variation of the areas covered by DDT and aldrine; both range from over 99 % in soil to over 50 % in air. This is a warning that these substances might have a high mobility; the experiences in the 1960's show that this is a well - founded warning. For less known substances such a diagnosis would call for a laboratory testing program for these substances.

With dieldrine, we also have a high insecurity regarding the Henry coefficient, but the graphic shows that in spite of this insecurity the tendency to adsorption is in any case near 99 %. For a first analysis the other two compartments and the insecurity in the value for the Henry coefficient can be neglected.

For lindane, the data insecurity has little effect on the water exposition; it will be important to study the behaviour in rivers and lakes. However, the tendency to sorption is not to be neglected, and especially the effect on sediments has to be closely watched.

5. Additional activities

5.1 Relations depending on the chemical structure

Estimation of time constants governing chemical reactions are not implemented in DTEST. Up to now, we have only found structure-dependent LFER, which could not be supported within E4CHEM. For example, [8,32] show estimation methods for hydrolysis and photolysis rates.

Our work now concentrates more and more upon single substances rather than processing large lists. This makes estimation methods relying on partial structure analysis and substance classes more interesting. For one of the most central data, the Henry coefficient, UNIFAC seems to be the most interesting method at hand.

For the estimation of K_{ow}, methods derived from algebraic graph theory have produced good results [33]. We already use these methods in individual cases, but not yet on such a large scale as we use DTEST. This is the scope of our coming activities regarding data estimation.

5.2 Information system for environmental chemicals

Throughout this talk I have been talking of 'known' and 'unknown' data. But how can we know which data actually are 'known' ? During our work on E4CHEM and some connected projects we have accumulated some experience on data retrieval. We found that a lot of information could be found in handbooks and data bases, but it usually took a variety of these sources to find all the information needed, and that it took a lot of practice to get the information within a reasonable time at acceptable costs. The main questions for a potential user of databases are :

- Which kind of data can I find in which database ?
- What amount of data is stored in these sources, i.e. what is the probability to actually find the information I need ?
- How do I get optimal access to the required information (retrieval language, search profiles, logon procedures) ?
- How much will I have to pay for a search ? Are there less expensive alternatives ?

These questions were the basis for the information system for environmental chemicals, a project sponsored by the Bavarian ministry of land development and environmental protection. This is not another database on chemicals but a PC - based retrieval system accessing a multitude of international public and private databases. The user interface is independent of the different retrieval languages, operating systems and logon procedures of the various databases. The system includes strategies to select the databases with the highest probability to find the specific information wanted and the lowest costs of getting this information. The system contains standard search profiles, supports the generation of new profiles and adapts these search profiles to the different retrieval languages.

For the details of this program see [7]. In this talk I just want to add that data retrieval, data estimation and modelling are no isolated solutions, but the equally important corners of an equilateral triangle. Each corner is needed to stabilize the other two by frequent validation and cross-checking. Only the combination of these can improve our knowledge and give rise to new questions. But in this workshop full of theory we must not forget that the most beautiful triangle is rather flat. To get a solid tetrahedron we need a fourth corner, representing measurements in the laboratory as well as on location (figure 5).

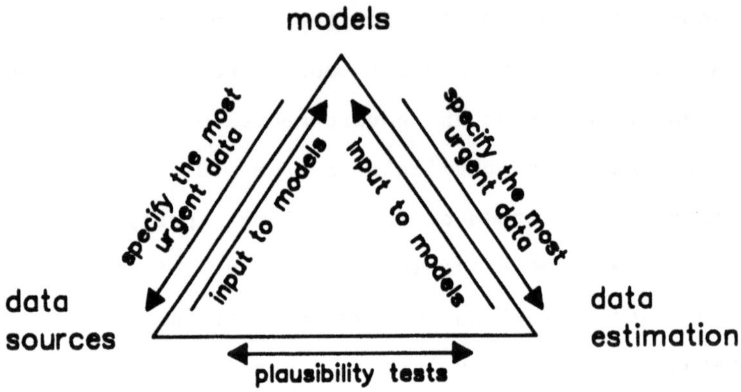

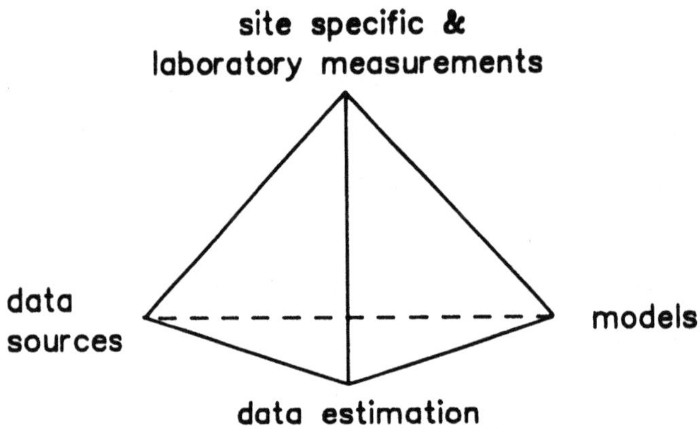

Figure 5 : schematic relations : models - data sources - data estimation - experiments

6. Literature

[1] H. Rohleder, M. Matthies, J. Benz, R. Brüggemann, B. Münzer, R. Trenkle, K. Voigt
"Umweltmodelle und rechnergestützte Entscheidungshilfen für die vergleichende Bewertung und Prioritätensetzung bei Umweltchemikalien"
Projektgruppe Umweltgefährdungspotentiale von Chemikalien
GSF-Bericht 42/86, München-Neuherberg

[2] M. Matthies, R. Brüggemann and R. Trenkle
Multimedia Modelling Approach for Comparing the Environmental Fate of Chemicals
In: "Environmental Modelling for Priority Setting among Existing Chemicals"
Proceedings International Workshop 11. - 13.11.1985
Projektgruppe Umweltgefährdungspotentiale von Chemikalien (Ed.)
GSF München-Neuherberg, Ecomed-Verlag Landsberg, 1986

[3] H. Rohleder, B. Münzer, K. Voigt
"E4CHEM (Exposure and Ecotoxicity Estimation for Environmental Chemicals) - A Computerized Aid for Priority Setting", in: Proceedings Environmental Modelling for Priority Setting among Existing Chemicals
Projektgruppe Umweltgefährdungspotentiale von Chemikalien
Gesellschaft für Strahlen-und Umweltforschung
München, Ecomed Verlag, November 1985

[4] M. Matthies, R. Brüggemann, B. Münzer, G. Schernewski and St. Trapp
"Exposure and Ecotoxicity Estimation for Environmental Chemicals (E4CHEM): Application of Fate Models for Surface Water and Soil", April 1988, to be published
in: Ecological Modelling

[5] R. Brüggemann, B. Münzer
"EXWAT Multikompartiment-Modell für den Transport von Stoffen in Oberflächengewässern"
Projektgruppe Umweltgefährdungspotentiale von Chemikalien
GSF-Bericht 33/87, München-Neuherberg

[6] St. Trapp, R. Brüggemann
"Untersuchung der Ausgasung leichtflüchtiger Substanzen aus mitteleuropäischen Fließgewässern mit dem Fließgewässermodell EXWAT"
Gesellschaft für Strahlen-und Umweltforschung München, February 1988, accepted for publication in Deutsche Gewässerkundliche Mitteilungen

[7] J. Benz, K. Voigt
"Konzeption rechnergestützter Suchhilfen für die Beschaffung von Chemikaliendaten", 2. Symposium "Informatikanwendungen im Umweltbereich"
KFK Karlsruhe, 9. - 10.11.1987, A.Jeschke, B. Page (Eds.)
Springer-Verlag Berlin, 1988

[8] W.J. Lyman, W.F. Reehl and D.H. Rosenblatt
"Handbook of Chemical Property Estimation Methods"
Mc Graw Hill Book Company, New York, 1981

[9] W.J. Lyman, R.G. Potts, G.C. Magil
"CHEMEST: User's Guide; A Program for Chemical Property Estimation"
A.D. Little, Inc., Acorn Park, Cambridge, March 1983

[10] E.E. Kenaga, C.A.I. Goring "Relationship between Water Solubility, Soil Sorption,
Octanol-Water Paritioning and Concentration of Chemicals in Biota"
ASTM Spec. Tech. Publ. 1980, SIP 707, 78 - 115

[11] G.J. Pierotti, C.H. Deal, E.L. Derr
"Activity Coefficients and Molecular Structure"
Industrial and Engineering Chemistry 51, 95 - 102, 1959

[12] W.B. Arbuckle
"Estimating Activity Coefficients for Use in Calculating Environmental Parameters"
Environ. Sci. Technol. 17, 537 - 542, 1983

[13] J. Gmehling, P. Rasmussen, A. Fredenslund
"Eine Übersicht zur Berechnung von Phasengleichgewichten mit Hilfe der UNIFAC-Methode"
Chem.-Ing.-Tech. 52, 724, 1980

[14] K. Stephan.
"Anwendung der Mischphasen-Thermodynamik auf die Berechnung von thermischen Stofftrennverfahren"
Chem.-Ing.-Tech. 52, 209, 1980

[15] J. Gmehling
UNIFAC - ein wichtiges Werkzeug für die chemische Industrie
S. 137 - 151, in: Software-Entwicklung in der Chemie 1
Proceedings: Computer in der Chemie, Hochfilzen/Tirol, 19. - 21. November 1986, J. Gasteiger (Ed.), Springer-Verlag Berlin 1987

[16] M.M. Miller, S.P. Wasik, G.-L. Huang, W.-Y. Shiu, D. Mackay
"Relationships between Octanol-Water Partition Coefficient and Aqueous Solubility"
Environ. Sci. Technol. 19, 522 - 528, 1985

[17] A. Leo and D. Weininger
Estimation of the n-octanol/water partition coefficient for organics in the TSCA industrial inventory, Pomona Medicinal Chemistry Project, Pomona College, Claremont, California (mimeo)
CLOGP Version 3.2, 1984

[18] C. Hansch, J.E. Quintan, G.L. Lawrence
"The linear Free-Energy Relationships between Partition Coefficients and the Aqueous Solubility of Organic Liquids", J. Org. Chem. 33, 347 - 350, 1968

[19] S.H. Yalkowsky, S.C. Valvani
"Solubilities and Partitioning: 2. Relationships between Aqueous Solubilities, Partition Coefficients and Molecular Surface Areas of Rigid Aromatic Hydrocarbones"
J. Chem. Eng. Data 24, 127 - 129, 1979

[20] D. Mackay, A. Bobra, W.-Y. Shiu, S.H. Yalkowsky
"Relationships between Aqueous Solubility and Octanol-Water Partition Coefficients"
Chemosphere 9, 701 - 711, 1980

[21] S.W. Karickhoff.
"Semi-Empirical Estimation of Sorption of Hydrophobic Polutants on Natural Sediments and Soils"
Chemosphere 10, 833 - 846, 1981

[22] D. Mackay, A. Bobra, D.W. Chan and W.Y. Shiu
Vapor Pressure Correlation for Low-Volatility Environmental Chemicals
Environ. Sci. Technol. 16, 645 - 649, 1982

[23] H. Fromherz
Physikalisch-chemisches Rechnen in Wissenschaft und Technik, Verlag Chemie, Weinheim, 1967

[24] K. Verschueren
"Handbook of Environmental Data on Organic Chemicals"
Second Edition, von Nostrand Reinhold Company, New York 1983

[25] D. Mackay, W.-Y. Shiu
"A Critical Review of Henry's Law Constants for Chemicals of Environmental Interest"
J. Phys. Chem. Ref. Data 10, 1175 - 1199, 1980

[26] R.C. Reid, J.M. Prausnitz, T.K. Sherwood
"The Properties of Gases and Liquids" 3rd Edition
Mc Graw Hill Book Company, New York 1979

[27] OECD Hazard Assessment Project
OECD Working Group on Exposure Analysis
"Collection fo Minimum Pre-Marketing Sets of Data including Environmental Residue Data on Existing Chemicals", Paris 1982

[28] W. Karcher et al.
Band II des "Spectral Atlas of PAC"
D. Reidel Publishing Company, Dordrecht

[29] D. Mackay, S. Paterson
"Calculating Fugacity"
Envir. Sci. Technol. 1981, 15, 1006 - 1014

[30] R. Brüggemann
Mackays Fugazitätsmodell mit Level I bis IV
Parameter, Kompartmentalisierung, Sensitivität
Projektgruppe Umweltgefährdungspotentiale von Chemikalien (PUC)
GSF-Bericht 43/86

[31] J.M. Prausnitz, R.N. Lichtenthaler, E.G. de Azevedo
"Molecular Thermodynamics of Fluid-Phase Equilibria" Second Edition
Prentice-Hall Inc., Englewood Cliffs, N.J. 07632, 1986

[32] L.A. Burns, D.M. Cline, R.R. Lassiter
"Exposure Analysis Modeling System (EXAMS): User Manual and System Documentation EPA-600/3-82-023
Environmental Protection Agency, Athens, Georgia, 1982

[33] A. Sabljic
"On the Prediction of Soil Sorption Coefficients of Organic Pollutants from Molecular Structure: Application of Molecular Topology Model"
Envir. Sci. Technol. 1987, 21, 358 - 366

Application of Molecular Topology for the Estimation of Physical Data for Environmental Chemicals

Aleksandar Sabljić

Institute Ruder Bošković, POB 1016, 41001 Zagreb, Croatia, Yugoslavia

1. INTRODUCTION

The chemical industry has grown enormously in recent decades. It provides us with numerous vital chemicals (fuels, antibiotics and other drugs, plastics, pesticides, fertilizers, etc.) without which our society cannot survive and preserve its present life style and high living standards. Many of these substances have little or no adverse environmental effects, but some may be harmful to human health and the natural environment. Usually these effects only become apparent after wide and prolonged usage and at that point authorities introduce control measures. Clearly, there is a need for an effective evaluation and testing program to identify, before their use, those chemicals or classes of chemicals that present a potential environmental hazard. Such evaluation procedure should trace the fate of chemicals from discharge and dispersal to subsequent effects on biota. The ecotoxicological profile of a chemical is based on a sequence of interactions and effects controlled by its physical, chemical, and biological properties. At the first stage, a chemical released into the environment is subject to physical distribution between the atmosphere (air), water, soils, and sediment depending on its physico-chemical properties. At the same time, it can be chemically modified and degraded by abiotic processes or more often by microorganisms in the environment. During the following stage organisms will be exposed to the chemical either in its original or in its degraded or transformed form. The uptake of the chemical and degradation products will occur. Organisms may react to such exposure by variety of negligible and sublethal effects or ultimately by death. Finally, the complex natural ecosystem of which the organisms are an integral part may react in a variety of ways to the effects on the component organisms. Thus the ecotoxicological profile of a chemical may be described as a sequential set of steps starting with the source and proceeding through to the ecosystem response.

Present laws and regulations about control of toxic substances (Toxic Substances Control Act, US Federal Pesticides Law, OECD-Guideline for Testing Chemicals and similar) require that all commercial chemicals must be assessed for their environmental behavior and hazards. The experimental determination of environmental parameters (e.g. soil sorption coefficients, bioconcentration factors, biodegradation and biotransformation, toxic effects, etc.) of those chemicals is a costly, time-consuming, and very tedious process. Since an estimated 100,000 chemicals are currently in common use [1] and new chemicals are registered at a rate of 1000/year it is obvious that our human and material resources are insufficient to obtain experimentally even a basic information about environmental fate for all those chemicals. Thus, it is necessary to develop quantitative models that will accurately and

C. Jochum M. G. Hicks J. Sunkel (Eds.)
Physical Property Prediction
in Organic Chemistry

rapidly predict environmental behavior for large sets of chemicals.

A lot of work is being done to find methods that will enable us to accurately and rapidly estimate environmental distribution and toxicity of organic pollutants. Empirical models [2-7], based on water solubility and n-octanol/water partition coefficients, have been proposed as accurate methods to estimate environmental distribution and toxicity of organic pollutants. Their analysis [8] has shown that they have serious shortcomings: (a) the low precision of water solubility and n-octanol/water partitioning data and (b) the large variety of quantitative models describing the relationship between the environmental distribution coefficients and above physico-chemical properties.

For these reasons, during the last decade, considerable efforts are also made to develop nonempirical models for environmental distribution and toxicity of organic pollutants [9,10]. Our long term interest in environmental QSAR (quantitative structure-activity relationships) is to develop general nonempirical model(s) for predicting environmental distributions and toxicity of organic pollutants based only on information encoded in their structural formulas. The ultimate goal is to develop the above quantitative models with the following desired characteristics: (a) to be highly accurate so that they can be reliably used, (b) to be sufficiently simple so as to be easily applicable by various profiles of scientist and other people involved in environmental problems (even layman) as daily routine either in the laboratory or in the field, and (c) to be short enough to be performed on the large samples in reasonable amount of time. Several years ago we have pioneered the idea that molecular topology [11,12] could be successfully applied in correlations between molecular structure and environmental distribution of organic pollutants [13-15]. Thus far, molecular topology has been shown to be the most successful structural property for describing and predicting their soil sorption coefficients [8,9,13,15-19], bioconcentration factors [9,14,16,17,20], biodegradation [21-23], and acute toxicity [9,24-30]. In the following sections we will describe the most successful nonempirical models, in terms of accuracy and range of applicability, for predicting soil sorption coefficients, bioconcentration factors, and fish acute toxicity of organic pollutants.

2. METHODS OF CALCULATIONS

The simplest way to represent a molecular structure is to assign to a structure a number or a set of numbers, termed indices. Indices generated by the application of the chemical graph theory [12] are called topological indices [11,12]. The concept of molecular connectivity indices was introduced by Randić [31] and further developed and extensively exploited by Kier and Hall [10] and many others [8-30,32-37]. Several extensive reviews of the theory and method of calculation of molecular connectivity indices have been published recently [10-12,36,37]. Thus only a brief description of the calculation of the topological indices used in the nonempirical models discussed in this review is given here.

The first-order ($^1\chi$) molecular connectivity index is calculated from the non-hydrogen part of the molecule. Each non-hydrogen atom is described by its atomic

δ value, which is equal to the number of adjacent non-hydrogen atoms. The $^1\chi$ is then calculated from the atomic δ values by equation 1 ,

$$^1\chi = \Sigma(\delta_i * \delta_j)^{-0.5} \qquad (1)$$

where i and j correspond to the pairs of adjacent non-hydrogen atoms and summation is over all bonds between non-hydrogen atoms.

The zero-order ($^0\chi^V$) and second-order ($^2\chi^V$) valence molecular connectivity indices are also calculated from the non-hydrogen part of the molecule. Each non-hydrogen atom is described by its atomic valence δ^V value which is calculated from the following equation

$$\delta^V = (Z^V - h)/(Z - Z^V - 1) \qquad (2)$$

where Z^V is the number of valence electrons in the atom, Z is its atomic number, and h is the number of hydrogen atoms bound to the same atom. The $^0\chi^V$ and $^2\chi^V$ are then calculated from the atomic δ^V values by equations 3 and 4, respectively,

$$^0\chi^V = \Sigma(\delta_i)^{-0.5} \qquad (3)$$

$$^2\chi^V = \Sigma(\delta_i * \delta_j * \delta_k)^{-0.5} \qquad (4)$$

where i, j, and k correspond to three consecutive non-hydrogen atoms and summations are over all non-hydrogen atoms or over all pairs of adjacent bonds between non-hydrogen atoms, respectively.

Molecular connectivity indices were calculated by the GRAPH III computer program on an IBM PC/XT personal computer [8,38]. Minimum hardware and software requirements for this program are IBM PC or compatible microcomputer, 256 KB of memory, 1 double sided/double density disk drive, and PC-DOS or MS-DOS operating system version 2.1 or higher. The use of mathematical coprocessor is highly recommended. GRAPH III can calculate the molecular connectivity indices up to tenth order for molecules with 35 non-hydrogen atoms or less. It is possible to extend the program to handle larger molecules if sufficient memory is available. To test the quality of the regression equations the following statistical parameters were used: the correlation coefficients (r), the standard error of the estimate (s), a test of null-hypothesis (F-test), and the amount of explained variance (EV). The majority of statistical calculations were also carried out on an IBM PC/XT microcomputer with the appropriate modules of statistical package SYSTAT.

The soil sorption coefficients are currently used as a quantitative measure of sorption of xenobiotic chemicals by soil from aqueous solutions [2]. They are defined as the ratios between the concentrations of a given chemical sorbed by the soil and dissolved in soil water. In order to be able to compare the soil sorption coefficients measure for different soils they have to be normalized either to the total organic carbon content of the soil (K_{oc}) or to the organic matter content of the soil (K_{om}). These two normalizing schemes are simply related by the factor 1.724, thus it is easy to convert coefficients reported on any basis. The bioconcentration factor, the ratio of concentrations in the biota and in the medium, is currently used to indicate the extend of bioconcentration process. The majority of bioconcentration tests are carried out on fish because of its obvious economic importance as human food. The main sources of soil sorption and bioconcentration data used in our studies are their large compilations [2-5,39-41] published recently.

All fish acute toxicity data are from the single source [42]. Toxicity tests were conducted with Fathead Minnows (*Pimephales Promelas*) under maximally controlled and uniform conditions. Mortality was recorded at 96 hours and their LC_{50} data were calculated and reported as g/L or mg/L. Those data were recalculated in molar units (mmol/L) and their negative logarithms are used as toxicity variables.

3. MODELLING SOIL SORPTION

Contamination of groundwater by pesticides and other agricultural chemicals, by hazardous chemicals from waste disposal sites, and by gasoline and chemicals from underground storage tanks is presenting a major environmental problem. Reports prepared by US Environmental Protection Agency show that there are up to 50 000 waste disposal sites in U.S.A. that may contain hazardous chemicals and groundwater systems close to many of these sites are being slowly degraded and the contamination often involves the presence of synthetic organic chemicals. To minimize the impact of man's activity on groundwater quality, mechanisms by which pollutants enter groundwater need to be better understood, and reliable techniques to either measure or predict the transport of contaminants within aquifers need to be developed. Since the large majority of synthetic organic chemicals are hydrophobic, their adsorption on soil and sediments plays a very important role in their transport and mobility in surface and subsurface systems. In addition, the adsorption of agricultural chemicals by soils from aqueous solution strongly influences their performance and residue problems. Thus far, topological approach which relies on molecular properties like the number and type of atoms and chemical bonds in the molecule to calculate soil sorption coefficients is found to be the most successful [8,9,19].

Pioneering study [13], in this area, was focused on polycyclic aromatic hydrocarbons (PAHs) since they constitute the major group of environmental hazard to all living species. Moreover, their soil sorption data are from the single source [43], measured under highly controlled and uniform conditions, and thus have the highest degree of internal consistency and comparability. Thus, this was an excellent starting point for our project since it gave us optimal control over the modelling process, good feeling for modelling environmental properties, and excellent chance to create sound model for predicting soil sorption coefficients. The first-order molecular connectivity index was found to correlate extremely well with the soil sorption coefficients of 8 PAHs as shown by eq. 5 and its statistics.

$$\log K_{om} = -0.10 + 0.63 * {}^{1}\chi \qquad (5)$$

$n = 8 \quad r = 0.986 \quad s = 0.202 \quad F^{1,6} = 205 \quad EV = 96.7\%$

Such gratifying result was a driving force to continue our investigation [15] in the same direction. The soil sorption coefficients were collected for 29 additional compounds, mainly halogenated hydrocarbons: chlorobenzenes, polychlorinated biphenyls (PCBs), and chlorinated or brominated alkanes and alkenes. Adding halogen atoms to hydrocarbon skeleton seemed to be a very small perturbation and it was reasonable to expect that resulting compounds will have similar environmental distribution patterns as parent compounds. We were fortunate to learn that our working hypothesis is correct. Quantitative model describing the soil sorption of hydrocarbons and their halogenated derivatives is given by eq. 6 and its statistics.

$$\log K_{om} = 0.42 + 0.53*{}^1\chi \qquad (6)$$

$$n = 37 \quad r = 0.976 \quad s = 0.300 \quad F^{1,35} = 704 \quad EV = 95.1\%$$

Statistically eq. 6 accounts for 95% of the variation in the *log* K_{om} data. This is as good as can be expected taking into account the accuracy of measured data.

Our next task was to expand molecular connectivity model (eq. 6), in an effort to define the whole range of its applicability and to test its level of accuracy for predicting soil sorption coefficients [8,9]. For this test we have used the similar set of 31 compounds (alkyl- and chlorobenzenes, heterocyclic and substituted PAHs, chlorinated alkanes and alkenes, and chlorinated phenols). Their ${}^1\chi$ indices were calculated and their soil sorption coefficients were predicted from the molecular connectivity model, eq. 6. Comparison of the observed and predicted soil sorption coefficients clearly demonstrates that the molecular connectivity model is very accurate in predicting the soil sorption coefficients. The average difference between predicted and observed soil sorption coefficients is only o.22 on the logarithmic scale (corresponding to a factor of 1.66) and more than 90% of coefficients are predicted within the two standard deviations.

The highly satisfactory performance of the molecular connectivity model, eq. 6, in predicting the soil sorption coefficients prompted us to combine both sets of compounds into a single regression model. The resulting molecular connectivity model for quantitative description of soil sorption coefficients of nonionic organic compounds is given in linear eq. 7.

$$\log K_{om} = 0.43 + 0.53*{}^1\chi \qquad (7)$$

$$n = 72 \quad r = 0.977 \quad s = 0.282 \quad F^{1,70} = 1478 \quad EV = 95.4\%$$

The statistical parameters parameters show that eqs. 6 and 7 are statistically significant above the 99% level and that both have similar levels of accuracy. Thus the range of applicability of the molecular connectivity model is now extended to alkylbenzenes, heterocyclic PAHs, substituted PAHs, and chlorophenols.

The molecular connectivity model (eq. 7) for predicting soil sorption coefficients is based on nonpolar and nonionic compounds. Thus, it was reasonable to assume that it may not be valid for the highly polar and ionic compounds. To check this assumption, our model (eq. 7) was tested on the following classes of compounds: anilines, acetanilides, nitrobenzenes, carbamates, substituted benzenes and pyridines, phenylureas, 3-phenyl-1-methylureas, 3-phenyl-1,1-dimethylureas, 3-phenyl-1-cycloalkylureas, alkyl-N-phenylcarbamates, triazines, uracils, organic acids, and organic phosphates. The list of 143 compounds sorted by functional groups and their soil sorption coefficients are given in reference 8. The first-order molecular connectivity indices were calculated for all those compounds and their soil sorption coefficients predicted from the molecular connectivity model (eq. 7). We expected the predicted soil sorption coefficients of such a large number of compounds to be distributed more or less randomly around the regression line defined by eq. 7. To our surprise, the predicted soil sorption coefficients of all compounds used in this analysis fall below the regression line. Such an unusual result reveals valuable information about the relation between molecular structure and soil sorption properties of organic compounds. First, hydrocarbons and halogenated hydrocarbons must have optimal geometric and electronic features for strong sorption on soils. Second, the introduction of any polar atom and/or substitutent will always decrease the soil sorption capability of the resulting compound. Although, the second relation was previously known to exist this is the first systematic and general proof for it and, as will be shown below, it can be described quantitatively.

Table I. The differences between observed soil sorption coefficients and those predicted from molecular connectivity model (eq. 7) for 143 polar and ionic organic compounds. These differences are are used as the polarity correction factors (P_f).

Chemical group	Number of compounds	P_f log K_{om} (calc-obs)	Average deviation
Substituted benzenes & pyridines	6	1.00	0.16
Nitrobenzenes	6	1.00	0.24
Anilines	12	1.05	0.20
Organic phosphates (group 1)	9	1.07	0.33
Carbamates	5	1.11	0.11
Phenylureas	15	1.71	0.30
Alkyl-N-phenylcarbamates	10	1.87	0.14
Acetanilides	17	1.87	0.26
3-Phenyl-1-methylureas	5	1.89	0.28
Triazines	11	1.95	0.16
3-Phenyl-1-methyl-1-methoxyureas	4	2.04	0.18
Uracils	3	2.04	0.24
3-Phenyl-1,1-dimethylureas	16	2.23	0.24
di-Nitrobenzenes	7	2.36	0.36
Organic acids	8	2.46	0.22
3-Phenyl-1-cycloalkylureas	4	2.57	0.14
Organic phosphates (group 2)	5	3.05	0.28

The detailed analysis of the calculated soil sorption coefficients of polar and ionic compounds shows that the absolute difference between the predicted and observed coefficients depends strongly on the type of polar functional group and that variation within the groups are small. This result is shown in Table I, where 143 polar and ionic compounds are arranged into 17 classes of compounds in descending order of their soil sorption coefficients. The average error within each class in Table I is very small and it is always less than the standard deviation of the molecular connectivity model (eq. 7). Such systematic behavior of polar organic compounds can be used to indicate the presence or absence of a polar functional group. In addition, the numerical differences between predicted and observed soil sorption coefficients (Table I, column 3) were used to determine the magnitude of a set of semiempirical variables (the polarity correction factors - p_f) that can be employed to accurately predict the soil sorption coefficients of polar and ionic organic compounds. The polarity correction factor is calculated for each class of polar compounds by eq. 8,

$$p_f = (1/n) * \Sigma(\mathit{log}\ K_{om}^{eq\,7} - \mathit{log}\ K_{om}^{obs}) \qquad (8)$$

where n is the number of compounds in particular class. The soil sorption coefficients were recalculated using the polarity correction factors by eq. 9.

$$\mathit{log}\ K_{om}^{adj} = \mathit{log}\ K_{om}^{eq\,7} - p_f \qquad (9)$$

The quantitative relationship between recalculated (adjusted) and observed soil sorption coefficients for a total of 215 compounds is described by the regression eq. 10 and its statistics.

$$\mathit{log}\ K_{om}^{obs} = 0.08 + 0.97 * \mathit{log}\ K_{om}^{adj} \qquad (10)$$

$$n = 215 \quad r = 0.969 \quad s = 0.279 \quad F^{1,213} = 3291 \quad EV = 93.9\%$$

It is now possible, with this simple model, to accurately predict the soil sorption coefficients for almost 95% of all organic chemicals whose soil sorption coefficients have been measured. No empirical or nonempirical model has ever been able to predict the soil sorption coefficients to such a high level of accuracy on such a broad selection of structurally diverse compounds.

Our future efforts will be primarily focused on finding the structural variable(s) (property) which can explain and quantify the soil sorption behavior of polar and ionic chemicals and to generate general soil sorption model for all classes of organic chemicals. Consequently, developed models for estimating sorption coefficients will be coupled with the models for calculating concentration profiles and transport of chemicals in groundwater aquifers. This will result in better assessment methods for predicting exposure concentration of organic pollutants in groundwater resources. Better methods to estimate concentrations of chemicals in groundwater aquifers will also make it possible to examine feasible methods of reclamation of contaminated aquifers prior to field evaluation, and to modify agricultural and waste disposal practices to minimize aquifer contamination.

4. MODELLING BIOCONCENTRATION

The bioconcentration of chemicals, by organisms, from the environment is of particular importance. This process controls the concentrations of contaminant chemicals in organisms and thus the harmful effects on biota. The atmosphere contains very low concentrations of chemicals compared with water (i.e. oceans, lakes, rivers, groundwater) and as a result the bioconcentration processes in aquatic organisms have received the most attention. The fish bioconcentration factors of up to

10^6 have been measured for certain classes of hydrophobic chemicals which means that even if water concentration of these contaminants is acceptable, dangerous concentration may be present in aquatic organisms. Once used as a human food they will also become threat to human health. In addition, the general trend is that the potentially most dangerous chemicals have the largest bioconcentration factors. Thus, it is very important to develop quantitative models that will accurately and rapidly estimate bioconcentration factors for large sets of chemicals.

Several years ago we have demonstrated [14] that molecular connectivity indices can be used very successfully in quantitative correlation and prediction of fish bioconcentration factors of chlorinated benzenes, PCBs, and chlorinated diphenyl oxides. The second-order valence molecular connectivity index (${}^2\chi^V$), was found to correlate extremely well with the bioconcentration factors (BCFs) of chlorinated hydrocarbons in fish as shown by eq. 11 and its statistics.

$$\log \text{BCF} = -2.32 + 2.22 * {}^2\chi^V - 0.17 * ({}^2\chi^V)^2 \qquad (11)$$

$$N = 20 \quad r = 0.971 \quad s = 0.277 \quad F^{2,17} = 139. \quad EV = 93.6\%$$

Consequently, this result was confirmed by two other laboratories [16,17]. Since that time an improved parametrization scheme was suggested [10] for the valence type molecular connectivity indices and some fish bioconcentration factors of studied compounds were remeasured [9,20,40,41]. Thus, we have recalculated our quantitative model using the new parametrization scheme for connectivity indices and the average values for the fish bioconcentration factors. The following relationship was established between the ${}^2\chi^V$ index and fish bioconcentration factors for 20 chlorinated hydrocarbons.

$$\log \text{BCF} = -2.36 + 2.28 * {}^2\chi^V - 0.17 * ({}^2\chi^V)^2 \qquad (12)$$

$$N = 20 \quad r = 0.969 \quad s = 0.293 \quad F^{2,17} = 130. \quad EV = 93.2\%$$

The predictive power of this molecular connectivity model (eq. 12), based on the new parametrization scheme, have been tested in an effort to define the whole range of its applicability and to test its level of accuracy for predicting fish bioconcentration factors [9,20]. For this test, the set of 67 new compounds was used: chlorinated hydrocarbons, other halogenated hydrocarbons, alkyl and alkenyl benzenes, polycyclic aromatic hydrocarbons, substituted phenols, and other structurally similar compounds. The ${}^2\chi^V$ indices of these test compounds were calculated and their fish bioconcentration factors were predicted from the molecular connectivity model, eq. 12. Comparison of the observed and predicted bioconcentration factors clearly demonstrates that the molecular connectivity model is very accurate in predicting bioconcentration factors [9,20]. The average difference between predicted and observed bioconcentration factors is only 0.36 on the logarithmic scale (corresponding to a factor of 2.3) and nearly 90% of BCFs are predicted within the two standard deviations. Only eight compounds are predicted outside the two standard deviation range. Five of them are marginal outliers. Only benzo(a)pyrene, 3,4',5-tribromobiphenyl, and 4-nitrophenol do not fit into our model. Benzo(a)pyrene measured bioconcentration factor is considerably lower than the one predicted from our model because it undergoes rapid biotransformation in the fish [44] thus it cannot reach its theoretical maximum in bioconcentration. In a view of a rapid biotransformation of benzo(a)pyrene in fish it is very comforting to see that its predicted bioconcentration factor is higher than the experimental value. Two other results of the test are particularly gratifying. First, during our initial study [14] the experimental bioconcentration factor of 215 for 4,4'-dichlorobiphenyl was found to be in strong disagreement with the predicted value of 7100. Our conclusion was that *"this observed bioconcentration factor needs additional experimental confirmation and has to be used with caution"*. Recently, a new experimental value of 12 000 was reported [16] for 4,4'-dichlorobiphenyl, which is in close agreement with our predicted value. Second, the close correlation between the observed and predicted bioconcentration factors of o,p'-DDT, aldrin, chlordane, heptachlorepoxide, decabromobiphenyl, octachlorodipropylether, and 2,4,6-tribromophenyl(2-methyl-2,3-dibromopropyl)ether strongly supports our nonlinear model because they are located on the descending side of a parabola represented by eq. 12. This experimental support for our nonlinear model is very important because our nonempirical model is in sharp contrast with exclusively linear empirical models based on n-octanol/water partition coefficients or water solubilities [2,5,6,40,45,46].

The highly satisfactory performance of the molecular connectivity model, eq. 12, in predicting the fish bioconcentration factors prompted us to combine all compounds, except the three outliers indicated above, into a single regression model. The resulting molecular connectivity model for the quantitative description of fish bioconcentration factors of 84 studied organic compounds is given by eq. 13.

$$\log \mathrm{BCF} = -2.13 + 2.12 * {}^{2}\chi^{v} - 0.16 * ({}^{2}\chi^{v})^{2} \qquad (13)$$

$$N = 84 \quad r = 0.966 \quad s = 0.345 \quad F^{2,81} = 559. \quad EV = 93.1\%$$

The statistical parameters show that eq. 13 is statistically significant above the 99% level and its standard error (s) is only slightly higher than the standard error for eq. 12. Thus the range of applicability of the molecular connectivity model is now extended to halogenated hydrocarbons, alkyl and alkenyl benzenes, PAHs, substituted phenols, and similar compounds without sacrificing its high accuracy.

Our future efforts in modelling fish bioconcentration factors will be primarily focused on testing present model on other classes of industrial chemicals, mainly pesticides and other agricultural chemicals, in an effort to develop a general quantitative model that will accurately and rapidly estimate bioconcentration factors for large sets and diverse classes of organic chemicals. In addition, we plan to start to develop quantitative models for bioconcentration factors in other aquatic organisms as well as in terrestrial animals, mainly those used as a human food.

5. MODELLING ACUTE TOXICITY

Environmental toxicology is principally concerned with the investigation of lethal effects (acute toxicity) of environmental contaminants. Numerous efforts [7,9,10,16,17,24-30,47-50] were made to quantitatively model acute toxicity (LC_{50}) of aquatic organisms since the need for LC_{50} data are enormous due to the regulatory requirements. Although some high correlation models were developed their application and usefulness are limited since they are based on nonstandard toxicity data for the small sets of compounds. The first standardized data-base of acute fish toxicity data for a large number of structurally diverse chemicals (hydrocarbons, alcohols, ethers, aldehydes, ketones, carboxylic acids, esters, amines, phenols, etc.) is available only very recently due to the joint efforts of the people in EPA (Duluth, Minnesota) and Center for Lake Superior Environmental Studies (University of Wisconsin-Superior). It is published in several volumes by the Center for Lake Superior Environmental Studies [42].

Our primary objective is to develop a general model for predicting fish acute toxicity of structurally diverse commercial chemicals. The molecular connectivity indices will be calculated for alcohols, ethers, aldehydes, ketones, nitriles, aliphatic and aromatic amines, halogenated aliphatic hydrocarbons, substituted benzenes, and phenols. At the first stage individual models will be developed for each class of chemicals. During the second stage we will try to combine individual models in an effort to describe the major structural factor(s) that controls toxic properties for those structurally diverse classes of chemicals. Since molecular connectivity indices describe and quantify the size, the shape, the degree of branching and cyclicity of molecules, this investigation will show whether and how these particular structural features influence the fish acute toxicity of industrial chemicals.

Table II. The single variable (${}^0\chi^v$) fish acute toxicity models for ten classes of commercial chemicals. N is the number of compounds in each chemical class while a and b are the regression coefficients for equation $log\ LC_{50} = a + b\ ({}^0\chi^v)$.

Chemical Class	N	a	b	r	s	F	EV
Alcohols	20	3.99	-.74	.98	.28	441	96
Ethers	10	4.09	-.72	.98	.25	239	96
Aldehydes	39	-.18	-.18	.49	.64	12	22
Ketones	23	4.28	-.74	.97	.32	357	94
Nitriles	11	2.26	-.49	.95	.32	91	90
Aliphatic Amines	9	2.96	-.60	.97	.35	121	94
Aromatic Amines	13	3.37	-.72	.95	.24	93	89
Halogenated Hydrocarbons	15	2.75	-.66	.93	.36	83	85
Benzenes	12	2.02	-.57	.95	.26	96	90
Phenols	32	1.34	-.41	.92	.36	171	85

Fish toxicity data for ten groups of industrial chemicals, alcohols, ethers, aldehydes, ketones, nitriles, aliphatic and aromatic amines, halogenated aliphatic hydrocarbons, substituted benzenes, and phenols, were analyzed and correlated with their molecular connectivity indices. For each group of chemicals the single variable regression equations were calculated with the zero-, first-, and second-order molecular connectivity indices. The best one-variable equations were obtained for the valence zero-order molecular connectivity index, ${}^0\chi^v$. Results are shown in Table II, where regression coefficients, correlation coefficients, and other statistical tests are reported. The majority of one-variable equations account from 85% to 95% of the variation in toxicity data. This is probably the best that can be expected taking into account the purity of chemicals and the reproducibility of toxicity data. Aldehydes gave very poor correlations with all tested molecular connectivity indices. We will continue to study aldehydes because this lack of correlation is puzzling. Combination of the valence zero-order index with the higher order indices and/or cluster, path/cluster, and chain type indices did not significantly improve the correlations with toxicity data. The ${}^0\chi^v$ index describes the size of molecule. It seems that the most important structural factor which influences the level of toxicity of studied industrial chemicals is their size while the other topological properties like degree of branching and cyclicity have negligible contribution.

The analysis of results from Table II shows that the regression coefficients for the majority of chemical classes are very similar. This result prompted us to combine all compounds, except aldehydes and phenols, into a single regression model. The resulting molecular connectivity model for the quantitative description of fish acute toxicity of commercial organic chemicals is given by linear eq. 14.

$$\log LC_{50} = 3.43 - 0.68 * {}^0\chi^v \qquad (14)$$

$$N = 113 \quad r = 0.915 \quad s = 0.541 \quad F^{1,111} = 568 \quad E.V. = 84\%$$

The standard error (s) for eq. 14 is only 8% of the range for the LC_{50} data which is 6.6 logarithmic units. The statistical parameters for eq. 14 clearly show that a simple and reasonably accurate quantitative model is developed based only on topological properties of molecules. This model is applicable for organic chemicals of diverse structures and LC_{50} data. None of the other molecular connectivity indices was able to improve correlation leading to the earlier conclusion that the size of molecule is the most important structural property responsible for the level of fish acute toxicity of organic chemicals. The size of molecule is directly proportional to the acute toxicity of industrial chemicals. This means that the larger molecules are generally more toxic while the smaller molecules are less toxic.

Quantitative model for phenols (Table II) considerable differ from the models for all other classes of chemicals. This may be an indication that the mechanism of toxicity of phenols is different from the mechanism of other chemicals. In fact, it is published that substituted phenols cause toxicity by specific mode of action, disrupting metabolism by inhibition of oxidative phosphorylation [51]. The other classes of chemicals studied here are considered to be narcotics [7], having nonspecific mode of action. This nonspecific mode of toxic action is now recognized and visualized [7,47,48,50,52-54] as the disruption of the cell membranes structure by the simple physical presence of these chemicals. Thus, the conclusion that the size of molecule is directly proportional to the extend of its toxic activity and that the larger molecules will cause more disruption and damage to the structure of cell membranes (e.g. being more toxic) is in complete agreement with the proposed nonspecific mode of action for narcotic chemicals. It is very gratifying that the distinction between two modes of action comes direct from our models. Finally, it is reasonable to conclude that our quantitative model (eq. 14) is a solid starting point in developing global model for predicting the fish acute toxicity that will be based on topological and possibly other structural properties.

This model is in its early developing stage and our future efforts in modelling fish acute toxicity will be focused on its validation and improvement and on expanding it to new classes of chemicals. However, the rate limiting step for our efforts in this area will be the progress on fish acute toxicity data-base described above.

6. SUMMARY AND PERSPECTIVES

All results described above strongly support our idea that molecular topology can be used as a predictive tool for ranking potentially hazardous chemicals and for creating priority lists for testing them so that experimental efforts can be focused on the potentially most dangerous chemicals. The direct benefits of this research will be the general nonempirical models for estimating environmental fate and adverse effects of organic pollutants based on their topological properties. These models will enable government agencies, industry, and public groups to make fast and accurate assessments on environmental fate of proven or potentially hazardous chemicals. In addition, they will enable the manufacturers of pesticides and other classes of organic pollutants to accurately predict environmental distribution, and consequently the potential environmental hazard, of their future products, even before such compounds are synthesized. The high predictive power of these nonempirical models indicates that they will very accurately estimate environmental distribution coefficients and acute toxicity of chemicals whose environmental properties have not been measured. In the future, these nonempirical quantitative models will facilitate development of new global ecological models that will give more insight into the expected distribution pattern and behavior of pollutants in the environment. In ecological models, different distribution coefficients are combined to describe a type of ecosystem. It would be extremely convenient to have quantitative models describing various environmental distribution coefficients of commercial chemicals based on the same of similar concept(s). This will facilitate development of global, complex environmental models since they can be then easily built up as combinations of modules, each module representing an individual environmental property. Since the molecular connectivity indices were found to correlate with the soil sorption coefficients, bioconcentration factors, biodegradability, and acute toxicity of xenobiotic chemicals, it should be feasible from molecular topology alone to evaluate the overall distribution pattern and possible adverse effects of chemicals in ecosystems.

REFERENCES

1. V.H. Freed, in *Dynamics, Exposure and Hazard Assesment of Toxic Chemicals,* R. Hague, Ed., Ann Arbor Science, Ann Arbor, 1980.
2. E.E. Kenaga and C.A.I. Goring, in *Aquatic Toxicology,* J.C. Eaton, P.R. Parrish, and A.C. Hendricks, Eds., American Society for Testing and Materials, Philadelphia (PA) 1980, pp. 78-115.
3. G.G. Briggs, J. Agric. Food Chem. **29** (1981) 1050-1059.
4. J.C. Means, S.G. Wood, J.J. Hassett, and W.L. Banwart, Environ. Sci. Technol. **16** (1982) 93-98.
5. D. Mackay, Environ. Sci. Technol. **16** (1982) 274-278.
6. G.D. Veith, D.L. DeFoe, and B.V. Bergstedt, J. Fish. Res. Board Can. **36** (1979) 1040-1048.
7. G.D. Veith, D.J. Call, and L.T. Brooke, Can. J. Fish. Aquat. Sci. **40** (1983) 743-748.
8. A. Sabljić, Environ. Sci. Technol. **21** (1987) 358-366.
9. A. Sabljić, in *QSAR in Environmental Toxicology - II,* Ed. K.L.E. Kaiser, D. Reidel Publishing Co., Dordrecht, Holland, 1987, pp. 309-332 and references therein.
10. L.B. Kier and L.H. Hall, *Molecular Connectivity in Structure-Activity Analysis,* Research Studies Press, Chichester (UK), 1986 and references therein.
11. A. Sabljić and N. Trinajstić, Acta Pharm. Jugosl. **31** (1981) 189-214.
12. N. Trinajstić, *Chemical Graph Theory,* CRC Press: Boca Raton, Florida, 1983.
13. A. Sabljić and M. Protić, Bull. Environ. Contam. Toxicol. **28** (1982) 162-165.
14. A. Sabljić and M. Protić, Chem.-Biol. Interact. **42** (1982) 301-310.
15. A. Sabljić, J. Agric. Food Chem. **32** (1984) 243-246.
16. R. Koch, Toxicol. Environ. Chem. **6** (1983) 87-96.
17. H. Govers, C. Ruepert, and H. Aiking, Chemosphere **13** (1984) 227-236.
18. Z. Gerstl and Ch. S. Helling, J. Environ. Sci. Health **B22** (1987) 55-69.
19. A. Sabljić, Environ. Health. Perspect. (1988) in press.
20. A. Sabljić, Zeitschrift für Gesamt Hygiene **33** (1987) 493-496.
21. R.S. Boethling, Environ. Toxicol. Chem. **5** (1986) 797-806.
22. R.S. Boethling, B. Gregg, R. Frederick, N.W. Gabel, S.E. Campbell, and A. Sabljić, Environ. Sci. Technol. (1988) submitted.
23. R.S. Boethling and A. Sabljić, Environ. Sci. Technol. (1988) submitted.
24. A. Sabljić, Bull. Environ. Contam. Toxicol. **30** (1983) 80-83.
25. L.H. Hall and L.B. Kier, Bull Environ. Contam. Toxicol. **32** (1984) 354-362.
26. L.B. Kier and L.H. Hall, Bull Environ. Contam. Toxicol. **29** (1982) 121-126.
27. M. Vighi and D. Calamari, Chemosphere **14** (1985) 1925-1932.
28. J.L. Newsted and J.P. Giesy, Environ. Toxicol. Chem. **6** (1987) 445-461.
29. M. Protić-Sabljić and A. Sabljić, Aquat. Toxicol. (1988) submitted.
30. Y. Yoshioka, T. Mizuno, Y. Ose, and T. Sato, Chemosphere **15** (1986) 195-203.
31. M. Randić, J. Amer. Chem. Soc. **97** (1975) 6609-6615.
32. A. Sabljić, J. Chromatogr. **314** (1984) 1-12.

33. A. Sabljić, J. Chromatogr. **319** (1985) 1-8.
34. A. Sabljić and M. Protić-Sabljić, Mol. Pharmacol. **23** (1983) 213-218.
35. M. Šoškić and A. Sabljić, Croat. Chim. Acta **60** (1987) 755-764.
36. P.G. Seybold, M. May, and U.A. Bagal, J. Chem. Educ. **64** (1987) 575-581.
37. A.T. Balaban, I. Motoc, D. Bonchev, and O. Mekenyan, Top. Curr. Chem. **114** (1983) 21-55.
38. Program GRAPH III is now fully operational and it is available for distribution. More details about this program and conditions for its distribution are available on request from the author.
39. C.T. Chiou, P.E. Porter, and D.W. Schmedding, Environ. Sci. Technol. **17** (1983) 227-231.
40. R.P. Davis and A.J. Dobbs, Water Res. **18** (1984) 1253-1262.
41. D.N. Brooke, "A comparison of four methods for the prediction of fish bioconcentration factors", Building Research Establishment, Note No. 163/84, Department of Environment, UK, 1984.
42. D.L. Geiger, C.E. Northcott, D.J. Call, and L.T. Brooke, (Editors), 1984/85. Acute toxicities of organic chemicals to fathead minnows (*Pimephales promelas*). Vol. I and II, Center for Lake Superior Environmental Studies, University of Wisconsin-Superior, Superior, Wisconsin, 414 and 326 pp.
43. S.W. Karickhoff, D.S. Brown, and T.A. Scott, Water Res. **13** (1979) 241-248.
44. A. Spacie, P.F. Landrum, and G.J. Leversee, Ecotoxicol. Environ. Safety **7** (1983) 330-341.
45. W.B. Neely, D.R. Branson, and G.E. Blau, Environ. Sci. Technol. **8** (1974) 1113-1115.
46. M. Ogata, K. Fujisawa, Y. Ogino, and E. Mano, Bull. Environ. Contam. Toxicol. **33** (1984) 561-567.
47. R.L. Lipnick, D.E. Johnson, J.H. Gilford, C.K. Bickings, and L.D. Newsome, Environ. Toxicol. Chem. **4** (1985) 281-296.
48. H. Konemann, Toxicology 19 (1981) 209-221.
49. L.S. McCarty, P. Hodson, G. Craig, and K. Kaiser, Environ. Toxicol. Chem. **4** (1985) 595-606.
50. J.C. McGowan and A. Mellors, Bull. Environ. Contam. Toxicol. **36** (1986) 881-887.
51. V. Kozak, G. Simsiman, G. Chesters, D. Stensby, and J. Harkin, Review of the environmental effects of pollutants XI: Chlorophenols. EPA-600/1-79-012. U.S. Environmental Protection Agency, Cincinnati, OH, 1979, 228 pp.
52. L.J. Mullins, Chem. Rev. **54** (1954) 289-323.
53. G.M. Omann and J.R. Lakowicz, Biochem. Biophys. Acta **684** (1982) 83-95.
54. S.H. Roth, Fed. Proc. Fed. Am. Soc. Exp. Biol. **39** (1980) 1595-1599.

Industrial Use of Group Contribution Methods for Estimation of Physical Properties

T.W. Copeman, P.M. Mathias[1], and H.C. Klotz

[1]Air Products and Chemicals, Inc., Management Information Services, P.O. Box 538, Allentown, PA 18105, USA

Estimation methods for physical properties are widely used in industry. The number of chemical substances is enormous (several million) and thus complete or even partial experimental data are rare. In addition, experimental measurements are expensive and thus screening studies and preliminary design often must be done without measured properties.

In this paper, we analyze a number of existing methods for property prediction ranging from simple group additivity to molecular modelling methods. The properties we consider include vapor pressure, heats of formation and phase equilibria.

Simple additivity group contribution methods work surprisingly well. We present examples of vapor-pressure prediction and discuss the applicability of the simple methods to complex compounds. Often the applicability is considerably improved when the simple methods are confined to a single family of compounds.

An obvious way to improve the simple methods is to account for higher-order structure and molecular connectivity. However, these corrections are empirical and only provide marginal improvement in correlative capability.

Molecular modelling methods offer promise to provide fundamentally-based predictions. We discuss molecular mechanics which currently can be used to predict relative conformational energies, ideal-gas heats of formation and dipole moments. We discuss the trade-off between higher complexity and improved predictive power. Further work is needed to extend the application of these methods.

C. Jochum M.G. Hicks J. Sunkel (Eds.)
Physical Property Prediction
in Organic Chemistry

INTRODUCTION

Estimation methods for physical properties are widely used in industry. Experimental measurements are, of course, preferable to estimated values, but these are unlikely to be available due to the enormous number of compounds (several million) and the ranges of temperatures, pressures and mixture compositions at which data are required. In addition, experimental measurements are expensive and time-consuming and thus screening studies and preliminary design often must be done without measurement of properties.

Even when experimental measurements are being made, estimation methods are useful since they anticipate the approximate range of the experimental results. Somewhat similarly, estimation methods are a valuable ally to experimental research programs since they help select a small subset of potential systems for experimental investigation.

Broad reviews of estimation methods for physical properties are available (for example, see Reid, Prausnitz and Poling, 1987; and Lyman, Reehl and Rosenblatt, 1982). In this paper, we limit our discussion to three important properties (heats of formation, vapor pressure and aqueous solubilities) in order to focus on the capabilities - and deficiencies - of existing methods. Other properties that are important to the chemical process industry include: density, heat capacity, enthalpy of vaporization, free energy of formation, viscosity, thermal conductivity, surface tension and diffusion coefficients. Out of the vast area of multicomponent, multiphase equilibria, we have somewhat arbitrarily chosen to discuss aqueous solubilities. We believe that some general conclusions about the state of the art of estimation methods can be drawn despite the narrow focus of this paper.

In general, the models available for correlation of thermodynamic properties and phase equilibrium are good. For example, in the case of thermodynamic properties, Soave (1984) has shown that even the simple van der Waals equation of state, together with enhancements like extended temperature dependence and the "Volume-translation" concept (Peneleoux, 1982), provides an adequate correlation framework, with remarkable capability to interpolate and even extrapolate. More sophisticated models (for example using the perturbed-hard-chain approach, Donohue, 1988) generally provide better descriptive capability.

But all existing correlative equations of state require at least a few well-chosen data points (e.g., vapor pressure, density, enthalpy of vaporization, heat capacity) to establish the model constants. Thus, there will always be a need for experimental measurements and estimations.

Our discussion will review some of the popular methods in industry for predicting heats of formation, vapor pressure, and solubility of organic chemicals in water. The strengths and weaknesses of these methods are discussed and compared.

ESTIMATION METHODS

We have found it useful to divide estimation methods into four categories:

1. Simple Group Additivity
2. Higher-Order Group Additivity
3. Structural Methods
4. Fundamental Methods

The simple-additivity method assumes that a property or a model parameter is directly equal to the sum of the individual contributions of the fragments or groups. In other words, the contribution of a particular group is independent of the presence of other groups. Despite its crudeness, the simple-additivity method works surprisingly well, especially for properties such as molar volume. But there are problems for other properties. For example, Wu and Sandler (1988) have presented a study which shows that UNIFAC method (effectively a simple additivity approach) has difficulty in estimating activity coefficients of compounds with two ether groups, and the errors worsen as the distance between the ether groups decreases.

Higher-order corrections are an obvious way to improve the simple-additivity method. For example, Benson (1968) and Hansch and Leo (1979) have successfully applied higher-order corrections to obtain improved correlations of heats of formation and octanol-water partition coefficients, respectively.

Higher-order corrections have mostly been empirical, deduced in hindsight by studying the error patterns of the simple additivity method. An objective, and potentially valuable, approach is "molecular connectivity" proposed by Randic (1975) and popularized by Kier and Hall (1985). We study the correlation of vapor pressure to assess the value of molecular connectivity.

"Structural Methods" consider the detailed molecular structure and are considerably less empirical than the group-contribution approach. An example of a structural method is molecular mechanics (Burkert and Allinger, 1982), which we evaluate as an estimation method for ideal-gas enthalpy of formation.

We classify ab-initio quantum mechanical calculations as fundamental methods. These methods are highly complex, computationally intensive and are generally not yet useful in practical engineering applications. We do not discuss these methods further.

FIGURE 1

ALKANE HEATS OF FORMATION
JOBACK METHOD

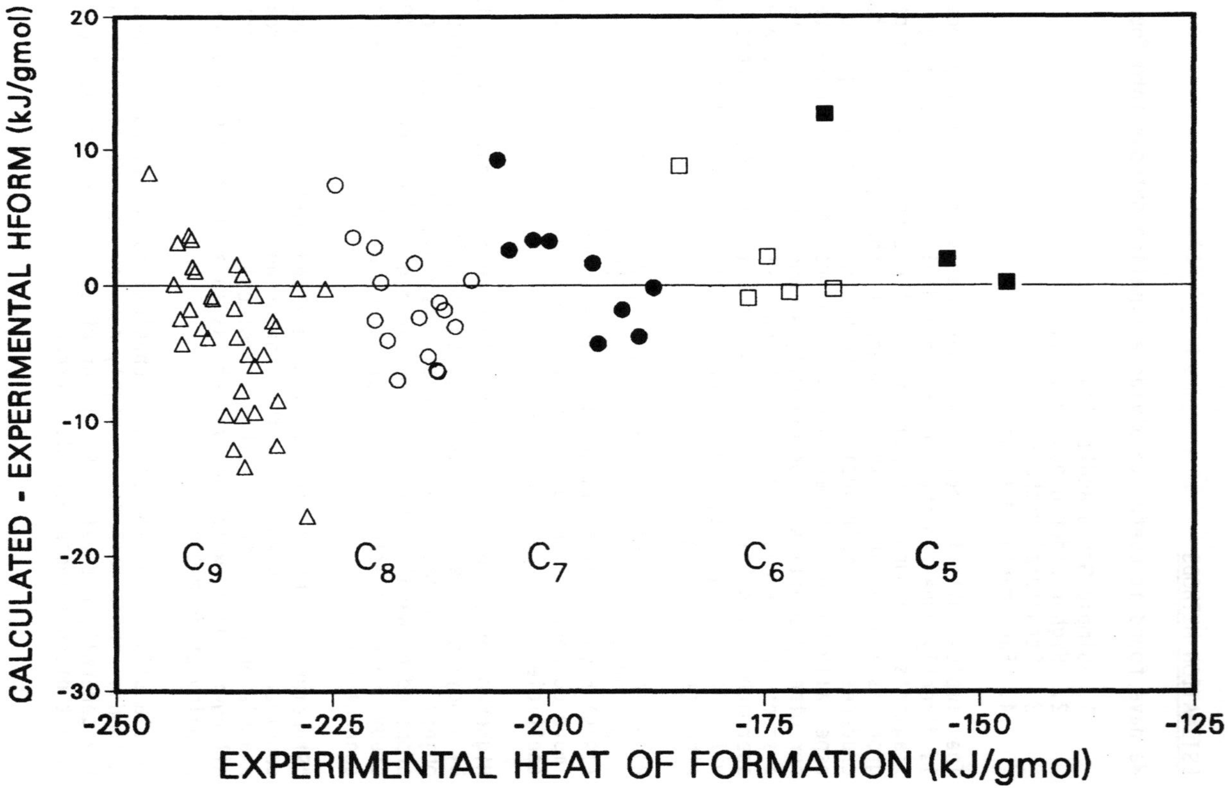

When faced with a property estimation problem, the properties of a similar substance can be used as a reference and structural corrections can be determined from the estimation method. The estimated property, therefore, is based on both experimental data and estimated values. For example, if we need to estimate the enthalpy of formation of substance A, we could locate a similar substance B, use the estimation method to predict the enthalpy of formation for both A and B, and then only use the estimation method to determine the (small) difference in heats of formation between A and B. In this study, we analyze groups of similar substances to evaluate this capability.

IDEAL GAS HEAT OF FORMATION

Since ideal gas properties are unimolecular, the estimation method need only account for a single molecule in its procedure. A number of estimation techniques of varying complexity have been applied to to ideal gas heats of formation. These methods range from chemical-bond group additivity methods to molecular orbital calculations. We present a brief evaluation of a cross-section of these methods for branched hydrocarbons and alcohols.

A. Simple Group Contribution: Joback's method

The simplest type of group contribution method that is moderately successful in predicting ideal gas heats of formation assigns contributions based on molecular groupings such as -CH3 and >CH2. An example of this method type is Joback's method (Joback, 1984). Joback's heat of formation method is based on the equation:

$$\Delta H_f^\circ(298K) = C + \sum_j n_j \Delta H_j^j \qquad (1)$$

where: $C = 68.29$ kJ/gmol

The groups used in this study were obtained from Reid et al. (1987) and are not reproduced here.

Figure 1 shows a comparison between heats of formation predicted by Joback's method and data for C_5 to C_9 alkanes. The experimental data were obtained from the TRC (1987a) and DIPPR (1987) data compilations. As shown, the heats of formation predicted by Joback method are generally within 10 kJ/gmol for this set of data. The method does not account for branching in a consistent way as evidenced by the random scatter both above and below the experimental values. Not surprisingly, the method is least accurate for the highly branched alkanes.

A comparison between estimations by Joback's method and experimental heats of formation for C_4 to C_8 alcohols is shown in Figure 3. The experimental data were obtained from TRC (1987b). The database for alcohol heats of formation is considerably smaller than the alkane database. Joback's method shows a large scatter for the alcohols with maximum errors of about 25 kJ/gmol for the tertiary alcohols tested.

FIGURE 2

ALKANE HEATS OF FORMATION
COMPARISON BETWEEN JOBACK AND BENSON METHODS

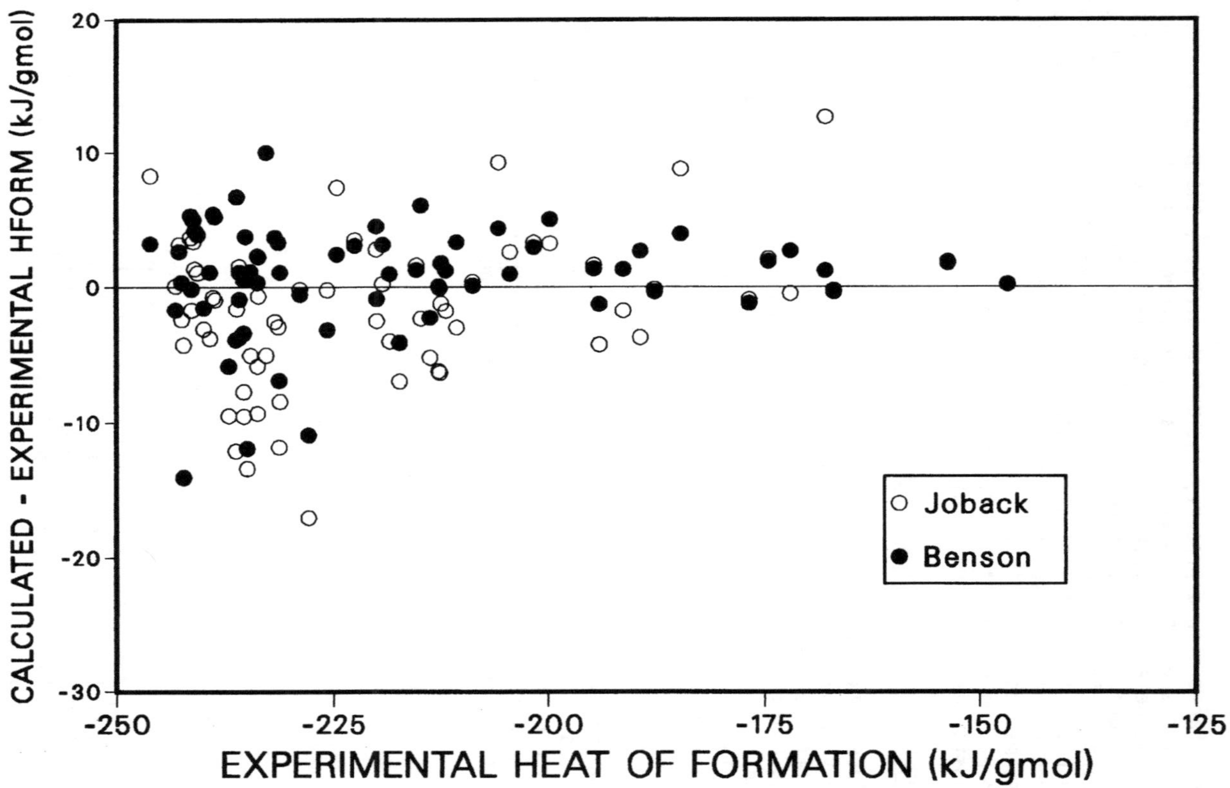

FIGURE 3

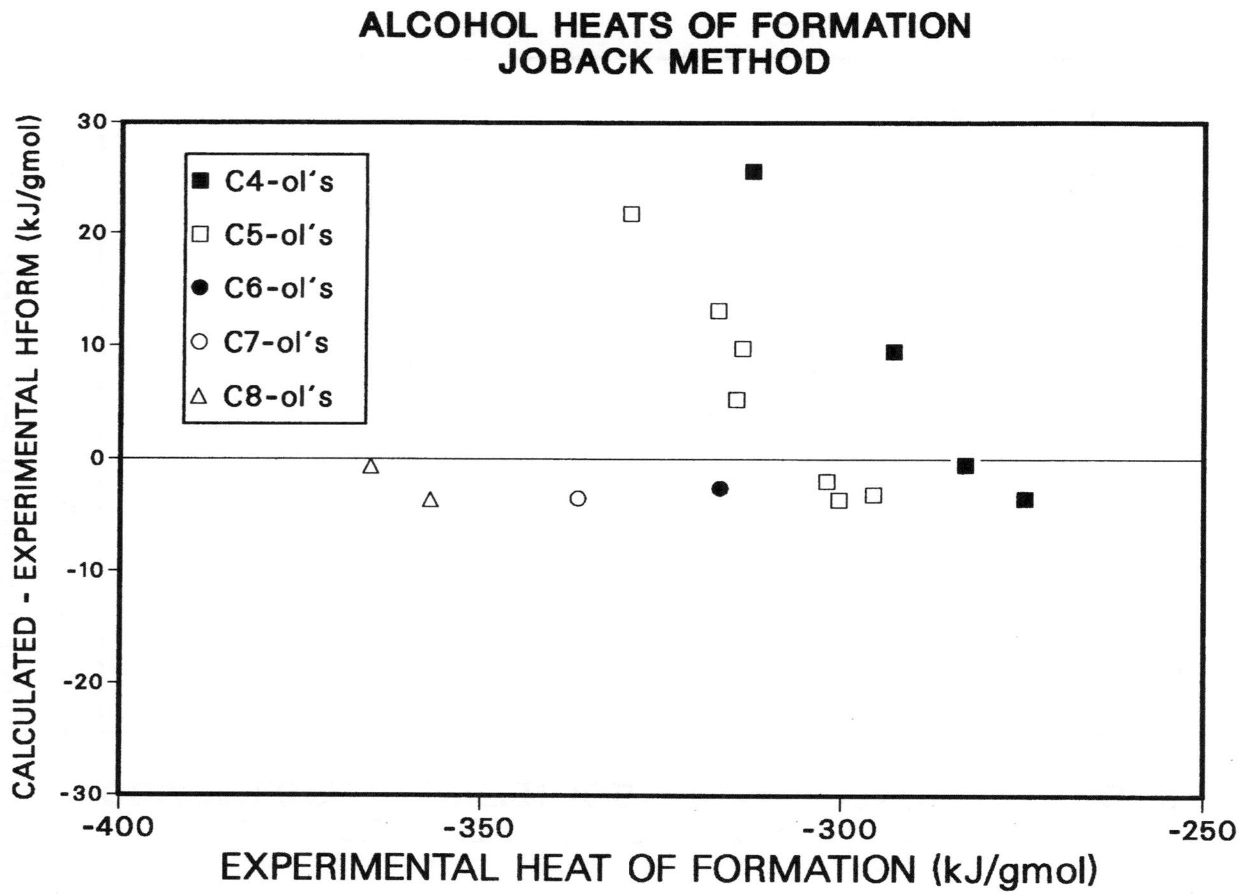
ALCOHOL HEATS OF FORMATION
JOBACK METHOD
C4-ol's
C5-ol's
C6-ol's
C7-ol's
C8-ol's
CALCULATED - EXPERIMENTAL HFORM (kJ/gmol)
30
20
10
0
-10
-20
-30
-400
-350
-300
-250
EXPERIMENTAL HEAT OF FORMATION (kJ/gmol)

FIGURE 4

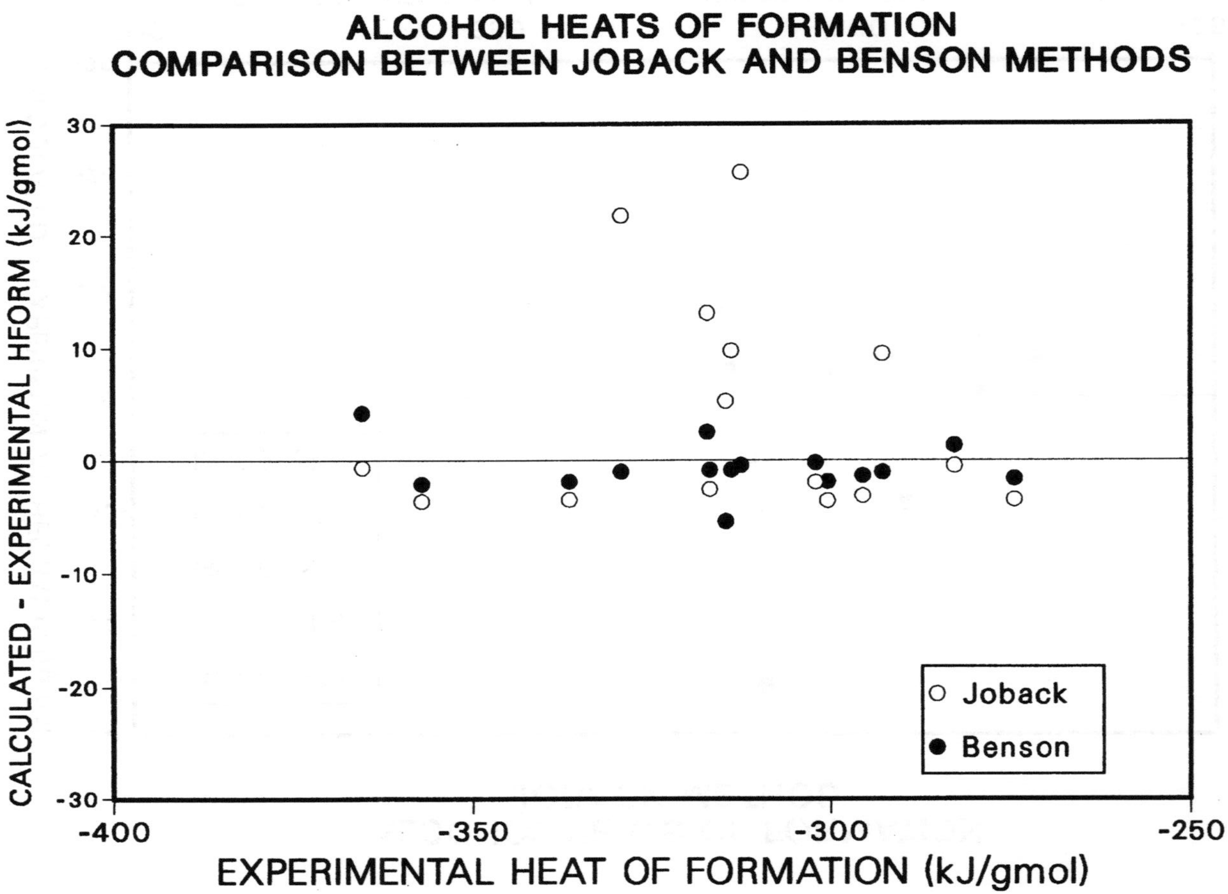
ALCOHOL HEATS OF FORMATION
COMPARISON BETWEEN JOBACK AND BENSON METHODS
CALCULATED - EXPERIMENTAL HFORM (kJ/gmol)
30
20
10
0
-10
-20
-30
-400
-350
-300
-250
EXPERIMENTAL HEAT OF FORMATION (kJ/gmol)
Joback
Benson

B. Higher Order Group Contribution: Benson's Method

A more sophisticated type of group-contribution method accounts for nearest neighbors to the atom or group. Benson's method (Benson, 1968) is an example of a method that falls into this category. For example, Benson's method includes a gauche correction term for alkanes which accounts for steric interactions of neighboring groups. For heats of formation, Benson's method uses a simple additivity approach given by:

$$\Delta H_f^\circ(298) = \Sigma n_j \Delta H_j^B \tag{2}$$

The group contributions used for this study were obtained from Reid et al. (1987) and are not presented here.

Figure 2 shows a comparison between Benson's and Joback's predictions and experimental heats of formation for a number of C_5 to C_9 hydrocarbons. Predicted heats of formation with Benson's method are slightly more accurate than with Joback's method. However, again the accuracy of Benson's method tends to decrease for highly branched compounds (e.g., tetramethylpentanes).

Benson's method was also evaluated against alcohols. Figure 4 shows a comparison between Benson's and Joback's methods and experimental data for a number of C_4 to C_8 alcohols. The agreement between predicted and experimental results with Benson's method is considerably better than Joback's method. This is primarily because Benson's method differentiates between primary, secondary and tertiary alcohols, whereas Joback's method does not.

These simple additivity methods can also be used in a perturbation mode (Seaton et al. 1974). The experimental heat of formation of a compound similar to the compound of interest is used as a reference value. Appropriate groups are added and/or subtracted from this reference molecule to obtain the molecule of interest. Table I shows results using Benson's method in a perturbation mode for several highly branched alkanes. Clearly, the heats of formation estimated using the perturbation analysis are more accurate than those based on the group contribution method alone.

C. Molecular Mechanics: Allinger's Method

Obviously, the methods described above take molecular structure into account in a semi-empirical way. Molecular mechanics and molecular orbital methods provide a more fundamental description of molecular structure. The MM2 molecular mechanics approach (Burkert and Allinger, 1982; Boyd and Lipkowitz, 1982)) offers a middle of the road approach to including the effect of molecular geometry in a reasonable way. The MM2 method can be used for the determination of unimolecular properties such as ideal gas heat of formation, dipole moment, conformational energies and transition state structures. With molecular mechanics, a molecule is assumed to consist of a collection of atoms held together by harmonic forces. These forces are described by a potential energy function of bond length and bond angles. The steric energy of a molecule arises from the difference in energy of a real molecule and a

TABLE I: Perturbation Analysis Using Benson's Method Ideal-Gas Heats of Formation

	ΔH_f^o, kJ/gmol		
	Experimental	Benson's Method	Perturbation Analysis
2,2,3,4-tetramethylhexame	-253.9	-264.3	-252.4*
2,2,3,4-tetramethylheptane	-275.0	-285.0	-273.1*
2,2,3,3,4-pentamethylpentane	-241.3	-265.7	-241.0**

* - 2,2,3,4-tetramethylpentane used as reference molecule.

** - 2,2,3,4,4-pentamethylpentane used as reference molecule.

hypothetical molecule with ideal structural values. The steric energy is approximated as a sum of contributions arising from bond stretching, bond bending, bond twisting and non-bonded interactions. The molecular geometry is optimized by minimizing the steric energy. Molecular mechanics therefore provides the molecular potential energy as a function of the movements of atoms within the molecule.

Predicted heats of formation using the MM2 approach were compared against experimental data for some C_5 to C_8 alkanes by Burkert and Allinger (1982). Table II shows a comparison of the accuracy of Joback's, Benson's and the MM2 method for these components. As illustrated, all three methods do well for the non-branched alkanes, however the MM2 method appears most accurate for the highly branched alkanes.

MM2 predicted heats of formation were also compared to experimental data for some C_4 to C_8 alcohols by Burkert and Allinger (Table II). The agreement between predicted and experimental results is similar to Benson's method.

The somewhat improved predictions obtained with the MM2 approach is not without cost. Clearly, the MM2 approach is the most difficult to apply of the three methods presented. In order to estimate heat of formation, the energies of all conformations must be calculated and averaged based on their relative amounts. For large alkanes, this could be a formidable task. However, even with these drawbacks, structural methods like the molecular mechanics approach offer a strong potential for accounting for molecular geometry in the prediction of unimolecular properties. These methods are generally not discussed in the chemical engineering literature. More attention should be given to the development and use of these methods.

VAPOR PRESSURE

The vapor pressure estimation method we focus on is based upon a simple-additivity group-contribution method applied to the AMP equation (Abrams, Massaldi and Prausnitz, 1974; Macknick, Winnick and Prausnitz, 1977). We first show the capability of the model for perfluorocarbon compounds. Next we turn our attention to branched hydrocarbons and alcohols in order to assess the capability of the method to capture the structural features of a compound. Finally, we analyze the capability to "interpolate" the vapor pressure of a given compound, if the vapor pressure of similar compounds are available.

We investigate the possibility of simple improvements to the method. An attempt has been made to improve the group-contribution AMP method by using the experimental molar volume as input. We also analyze the capability of the "molecular connectivity" approach (Kier and Hall, 1985) to capture the isomeric differences between the hydrocarbons.

TABLE II: Comparison of MM2, Benson and Joback Methods For Ideal Gas Heat of Formation

Component	Heat of Formation kJ/gmol			
	Experimental	Joback	Benson	MM2
Alkanes				
n-Pentane	-146.76	-146.53	-146.56	-146.19
n-Hexane	-166.92	-167.17	-167.28	-167.49
n-Heptane	-187.65	-187.81	-188.00	-188.74
n-Octane	-208.82	-208.45	-208.72	-210.04
i-Butane	-134.99	-131.17	-134.55	-134.60
i-Pentane	-153.70	-151.81	-151.92	-152.62
Neopentane	-167.92	-155.28	-166.71	-169.69
2,3-Dimethylbutane	-176.80	-177.73	-178.00	-177.74
2,2,3-Trimethylbutane	-204.43	-201.84	-203.46	-204.22
2,2,3,3-Tetramethylbutane	-225.73	-225.95	-228.92	-224.51
2,2,4,4-Tetramethylpentane	-242.30	-246.59	-256.34	-247.07
3,3-Diethylpentane	-232.80	-237.84	-222.79	-229.66
Alcohols				
1-Butanol	-274.60	-278.12	-276.23	-275.22
2-Butanol	-292.88	-283.40	-293.94	-292.75
2-Methyl-1-Propanol	-282.92	-283.40	-281.59	-283.76
2-Methyl-2-Propanol	-312.46	-286.87	-312.91	-312.80
1-Pentanol	-295.58	-298.76	-296.95	-296.44
1-Hexanol	-316.80	-319.40	-317.67	-317.61
1-Heptanol	-336.50	-340.04	-338.39	-338.78
1-Octanol	-357.00	-360.68	-359.11	-359.99

The AMP vapor pressure equation (Abrams, Massaldi and Prausnitz, 1974; Macknick, Winnick and Prausnitz, 1977) has the form of an empirical equation proposed by Miller (1964):

$$\ln P^s = A + B/T + C \ln T + DT + ET^2 \quad (3)$$

The original form was empirical, but Abrams et al. (1974) showed that the five constants could be expressed in terms of three parameters that have physical significance. Thus,

$$\ln P^s = f\,(T;\, V_w,\, S,\, E_o) \quad (4)$$

The functional form of equation (4) is shown by Macknick, Winnick and Prausnitz (1977). V_w is the van der Waals volume of the molecule; S is the number of equivalent oscillators of the molecule and is related to the size and shape of the molecule; and E_o is the enthalpy of vaporization of the hypothetical liquid at T=0 and is a measure of the attractive forces of the molecule.

Due to their physical significance, it is expected that the parameters in the AMP equation will be amenable to a group contribution treatment. Macknick and Prausnitz (1974), Edwards and Prausnitz (1981) and Kelly, Mathias and Schweighardt (1988) have used a simple-additivity group-contribution approach to estimate the parameters in the AMP equation.

$$V_w = \sum_i n_i V_{wi} \quad (5)$$

$$S = \sum_i n_i S_i \quad (6)$$

$$E_o = \Sigma\, n_i E_{oi} \quad (7)$$

where n_i is the number of groups of type i in the molecule, and V_{wi}, S_i and E_{oi} are the group-contribution parameters of that group. Equation (5) was originally used by Bondi (1968).

Kelly, Mathias and Schweighardt (1988) showed that the group-contribution AMP method successfully correlated the vapor pressures of perfluorocarbon compounds. The group parameters are shown in Table III. Figure 5 presents a comparison between model predictions and experimental data for several representative perfluorocarbons. In general, the agreement is excellent. With few exceptions, the predicted normal boiling points are within 2°C of the observed values.

In general, good results can be obtained with this simple method. Macknick and Prausnitz (1977) and Edwards and Prausnitz (1981) showed that the approach is applicable to hydrocarbons and heteroatomic compounds containing nitrogen or sulfur.

TABLE III: Constituent Groups and Best-Fit Group Contributions
AMP Vapor-Pressure Method

Name	V_{wi} (cm^3/gmol)	E_{oi}/R (°K)	S_i
PERHYDRO GROUPS			
$-CH_3$ (Terminal)	13.67	1257.5	2.6402
$>CH_2$ (Chain)	10.23	757.0	0.5999
>CH- (Branch Point)	6.78	27.9	-1.7126
>C< (Quaternary C in perhydro compound)	3.33	-1024.0	-4.6062
-OH(P) (Alcohol-primary)	8.36	5843.5	8.3589
-OH(S) (Alcohol-secondary)	7.90	5175.0	7.9027
-OH(T) (Alcohol-tertiary)	8.28	5122.6	8.2759
PERFLUORO GROUPS			
$-CF_3$ (Terminal)	21.33	1491.9	3.4104
$>CF_2$ (Chain)	15.33	802.2	0.7632
>CF- (Branch Point)	9.33	68.2	-2.0267
>C< (Quaternary C in perfluoro compound)	3.33	-310.9	-3.9573
Six-membered ring (saturated HC precursor)	-1.69	1121.4	4.9731
Five-membered ring (saturated HC precursor)	-1.80	1096.9	4.7807
Fusion of two rings (per fusion)	0.00	-579.2	-1.2055

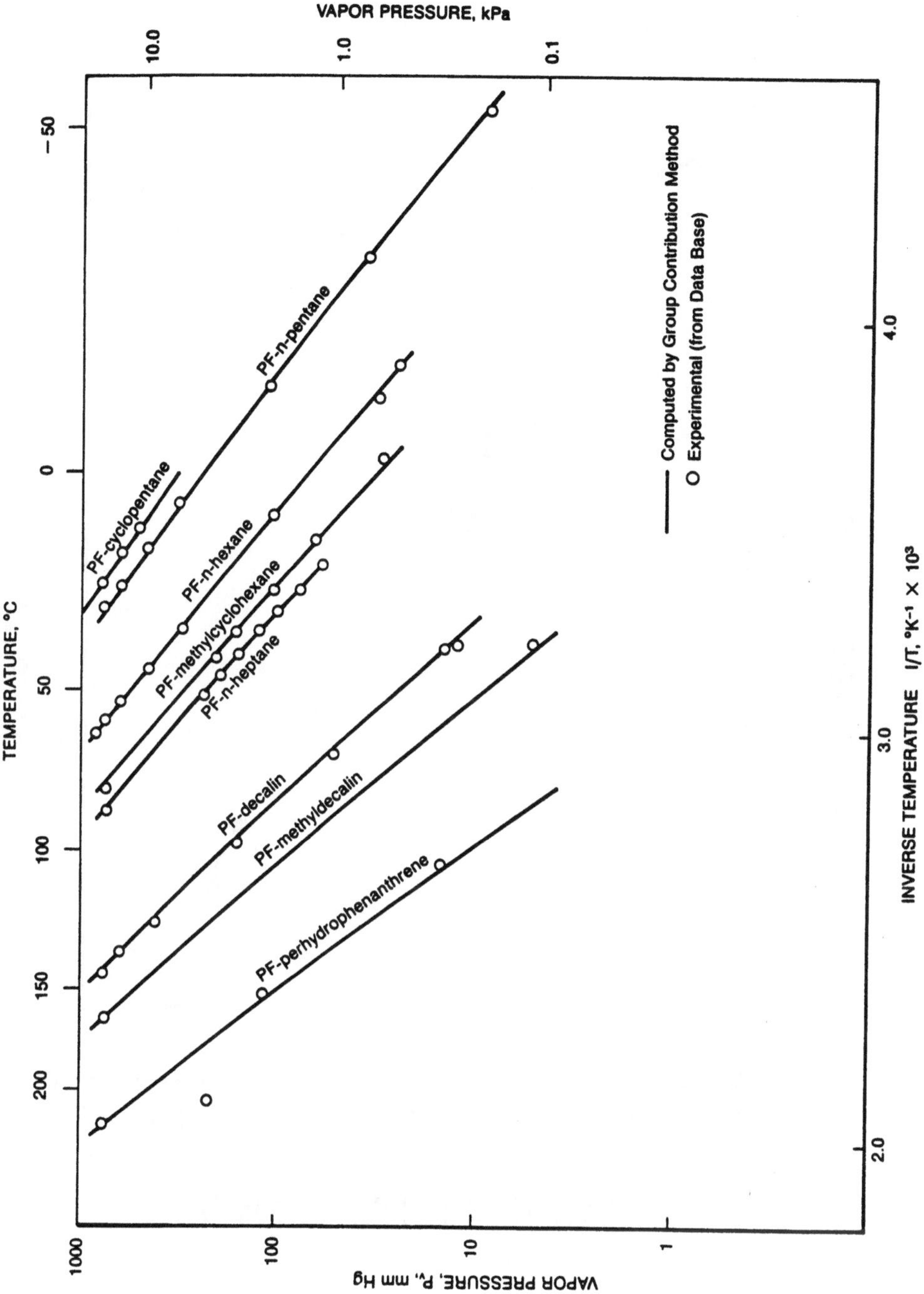
VAPOR PRESSURE, kPa
10.0
1.0
0.1
TEMPERATURE, °C
−50
0
50
100
150
200
PF-n-pentane
PF-cyclopentane
PF-n-hexane
PF-methylcyclohexane
PF-n-heptane
PF-decalin
PF-methyldecalin
PF-perhydrophenanthrene
Computed by Group Contribution Method
Experimental (from Data Base)
INVERSE TEMPERATURE 1/T, °K⁻¹ × 10³
4.0
3.0
2.0
VAPOR PRESSURE, P, mm Hg
1000
100
10
1

FIGURE 5

FIGURE 6

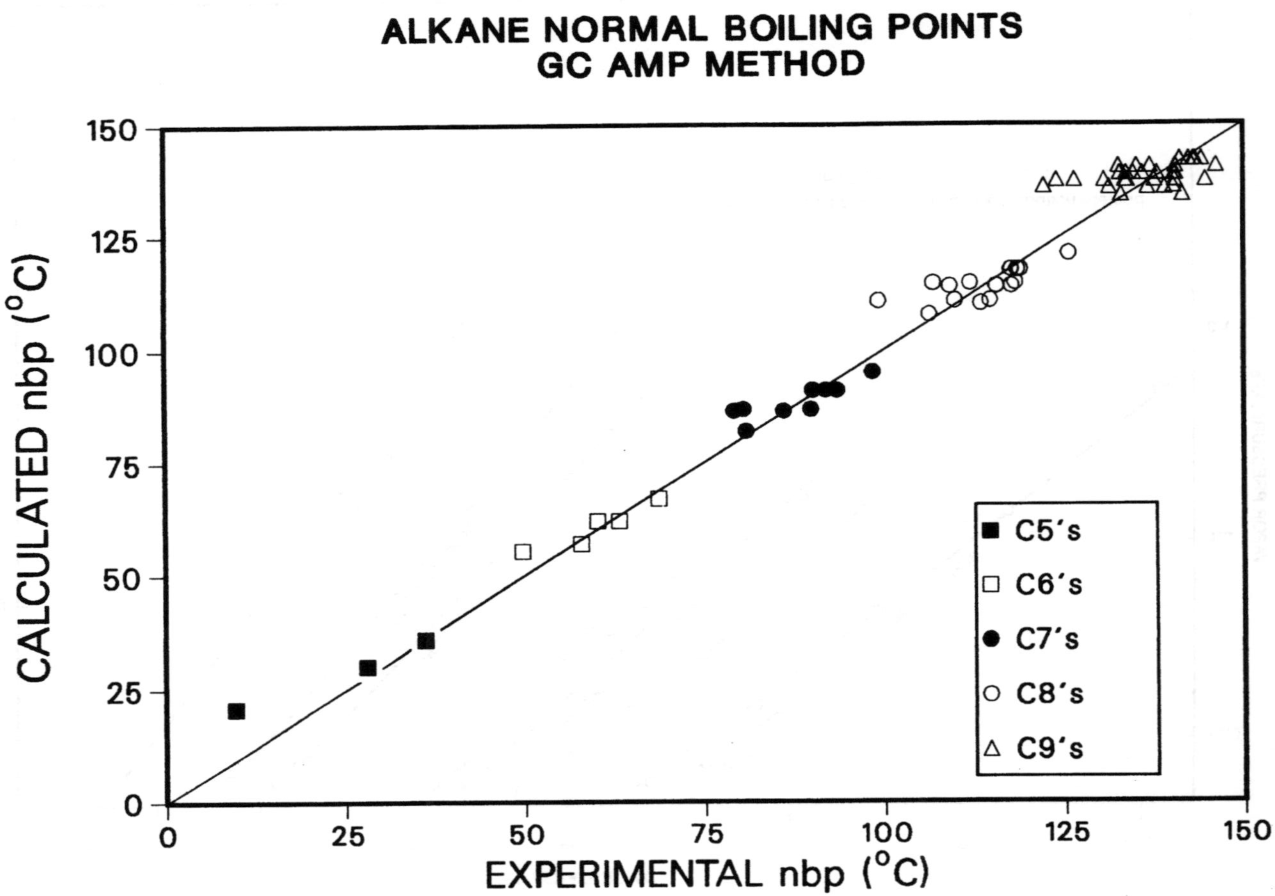

ALKANE NORMAL BOILING POINTS
GC AMP METHOD
CALCULATED nbp (°C)
EXPERIMENTAL nbp (°C)
0
25
50
75
100
125
150
C5's
C6's
C7's
C8's
C9's

It should be noted that the models are not entirely based upon simple additivity. Thus, for example, Edwards and Prausnitz (1981) allow different parameters for "aliphatic SH" and "aromatic SH" groups, and Kelly et al. (1988) apply corrections for 5-membered rings, 6-membered rings and fused rings (Table III). Of course, such "extended groups" increase the correlative power of the model, but at the loss of predictive capability. In the limit as entire molecules are treated as groups, the descriptive capability will be excellent, but the predictive power will vanish.

Models similar to the AMP group-contribution method can be developed, typically with equivalent capability. For example, Jensen, Fredenslund and Rasmussen (1981) developed a group-contribution method based, in part, on the UNIFAC method for vapor-liquid equilibria.

The simple methods apparently work well, but can they capture the subtle structural variations of molecules? We explored this question by studying two families of compounds for which extensive data compilations are available: alkanes and alcohols.

Figure 6 shows the comparison for the normal alkanes, C_5-C_9. The normal boiling points predicted by the AMP group-contribution method are plotted against the experimental values. Not surprisingly, the method is hardly able to capture the variation among the various components with the same carbon number. In the case of the C_9's, the points fall approximately on a horizontal line, indicating that a single value would provide a roughly equivalent correlation. The average absolute error for the 69 compounds in Figure 6 is 3.8°C and the maximum error is 13.2°C, for 2,3,5-trimethylhexane.

Similar results are presented in Figure 7 for the alcohols. The results are again bad, but surprisingly not considerably worse than for the alkanes. Thus, the effects of molecular branching are almost as difficult to describe as polarity. The points for the various alcohols of each carbon number appear to fall into three groups. This has been artificially imposed due to the fact that the primary, secondary and tertiary alcohol groups have been allowed to have different parameter values. The existence of the three groups and the tendency of the points of each of the three groups to fall along a horizontal line suggests that the empirical correlation does not capture the variation in the vapor pressures among the alcohol groups. For the 123 alcohols considered, the average absolute error is 4.6°C and the maximum error is 13.9°C, for 5-methyl-3-heptanol.

FIGURE 7

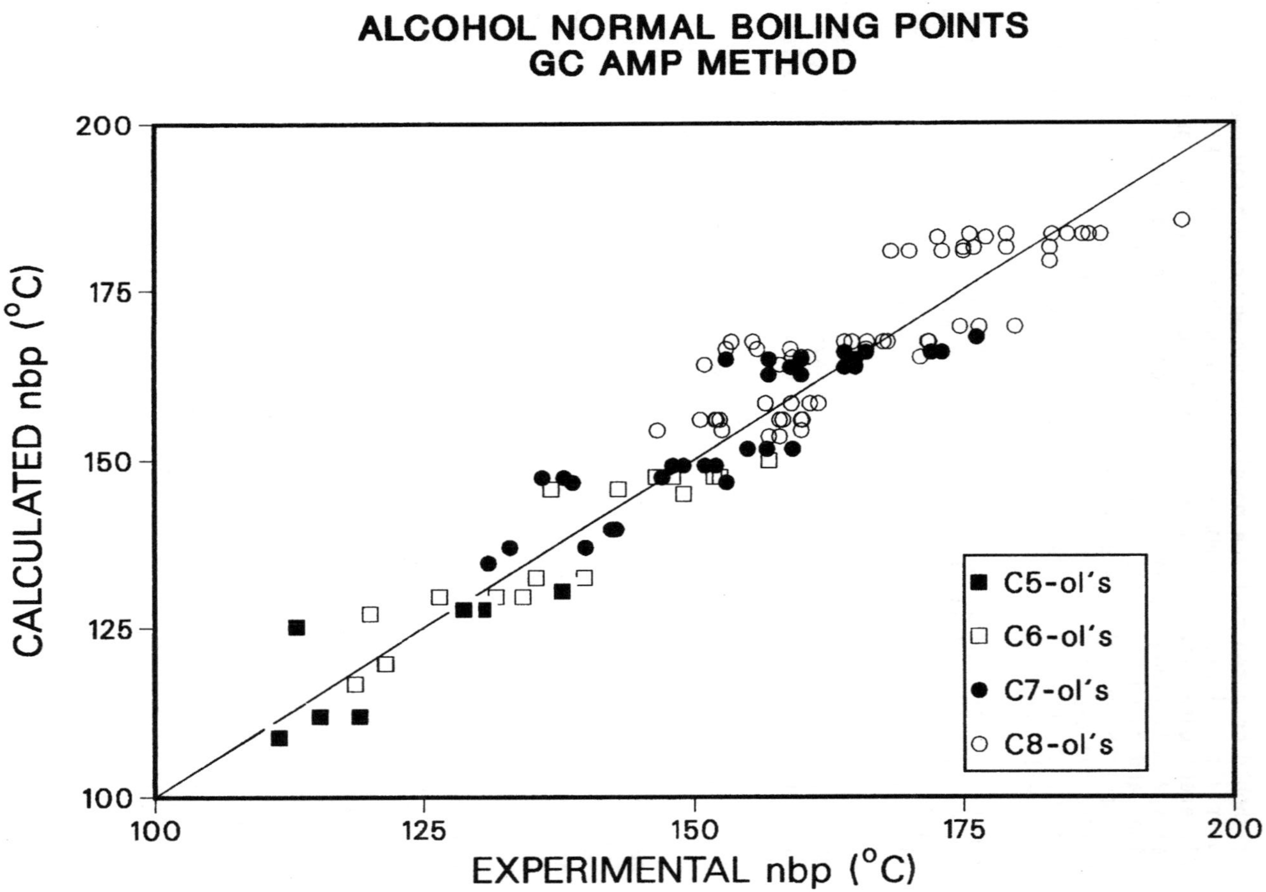
ALCOHOL NORMAL BOILING POINTS
GC AMP METHOD
CALCULATED nbp (°C)
EXPERIMENTAL nbp (°C)
100
125
150
175
200
C5-ol's
C6-ol's
C7-ol's
C8-ol's

FIGURE 8

VAPOR PRESSURE: 2,2,4-TRIMETHYL-PENTANE, -HEXANE
PERTURBATION ANALYSIS

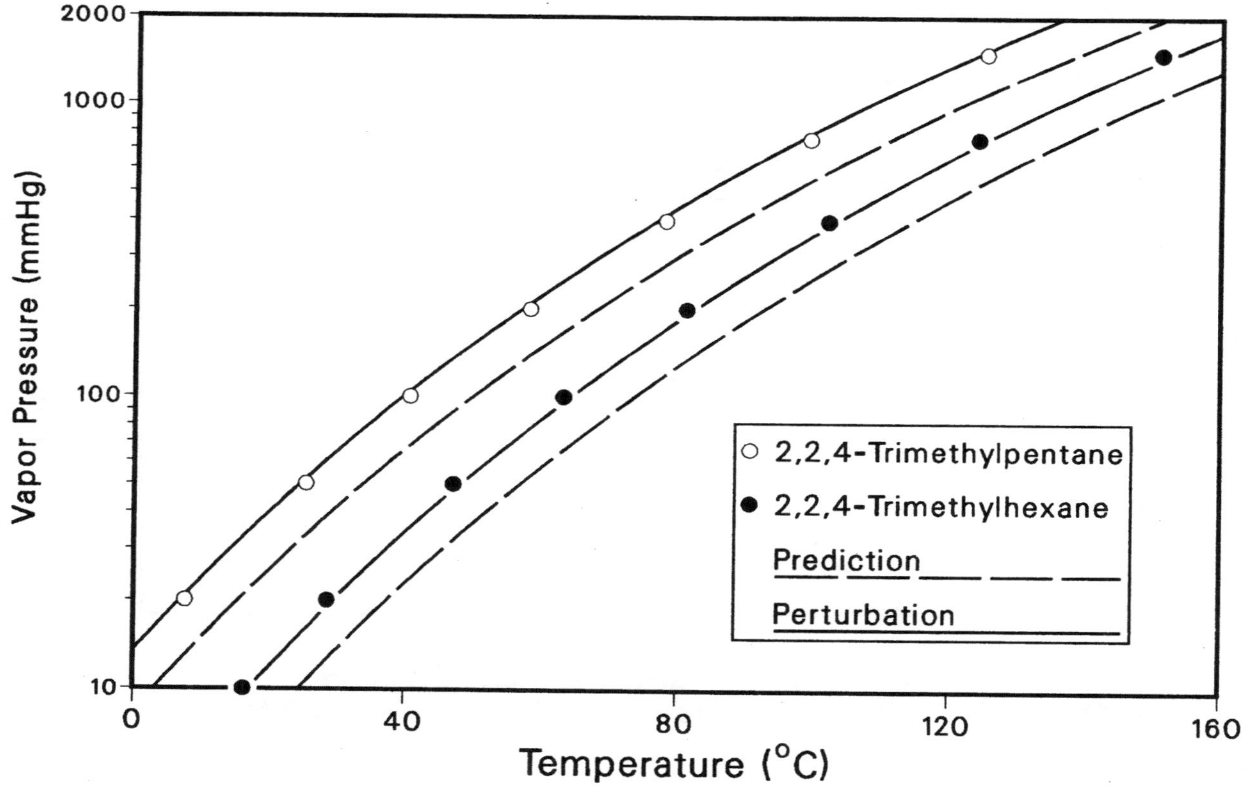

FIGURE 9

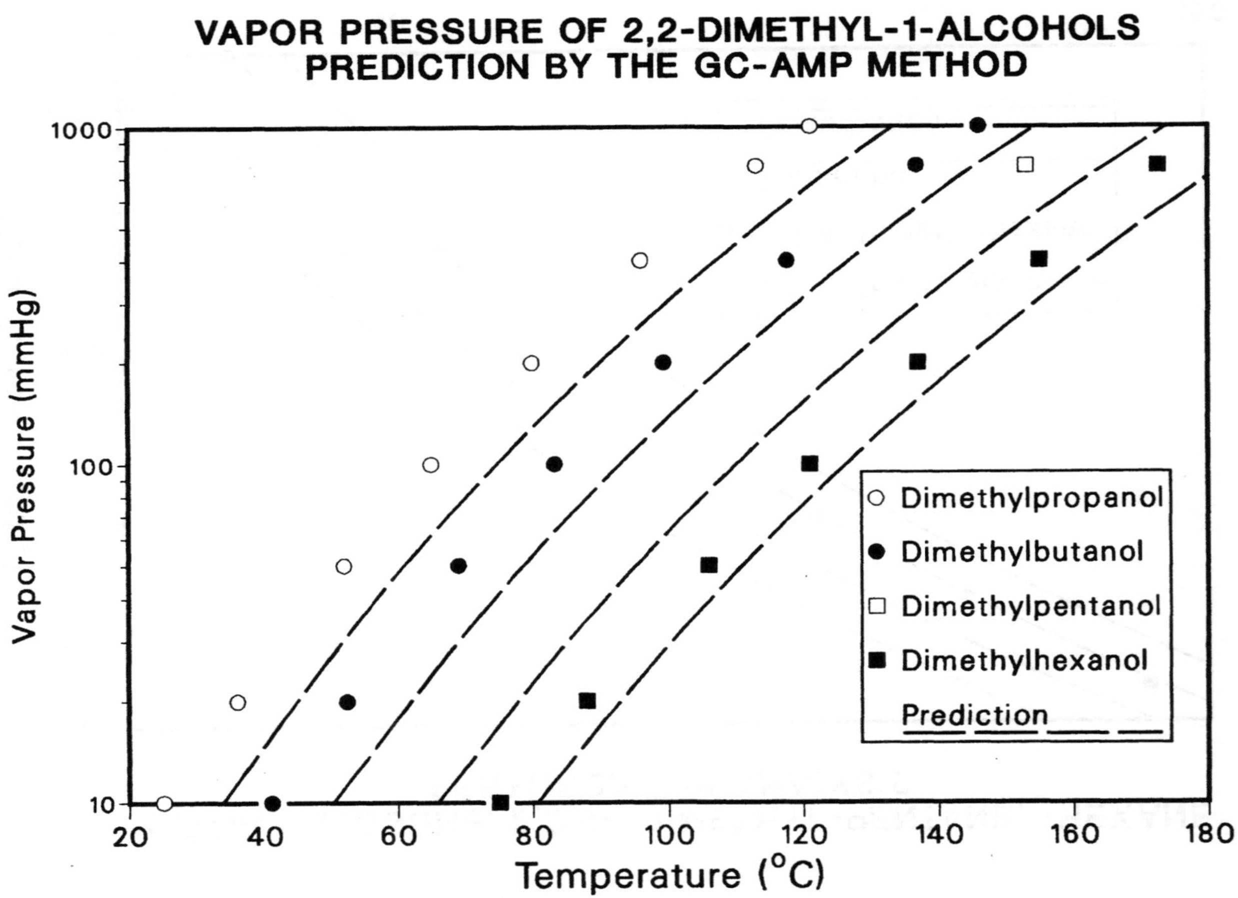
VAPOR PRESSURE OF 2,2-DIMETHYL-1-ALCOHOLS
PREDICTION BY THE GC-AMP METHOD
Vapor Pressure (mmHg)
1000
100
10
20
40
60
80
100
120
140
160
180
Temperature (°C)
Dimethylpropanol
Dimethylbutanol
Dimethylpentanol
Dimethylhexanol
Prediction

But this correlation, inadequate as it may be, can be used effectively in the "perturbation" mode. Figure 8 presents an analysis for 2,2,4-trimethylpentane and 2,2,4-trimethylhexane. In the perturbation method, we assume that the vapor pressure data for one of the substances is known. Now the AMP constants of the known substance are fit and the prediction method only estimates the difference in model constants between the reference and target molecules. As shown in Figure 8, the method works well. For the case of 2,2,4-trimethylhexane, the error in the predicted normal boiling point is reduced from 13.3°C to 1.3°C. This is a common-sense method, but in our experience it consistently provides a good alternative for vapor pressure estimation.

Figures 9 and 10 present a similar analysis for the 2,2-dimethyl-1-alcohols. Here 2,2-dimethyl-1-butanol has been arbitrarily chosen as the reference substance. It is worth noting that the perturbation method probably provides good estimates of the low-temperature vapor pressures of 2-2-dimethyl-1-pentanol for which no data exist.

We have made an attempt to improve the predictive power of the group-contribution AMP method. Analysis of the data for the alkanes indicates that the method provides good estimates for "E_o", but poor estimates for "S". This is reasonable since the dominant difference among the alkanes is branching (size and shape) rather than attractive forces, which would effect the energy. We therefore tried several modifications to equation 6. Surprisingly, a simple correction, using the molar volume at 25°C works well.

$$S = C\,V_{25} + \sum_i n_i S'_i \qquad (8)$$

In equation 8, "C" is a constant and V_{25} is the molar volume (cc/gmole) at 25°C. Figure 11 shows that the "volume-corrected" method provides a significantly improved correlation for the limited number of compounds analyzed. Here, the numerical value 'C' is 0.0500 and the 'S' parameters for the $-CH_3$, >CH- and >C< groups are 0.700, -0.231, -1.051 and -2.645, respectively. We do not offer this as a new method, but rather as an example of an approach where an easily-accessible property can be used to improve the estimation of a more difficult-to-measure property.

The "Molecular Connectivity" approach introduced by Randic (1975) and extensively applied and refined by Kier and Hall (1985) could potentially provide an improved description of molecular branching. Figure 12 shows a plot of the normal boiling point of the alkanes against the first-order connectivity index, $^1\chi$. There is a reasonable pattern, but the correlation is roughly equivalent to that provided by the group-contribution AMP method (Figure 6). We were unable to obtain improved correlations through the use of the higher-order connectivity indices recommended by Kier and Hall (1985).

FIGURE 10

VAPOR PRESSURE OF 2,2-DIMETHYL-1-ALCOHOLS
PERTURBATION ANALYSIS USING 2,2-DIMETHYL-1-BUTANOL

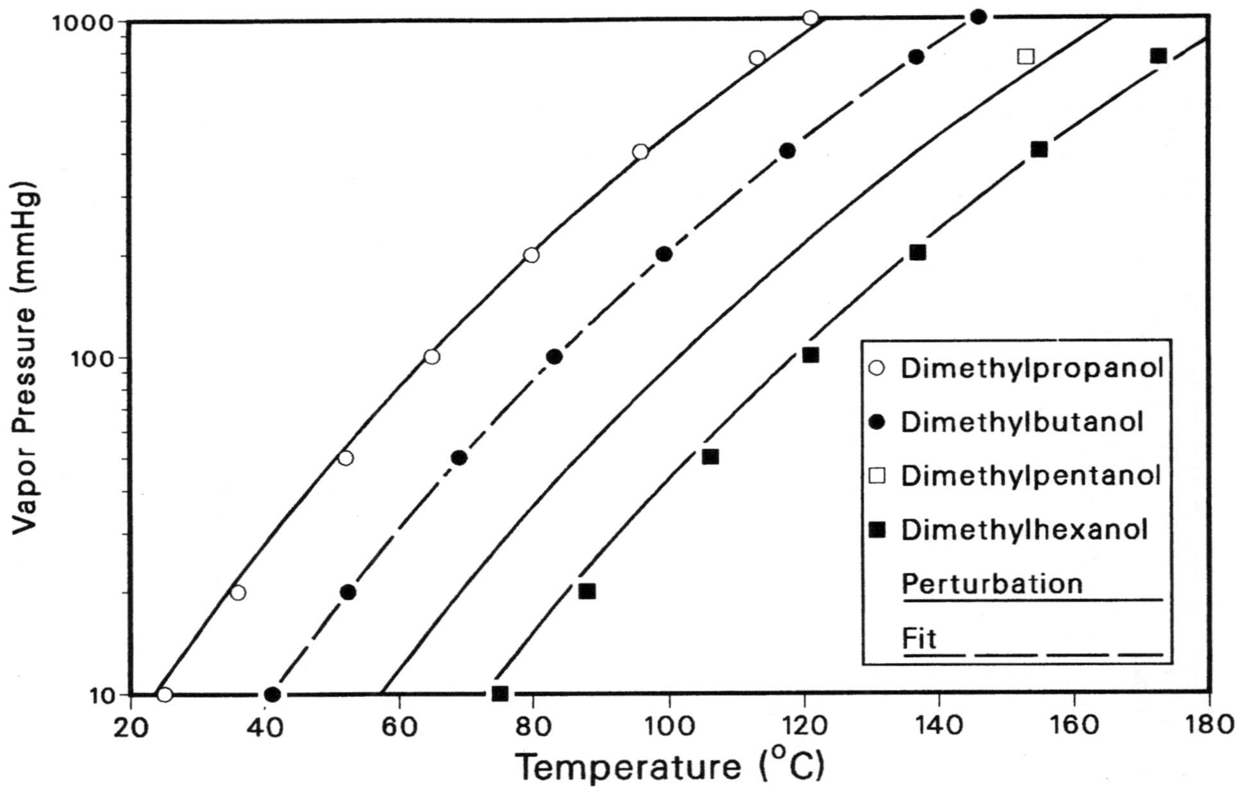

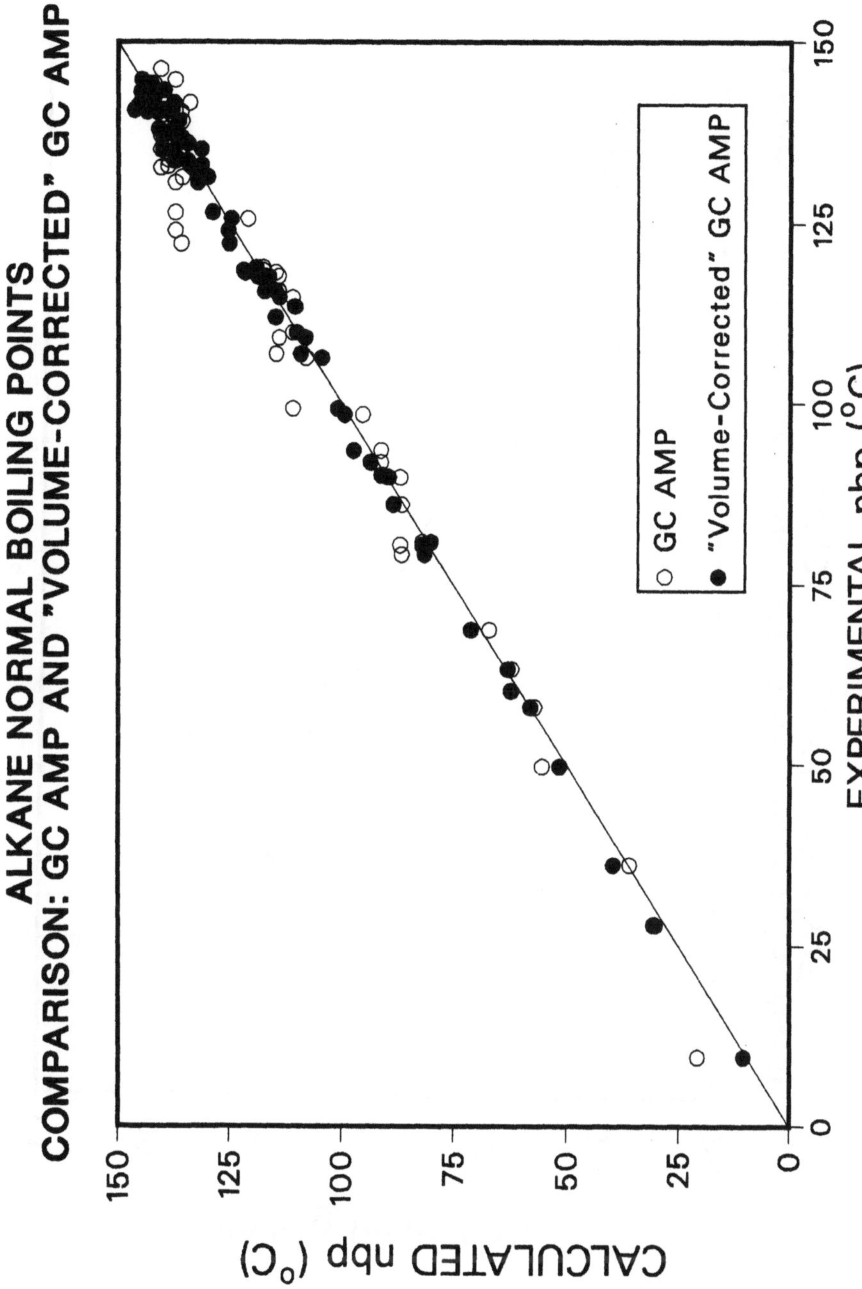
ALKANE NORMAL BOILING POINTS
COMPARISON: GC AMP AND "VOLUME-CORRECTED" GC AMP
GC AMP
"Volume-Corrected" GC AMP
CALCULATED nbp (°C)
EXPERIMENTAL nbp (°C)
0
25
50
75
100
125
150

FIGURE 11

FIGURE 12

BOILING POINT VS CONNECTIVITY INDEX, $^1\chi$
ISOMERS OF HEXANE, HEPTANE, OCTANE AND NONANE

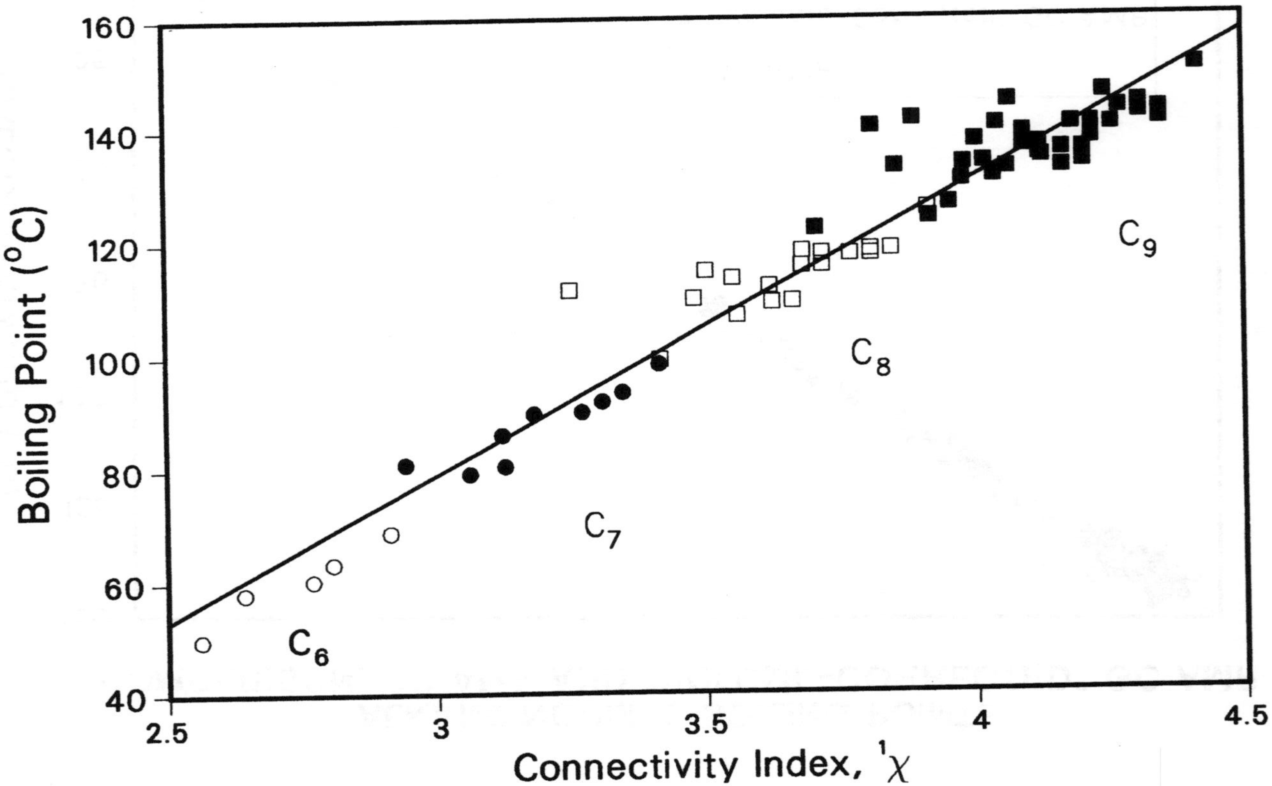

In summary, the simple-additivity methods provide only approximate estimates of the vapor pressure. The "perturbation" approach is effective, even with a relatively simple estimation method. Finally, the molecular connectivity does not appear to provide quantitative improvements.

PHASE EQUILIBRIA

Phase equilibria is of key importance to the chemical process industries. Design of diffusional separations, such as distillation and extraction, requires quantitative knowledge of the phase equilibria of fluid mixtures. Rather than attempt to review this broad subject, we focus our discussion on prediction of the solubility of organic chemicals in water. The fate of organic chemicals in the environment strongly depends on solubility in water. Chemicals with high solubility tend to have relatively low adsorption coefficients for soils and bioconcentration factors in aquatic life. Also, they tend to be more biodegradable by microorganisms (Lyman, Reehl and Rosenblatt, 1982).

Estimation of the solubility of organic chemicals in water can be divided into two broad categories: statistical correlations (e.g., linear free energy methods) and molecular-thermodynamic models (e.g., activity coefficient models and equations of state). Statistical correlations offer a high degree of simplicity while molecular thermodynamic models require more involved computations. Conversely, molecular-thermodynamic models offer the hope of higher accuracy in interpolation and extrapolation. Theoretical understanding of the molecular thermodynamics of liquid mixtures is still quite limited. The few available molecular thermodynamic correlations are, at best, semi-empirical. Methods in each category are discussed below.

The method of Irmann (1965) provides a useful estimation method for the solubility of hydrocarbons and halohydrocarbons in water based on structural information. The solubility of an organic liquid at 25°C is given by:

$$-\log W = A + \Sigma B_i N_i + \Sigma C_j N_j \qquad (9)$$

A is dependent on the compound type. B accounts for contributions of the various atom types. C accounts for various structural elements. N represents the frequencies of the contributions from atom types or structural elements, respectively in the molecule. For a material that is solid at 25°C, Irmann suggests:

$$-\log W_{solid} = -\log W + 0.0095\ (T_M - 25) \qquad (10)$$

The 0.0095 factor is based on an assumed entropy of fusion of 13 cal/gmole-°C.

Irmann derived the constants from a database of 200 hydrocarbons and halohydrocarbons. Estimates for nearly 90% of the compounds were within 15% of experimental values. The largest deviation was a factor of 1.6. Not surprisingly, these results demonstrate the applicability of a simple linear-free-energy group-contribution method to estimation of a property (low-dilution activity coefficients in this case) for a family of similar compounds.

TABLE IV: Comparison Between UNIFAC Predicted and "Experimental" Infinite Dilution Activity Coefficients for Higher Alcohols in Water at 25°C

Component	x' (Aqueous Phase)	x" (Alcohol Phase)	(x"/x')*	γ^{∞}(UNIFAC)
2-Methyl-1-Butanol	.006217	.6688	107.6	158.3
3-Methyl-1-Butanol	.00540	.6566	121.6	158.3
3-Methyl-2-Butanol	.01187	.6014	50.66	158.5
1-Pentanol	.00375	.6580	175.5	158.2
2-Pentanol	.00945	.6050	64.02	158.3
3-Pentanol	.01097	.6925	63.12	158.3
2,2-Dimethyl-1-Propanol	.007358	.6914	93.96	150.7
2,2-Dimethyl-1-Butanol	.001348	.90631	672.3	457.8
2,3-Dimethyl-2-Butanol	.007632	.5909	77.42	458.1
3,3-Dimethyl-2-Butanol	.004372	.6916	158.2	458.1
1-Hexanol	.00104	.7100	682.7	482.3
2-Hexanol	.002443	.7132	291.9	482.3
3-Hexanol	.002877	.7754	269.5	482.3
2-Methyl-2-Pentanol	.005869	.6121	104.3	457.7
3-Methyl-2-Pentanol	.003476	.7112	204.6	483.3
4-Methyl-2-Pentanol	.00286	.7650	267.5	483.3
2-Methyl-3-Pentanol	.003603	.7660	212.6	483.3
3-Methyl-3-Pentanol	.007784	.6040	77.6	457.7
1-Heptanol	.000264	.75	2840	1480
2-Methyl-2-Hexanol	.001516	.6957	458.9	1400
3-Methyl-3-Hexanol	.001863	.7367	395.4	1401
2,3-Dimethyl-2-Pentanol	.002419	.6971	288.2	1402
2,4-Dimethyl-2-Pentanol	.002101	.6897	328.3	1402
2,2-Dimethyl-3-Pentanol	.001280	.8308	649.1	1402
2,3-Dimethyl-3-Pentanol	.002578	.7128	276.5	1402
2,4-Dimethyl-3-Pentanol	.001092	.8187	749.7	1485
3-Ethyl-3-Pentanol	.002642	.7172	271.5	1400
1-Octanol	.0000703	.793	11280	4562
2,2,3-Trimethyl-3-Pentanol	.0009602	.8709	907.0	4087
1-Nonanol	.0000162	.736	45432	14109

* - x"/x' is approximately the experimental infinite dilution activity coefficient since the solubility of alcohol in the aqueous phase is very low and the alcohol activity coefficient in the alcohol phase is near unity. The actual alcohol infinite dilution activity coefficient will be up to 10% higher than x"/x'.

Kamlet, Taft and co-workers (1985) have applied linear-solvation energy relationships (LSER) to a number of applications. The LSER method loosely parameterizes solvent and solute interactions by dipolarity/polarizability, hydrogen bond donor/acceptor strength and molecular size. Linear correlations between properties of dilute solutes in solvents are developed. An example of an LSER application is prediction of the solubility of a variety of solutes in water.

Thomas and Eckert (1984) proposed a model for prediction of infinite-dilution activity coefficients based on an extension of regular solution theory to polar and association systems. The model is based on the assumption that the contributions from dispersion, induction, orientation and hydrogen-bonding forces to the cohesive energy density are additive.

The equation for infinite dilution activity coefficients is

$$\ln\gamma_2^\infty = v_2/(RT)((\lambda_1-\lambda_2)^2 + q_1^2 q_2^2(\tau_1-\tau_2)^2/\tau_1+(\alpha_1-\alpha_2)(\beta_1-\beta_2)/\varepsilon_1 + d_{12}) \qquad (11)$$

where λ is the dispersion parameter (based on the refractive index), q is the induction parameter, τ is the polar parameter, α is the acidity parameter, β is the basicity parameter and d_{12} is determined from a Flory-Huggins combinatorial term. The model applies to a large number of binary mixtures. An average error of 9.1% (with few errors larger than 30%) resulted for prediction of 3357 infinite dilution activity coefficients. Unlike the method of Irmann, the parameters of the Thomas and Eckert model have a physical interpretation. Unfortunately, predictions are poor for systems where steric interactions are important, aqueous systems and systems with activity coefficients larger than about 100.

The UNIFAC group-contribution activity-coefficient model (Fredenslund et al., 1975) predicts activity coefficients from structural information and group interaction parameters. Unlike the previously discussed methods, UNIFAC is based on an excess free energy model for multicomponent solutions and can be applied to multicomponent mixtures over the entire composition range. This model has a wider range of applicability than the previously discussed methods and is in widespread use throughout industry. The expression for activity coefficients is summarized as

$$\ln\gamma_i = \ln\gamma_i \text{ (combinatorial)} + \ln\gamma_i \text{ (residual)} \qquad (12)$$

The combinatorial contribution is based on molecular van der Waals volumes and areas. The residual contribution is based on group interaction parameters obtained from regression of experimental phase equilibrium data.

Fredenslund and coworkers have continuously improved and extended the group interaction parameters (Skjold-Jorgensen et al., 1979; Magnussen, et al., 1981; and Tiegs et al., 1987). The model has essentially been used as simple group additivity method. In some cases, more complex interactions have been accounted for by defining and adding larger groups to the database (e.g., $CON(CH_2)_2$).

Magnussen et al. (1981) developed a UNIFAC group-interaction parameter table particularly suited to prediction of liquid-liquid equilibria. A total of 512 binary interaction parameters for 32 different groups were determined from an extensive liquid-liquid equilibrium data base. In general, predictions from the model are good in the range of 10°C to 40°C. Limitations do exist, however, arising from the phenomenological basis of the model. As an example, Magnussen discusses the "propanol problem" where the two propanols had to be treated differently from all of the other alcohols.

Gupte and Danner (1987) evaluated the liquid-liquid equilibrium predicted by UNIFAC (Magnussen parameter table) using a large data base. Of particular interest, ternary system predictions were evaluated for systems not used in the parameter determination, in addition to systems that were used. Gupte concluded that UNIFAC predicts liquid-liquid equilibrium for both sets of systems with similar accuracy. However, as Gupte reports, the error in the distribution ratio of the solute can be large (varying from 25% to 1600% in Gupte's evaluation). The solute distribution ratio is of key importance to the solubility of organic chemicals in water.

Molecular-thermodynamic models can predict phase equilibrium for a wide range of molecules and conditions while statistically-based linear-free energy relations are more limited to specific groups of molecules and a narrow range of conditions. However, linear-free energy methods can be more accurate in specific situations. As an example, we used UNIFAC and LSER (Kamlet and Taft) to predict solubilities of alcohols in water at 25°C. UNIFAC does not capture the subtle structural effects of the groups (i.e., primary, secondary, tertiary alcohol) and the accuracy of predictions varies widely (see Figure 13). LSER, specifically correlated to molar volume of alcohol and simple structural features, predicts solubilities to a higher degree of accuracy in this case (see Figure 14). The UNIFAC alcohol solubility predictions could be improved by differentiating between primary, secondary and tertiary alcohols.

The molecular thermodynamic approach is a good starting point from which to derive predictive models. We feel that more attention should be given to developing specific correlations based on simple properties (e.g. density, dipole moment and polarizability) and molecular structure within the molecular thermodynamic framework.

FIGURE 13

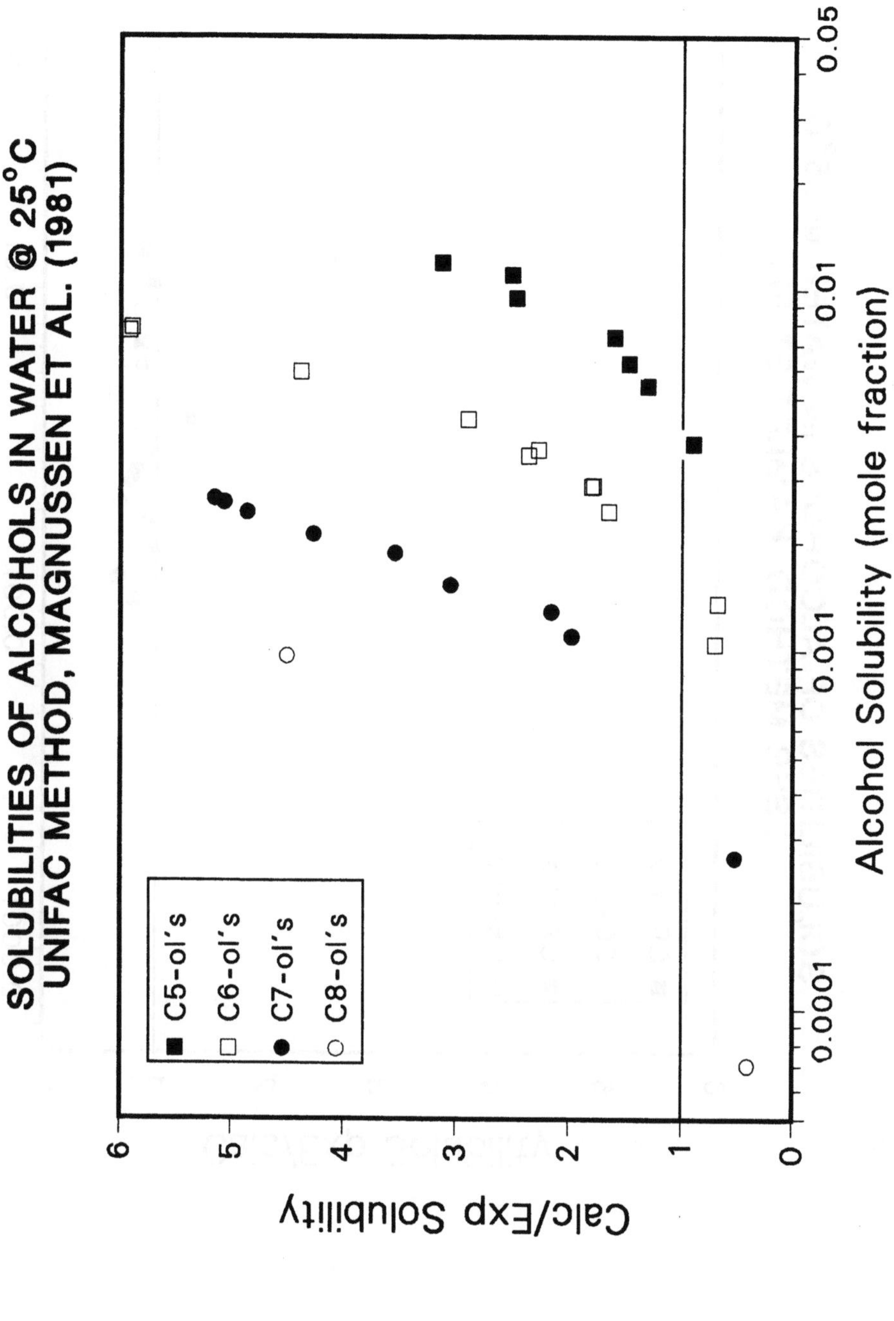
SOLUBILITIES OF ALCOHOLS IN WATER @ 25°C
UNIFAC METHOD, MAGNUSSEN ET AL. (1981)
C5-ol's
C6-ol's
C7-ol's
C8-ol's
Calc/Exp Solubility
0
1
2
3
4
5
6
0.0001
0.001
0.01
0.05
Alcohol Solubility (mole fraction)

FIGURE 14

SOLUBILITIES OF ALCOHOLS IN WATER @ 25°C
LSER METHOD, KAMLET ET AL.

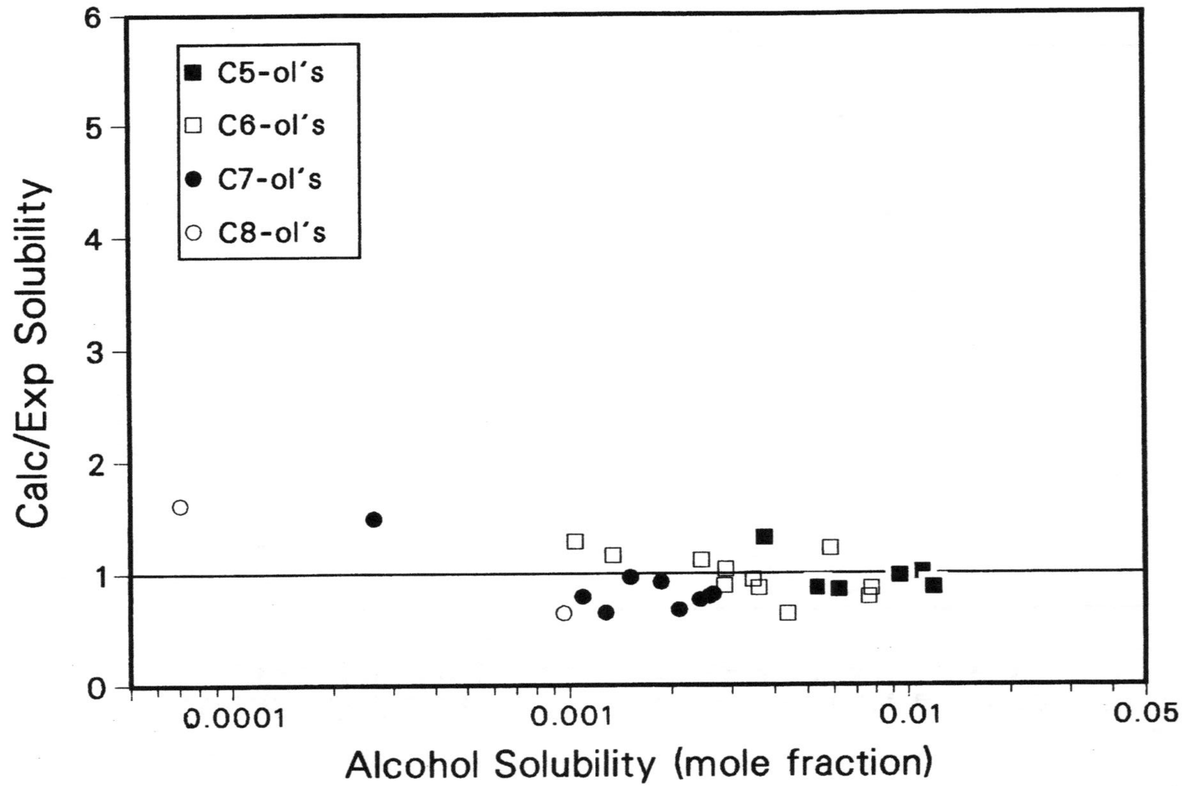

CONCLUSIONS

In conclusion, we offer the following comments:

- Simple group additivity can work surprisingly well and is the basis of most estimation methods.

- The simple additivity methods fail to capture the differences between similar molecules, e.g., those due to branching. These differences can be significant in many industrial applications.

- The molecular connectivity method, which was developed specifically to describe molecular branching, does not appear to provide a significantly improved quantitative correlation over the simple additivity methods.

- Second-order methods (e.g., Benson's correlation for ideal-gas heats of formation) provide improved predictions but, in our experience, are not significantly better than the simple-additivity methods. The corrections have tended to be empirical and new ideas based on physical arguments are needed.

- Group contribution methods can frequently be used to determine structural perturbations about a reference molecule to accurately estimate properties. This, of course, requires that the group contribution method is capable of distinguishing the structural perturbation.

- Theoretically-based methods (e.g, molecular mechanics and those based on quantum mechanics). will ultimately provide the basis for improved estimation methods. Theoretical methods are currently underdeveloped and require more attention, particularly in the chemical engineering literature.

REFERENCES

Abrams, D. S., Massaldi, H. A., and Prausnitz, J. M., "Vapor Pressures of Liquids as a Function of Temperature. Two Parameter Equation Based on Kinetic Theory of Fluids", Ind. Eng. Chem., Fundam., 13, 259-262(1974).

AIChE Design Institute for Physical Property Data Project 801, Data Compilation Manual (1987).

Benson, S. W., "Thermochemical Kinetics", Chap. 2 Wiley, New York, (1968).

Bondi, A., "Physical Properties of Molecular crystals, Liquids and Glasses", John Wiley & Sons, Inc., New York, 1968.

Burket, U. and Allinger, N. L., "Molecular Mechanics", ACS Monograph 177, American Chemical Society, Washington, DC (1982).

Boyd, D. B., and Lipkowitz, K. B., "Molecular Mechanics - The Method and its Underlying Philosophy", J. Chem. Education, 59, 269-274(1982).

Donohue, M. D., and Vimelchand, P., "The Perturbed-Hard-Chain Theory, Extensions and Applications", Fluid Phase Equilibria, 40, 185-211(1988).

Edwards, D. R., and Prausnitz, J. M., "Estimation of Vapor Pressures of Heavy Liquid Hydrocarbons Containing Nitrogen on Sulfur by a Group-Contribution Method", Ind. Eng. Chem. Fundam., 20, 280-283(1981).

Fredenslund, A., Jones, R. L., and Prausnitz, J. M., "Group Contribution Estimation of Activity Coefficients in Nonideal Mixtures", AIChE J., 21, 1086-1099(1975).

Gupte, P. A., and Danner, R. P., "Prediction of Liquid-Liquid Equilibria with UNIFAC: A Critical Evaluation", Ind. Eng. Chem. Res., 26, 2036-2042(1987).

Hansch, C., and Leo, A., "Substituent Constants for Correlation Analysis in Chemistry and Biology", Wiley, New York (1979).

Irmann, F., "Eine einfache Korrelation zwischen Wasserloslichkelt und Struktur von Kohlenwasserstoffen und Halogenkohlenwasserstoffen", Chem. Inq. Tech., 37, 789-798(1965).

Jensen, T., Fredunslund, A., and Rasmussen, P., "Pure-Component Vapor Pressures Using UNIFAC Group Contribution", Ind. Eng. Chem. Fundam., 20, 239-246(1981).

Joback, K. G., S. M. Thesis, Massachusetts Institute of Technology, Cambridge, Mass., June 1984.

Kamlet, M. J., Doherty, R. M., Abboud, J. L. M., Abraham, M. H., and Taft R. W., "Linear Solvation Energy Relationships. 36. An Amphihydrogen Bonding Parameter, W, Which Allows Correlation and Prediction of Solubilities of Aliphatic Alcohols and Other Solutes in Water", submitted to J. Am. Chem. Soc.

Kier, L. B., and Hall, L. H., "Molecular Connectivity in Structure Activity Analysis", Research Studies Press (a division of John Wiley and Sons), Letchworth, Hertfordshire, England (1985).

Lyman, W. J., Reehl, W. F., and Rosenblatt, D. H., "Handbook of Chemical Property Estimation Methods. Environmental Behavior of Orangic Compounds', McGraw-Hill Book Company, 1982.

Macknick, A. B., Winnick, J., and Prausnitz, J. M., "Vapor Pressures of Liquids as a Function of Temperature. Two-Parameter Equation Based on Kinetic Theory", Ind. Eng. Chem. Fundam., 16, 392(1977).

Magnussen, T., Rasmussen, P., and Fredenslund, A., "UNIFAC Parameter Table for Prediction of Liquid-Liquid Equilibria", Ind. Eng. Chem. Process Des. Dev., 20, 331-339(1981).

Miller, D. G., "Derivation of Two Equations for the Estimation of Vapor Pressures", J. Phys. Chem., 68, 1399-1408(1964).

Moelwyn-Hughes, E. A., "Physical Chemistry", 2nd Ed., Pergammon Press, Oxford, 1961.

Peneloux, A., and Rauzy, E., "A Consistent Correction for Redlich-Kwang-Soave Volumes", 8, 7-23(1982).

Randic, M., "On Characterization of Molecular Branching", J. Am. Chem. Soc., 97, 6609-15(1975).

Reid, R. C., Prausnitz, J. M., and Poling, B. E., "The Properties of Gases and Liquids", 4th Ed., McGraw-Hill Book Company (1987).

Seaton, W. H., Freedman, E., and Treweek, D. N., "CHETAH - The ASTM Chemical Thermodynamic and Energy Release Program", ASTM Data Series Publication DS 51, Philadelphia (1974).

Skjold-Jorgensen, S., Kolbe, B., Gmehling, J., and Rasmussen, P., "Vapor-Liquid Equilibria by UNIFAC Group Contribution - Revision and Extension", Ind. Eng. Chem. Process Des. Dev., 18, 714-722(1979).

Soave, G., "Improvements to the van der Waals Equation of State", Chem. Eng. Sci., 39, 357-369(1986).

Taft, R. W., Abboud, J. M., Kamlet, M. J., and Abraham, M. H., "Linear Solvation Energy Relationships", J. Sol. Chem., 14, 153-186(1985).

Tiegs, D., Gmehling, J., Rasmussen, P., and Fredenslund, P., "Vapor-Liquid Equilibria by UNIFAC Group Contribution. 4. Revision and Extension", Ind. Eng. Chem. Res., 26, 159-161(1987).

Thermodynamics Research Center, "Selected Values of Properties of Hydrocarbons," Data Project, Texas A&M University, College Station, Texas (1987a).

Thermodynamics Research Center, "Selected Values of Properties of Non-Hydrocarbons", Data Project, Texas A&M University, College Station, Texas (1987b).

Thomas, E. R., and Eckert, C. A., "Prediction of Limiting Activity Coefficients by a Modified Separation of Cohesive Energy Density Model and UNIFAC", Ind. Eng. Chem. Process Design Develop., 23, 194-209(1984).

Wu, H. S., and Sandler, S. I., "Proximity Effects of the UNIFAC Model - I. Ethers", submitted for publication, 1988.

Experience with the Development of a Group-Contribution Equation of State for the Prediction of Physical Properties for Process Engineering Purposes

Heiner W. Landeck and Hans F. Kistenmacher

LINDE AG, Division TVT Munich, 8023 Hoellriegelskreuth,
Federal Republic of Germany

Summary

The PFGC group-contribution equation of state as an example of a promising method to solve industrial application problems for complex mixtures containing non-polar, polar, associating, sub- and supercritical components, has been used to demonstrate the following aspects of general interest:

- The industrial requirements for improved physical properties prediction methods, and the range of molecular complexity of mixtures encountered in gas purification processes, are discussed.
- A general strategy for the preselection of physical property prediction methods, and especially those for testing equations of state or group-contribution equations of state, has been developed. The industrial boundary conditions which determine success or failure are given.
- A general multiproperty-multicomponent-multiphase strategy to be used for the fit of the parameters of equations of state or group-contribution equations is proposed.

For the PFGC group-contribution equation of state the investigations have the following specific results:

- The PFGC method has many fundamental weaknesses.
- The original PFGC method does not work for industrial applications with sufficient accuracy.

From an industrial point of view, a group-contribution equation of state is still required for all types of molecular interactions and all areas of applications.

C. Jochum M. G. Hicks J. Sunkel (Eds.)
Physical Property Prediction
in Organic Chemistry
© Springer-Verlag Berlin Heidelberg 1988

Introduction and Scope

Thermophysical properties are required for the design and optimization of thermal separation processes.

For an accurate design usually very complicated multicomponent mixtures have to be handled in addition to pure components. The various components may belong to a wide spectrum of organic and inorganic classes of components. The regions of components, temperature, and pressure may also be very wide. The components may be sub- or supercritical.

At the present time, there is no unique predition method available that is able to describe all the physical properties required over the very wide regions of interest with sufficient accuracy /1,2/.

In practice it is therefore necessary to locally fit the parameters of suitable literature methods to measurements in the specific regions of interest.

Equations of state have the advantage that not only phase equilibria, but also thermal properties and densities may be calculated for mixtures as well as for pure components, and for sub- and supercritical systems. Furthermore, these properties are thermodynamically consistent /1/.

The effort required to develop a prediction method suitable for a particular application can be minimized by the use of "group-contribution" methods in which the properties of a pure component or a mixture are predicted from the parameters of a limited number of structural groups /3,4,5/.

The PFGC (*P*arameters *f*rom *G*roup-*C*ontribution) group-contribution equation of state developed by Cunningham and Wilson /6/, modified, extended /7/ and improved /8/ by Erbar and co-workers appeared at the beginning of our investigations in 1979, as the most promising method from an industrial point of view to handle complex mixtures of the types mentioned above.

For the PFGC group-contribution method the following points have been studied in detail:

- The definition and discussion of industrial requirements for improved physical properties prediction methods, the thermodynamic complexity of the applications resulting from the different intermolecular interactions.
- The industrial boundary conditions for the development of calculation methods with indications of the factors which determine success or failure.
- The preselection of the most promising method and the strategy for testing the method and its improvements.
- The strategy to be used to fit the various parameters of multiparameter equations of state or group-contribution equations of state using multiproperty-multicomponent-multiphase regression.

Industrial Requirements

The requirements for the physical properties prediction methods considered in this work are those occuring in an engineering company. As typical examples gas treating processes are studied in detail.

Physical Properties Required and their Importance

Table 1 shows the physical properties which are necessary for the design of thermal separation processes and their relative importance for the design of different unit operations /9/ or for safety analyses.

unit operations or areas of application	densities PVT	phase equil. chem. phys.	caloric properties enth/entr.	cp	transport prop.	reaction rates	other prop. e.g. dielect
fractionators	b	a	b	d	b	s	s
extractors	b	a	c	d	b	s	s
flash drums	b	a	a	–	b	s	s
heat exchangers	c	a	a	c	b	s	s
compressors	a	a	a	a	d	s	s
reactors	c	b	b	b	b	a	s
safety	b	a	a	–	c	a	s

a most important c important s important in special cases
b very important d less important

Beilstein Workshop May 88	Industrial Requirements : Physical Properties required and their Importance	Table 1

Components and Mixtures of Interest

Typical components and mixtures of industrial interest, e.g. for gas purification processes, are shown in table 2. These mixtures contain non-polar, polar, associating, sub- and supercritical components. Furthermore on a molar basis, supercritical components may constitute substantial portions of the liquids under some process conditions /10, 11/.

Industrial Requirements :

Pure Components and Mixtures

Examples of typical mixtures encountered in industrial processes

1. **Components and liquid concentrations (mol-%) at the bottom of a RECTISOL wash column:**

	H_2	N_2	$CO/Ar/CH_4$	CO_2	H_2S	COS	CH_3OH	H_2O
%	.9	.01	.28	33.3	1.4	.01	63.9	.2

Operating conditions: 223 < T/K < 383, 20 < P/bar < 80

2. **Components and liquid concentrations (mol-%) at the bottom of a poly-glycol ether wash column**

	CH_4	C_2H_6	CO_2	H_2S	COS	RSH	H_2O	PGE
%	18.	.18	8.	18.7	.01	.001	19.	35.1

Operating conditions: 270 < T/K < 390, 20 < p/bar < 80

3. **Components of a glycol dehydration plant**

N_2, CO_2, H_2S, CH_4, C_2H_6 to $C_{12}H_{26}$, EG, DEG, TEG, H_2O

Beilstein Workshop May 88		Table 2

Molecular Thermodynamic Complexity

It is necessary to extend the picture of Donohue and Prausnitz from PHCT /12/ to a third dimension in addition to molecular size and density. As a third dimension the complexity of the molecular interactions must be taken into consideration to illustrate the actual molecular thermodynamic situation. Figure 1 also shows the molecular thermodynamic regions of interest for this work.

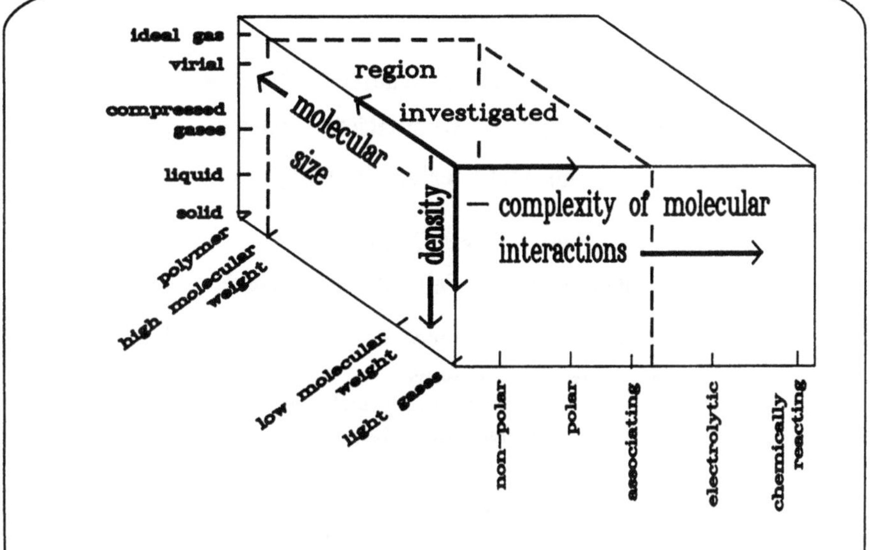

Beilstein Workshop May 88	Thermodynamic Complexity: Molecular Situation	Figure 1

Industrial Boundery Conditions

The boundary conditions for the development and implementation of physical property prediction methods differs significantly in industrial and university research. Table 3 summarize the critical success factors from an industrial point of view.

It is useful to give the following additional explantations to table 3:

When low accuracy is necessary, very rough estimation methods can be used. For final process design and for guarantees the methods used must be very accurate. E.g. in gas scrubbing processes the size of a plant is directly proportional to the K-value of the key component to be scrubbed; furthermore the compression work depends on the K-values of the coabsorbed gases. Thus for an incurracy of 10 percents in the size of the equipment, only 10 percent inaccuracy in the K-values can be tolerated. For other processes reliable estimations on the accuracy for physical estimation methods are reported by Streich and Kistenmacher /13/.

Industrial Boundary Conditions for the Development of Calculation Methods in an Engineering Company

Critical Success Factors for Physical Property Calculation Methods

*** available as fast as possible

*** as accurate as necessary
- low for feasabilitiy studies and control strategy design
- high for final design and for guarantees

*** as cheap as possible
considering: costs of experiments
software costs
CPU-costs

*** as widely applicable as possible

with respect to: T, P, components – traces

*** as consistent as necessary (properties)

Beilstein Workshop May 88	Industrial Boundery Conditions : Critical Success Factors	Table 3

Preselection and Strategy for Testing the most Promising Method

Preselection and Results

When the investigation was started in 1979 in a preliminary evaluation the following methods were taken into consideration:

- Extending the UNIFAC/UNIQUAC activity coefficient models /1a, 5/ to supercriticals, analogous to the Chao-Seader concept /14/. This possibility failed due to insufficient accuracy /15/.
- The DCFI method of Mathias and O'Connell /16/ based on a Henry coefficient approach failed because problems arose with the transition from the subcritical to the supercritical state. Furthermore the concept has not been tested for multicomponent solutes and large contents of supercritical components in the liquid phase.
- The widely used cubic equations of state /17,18,19/ failed also because of insufficient accuracy.
- The PFGC group-contribution of state developed by Cunningham and Wilson /6/, modified and extended by Erbar et. al. (PFGC-MES) /7/ appeared according the published results and test calculations for very complex systems to be the most promising method to meet the requirements outlined above.

The figures 2 summarizes the advantages of PFGC in contrast to UNIQUAC and SRK for the prediction of the complex ternary systems poly-glycol ether - water - CH_4 or CO_2 from binary information. Since the needed accuracy is in the order of 10 percents in the K-values, the diagram shows acceptable performance only for the PFGC method.

Table 4 shows that our industrial boundary conditions are fullfilled in principle by PFGC. Therefore PFGC was tested in detail.

The derivation of PFGC has been described by Cunningham /6a/ in detail; the modifications are reported by Erbar and co-workers /7/.

PFGC uses a Flory-Huggins approach as repulsive part /1a/ and a simplified Wilson mixing rule /20/ for the attractive term. The parameter requirements and the basic equations of PFGC are summarized in the appendices.

Result of Preselection (1979) :

Group Contribution Equation of State PFGC (Parameters from Group Contributions):

- developed by Cunningham and Wilson in 1974
- modified and extended by Erbar et. al. in 1979-86

*** handles high molecular and polar species because of its group contribution concept

*** handles mixtures of sub- and supercritical components (in contrast to activity coefficient models like UNIQUAC/UNIFAC)

*** handles components which in a single process are both sub- and supercrital (difficulties with the use of Henry's constants)

*** handles complex mixtures in contrast to widely used cubic equations of state

*** handles the properties fugacity, enthalpy, entropy, specific heat and density in a thermodynamically consistent way

Beilstein Workshop May 88	Preselection of Methods : Find most promising Method	Table 4

Figures 2
Predictions of gas solubilities in ternary mixtures from binary information according to different calculation methods

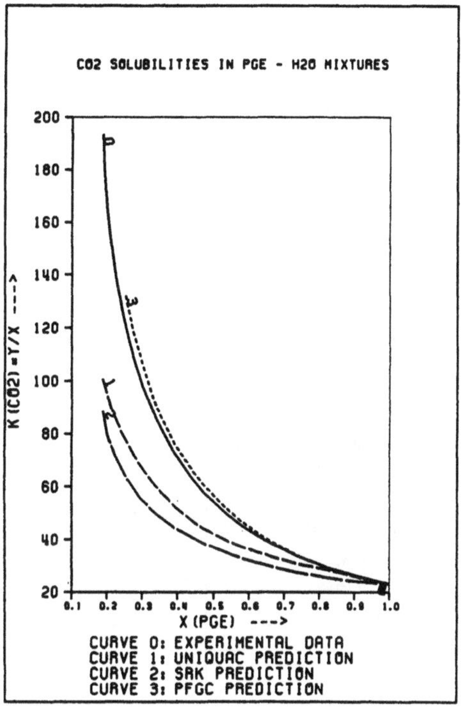

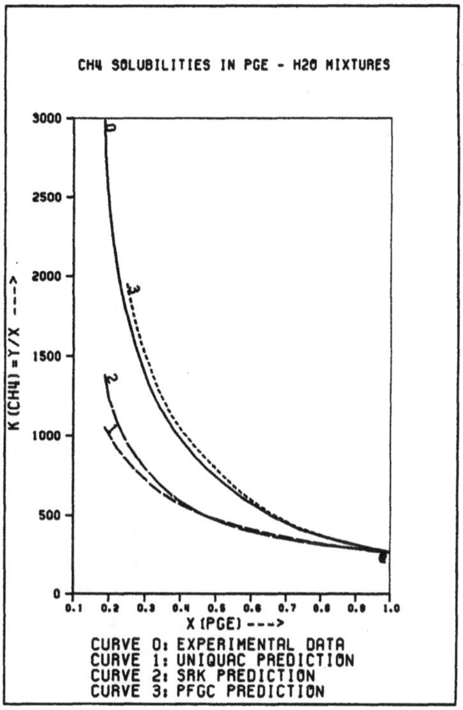

Strategy for a Systematic Test of PFGC

The systematic test of the original PFGC equation of state was organized into the following three steps:

- Test of the prediction accuracy without resorting to the group concept.
- Test of the group concept.
- If necessary: Improvements by means of parameter regression and analytical investigations.

The first step should demonstrate the principal advantages and weaknesses of PFGC when groups are equal to molecules. The different tests, which are nessecary in this step and the results for PFGC, are summarized in part I of table 5.

In figure 3 the predictions of PFGC-MES /7/ have been compared with experimental /21/ data for methanol. This figure illustrates the systematic deviations of PFGC for all components at temperatures below a reduced temperature Tr = 0.7 for vapour pressures, heats of vapourization, specific heats and densities.

In industrial applications the region below Tr = 0.7 is important especially for solvents. Therefore all parameters of PFGC-MES cannot be used for industrial applications and a refit is necessary.

Testing most promising Method

Experience with PFGC:
Critical Failure Factors

I. Test of eos without group concept

1. **Properties of pure components at saturation**
   P, Hv, dens at Tr > .7 (fit PFGC-MES) ok
   nessecary to refit for Tr < .7, Cp - not ok
2. **Critical region**
   errors in Tc < 5 %, Pc < 20 %, Vc < 20 % (ok)
3. **Desc. and prediction VLE of complex mixtures**
   PGE-H2O-CH4 or CO2, CH3OH-H2O-gases ok
4. **Multiphase predictions**
   method is able to predict VLLE ok
5. **Binary default value prediction**
   shows physical unrealistic behaviour - not ok

II. Test of group concept

1. **Properties pure components of homologous series**
   hc ok for Tr>.7, glycols present some problems(ok)
2. **Mixtures containing very different groups**
   many good results, problems with glycol-H2O (ok)

Beilstein Workshop May 88		Table 5

In the second step the group concept of PFGC has been tested. The results are summarized in part II of table 5.

The tests of the group concept of PFGC showed no severe problems.

As a consequence of the problems mentioned, a third step was undertaken in order to remove these problems. The results are given in table 6.

No satisfactory solution was found for a generally applicable method. As a consequence it was concluded, that PFGC can only be applied with sufficient accuracy in a very limited area of applications.

Improvements to PFGC by means of Parameter Regression and Analytical Investigations

I. New regression of eos parameters

1. Development of a multicomponent-multiproperty-multiphase regression program which may also be used for other methods
2. Refit of the PFGC parameters

Results :

a. PFGC cannot describe Cp behaviour accurately
b. PFGC needs temperature dependant binary parameters

II. Analytical investigations

1. The temperatur dependency of the energy parameters is incorrect at low temperatures :
   - Problems with Cp
   - No liquid density solution at low temperatures
2. The combinational rules must be improved

Beilstein Workshop May 88	Improvements to PFGC and Pricipal Weaknesses	Table 6

Figure 3
Systematic deviations of PFGC-MES predictions (dashed lines) and experimental data (solid lines) for methanol as an example

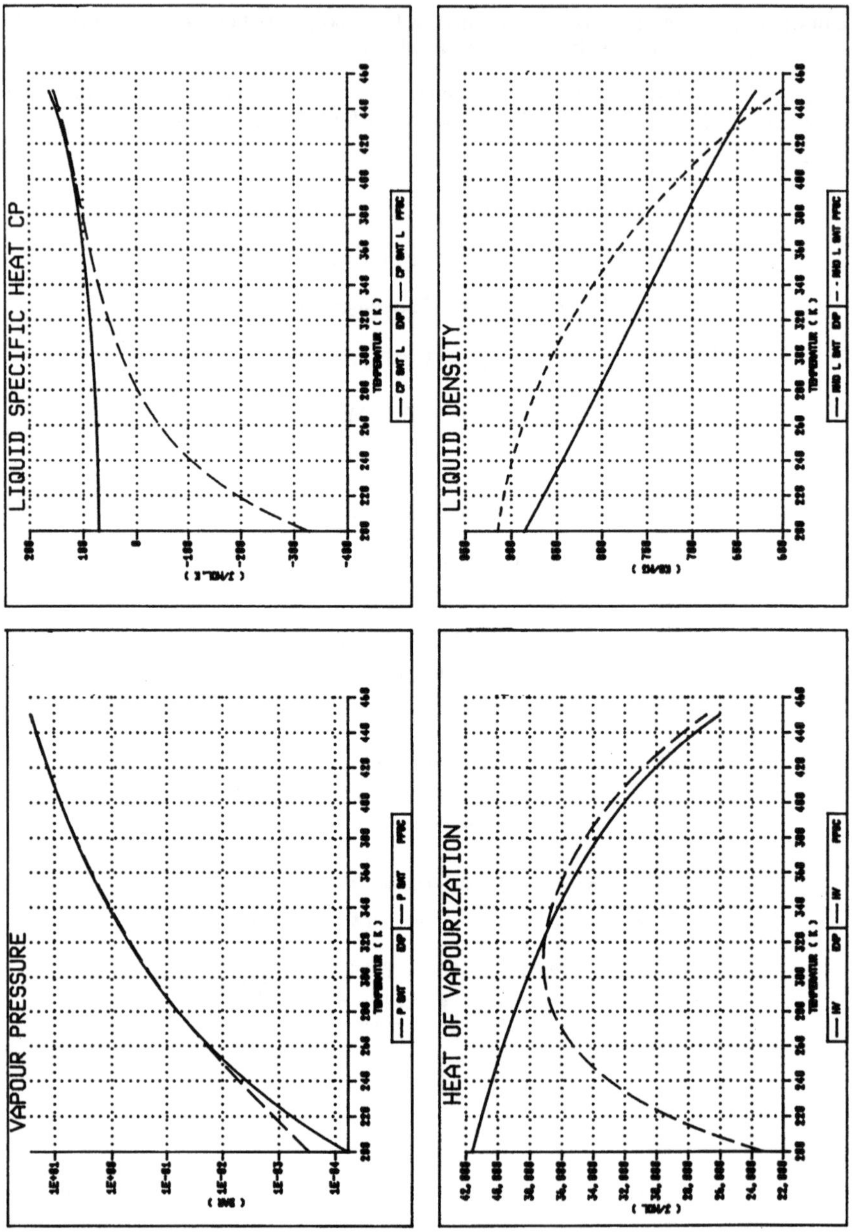

Figure 4
Fundamental weaknesses of the PFGC method:
Impossiblity to fit liquid specific heats.

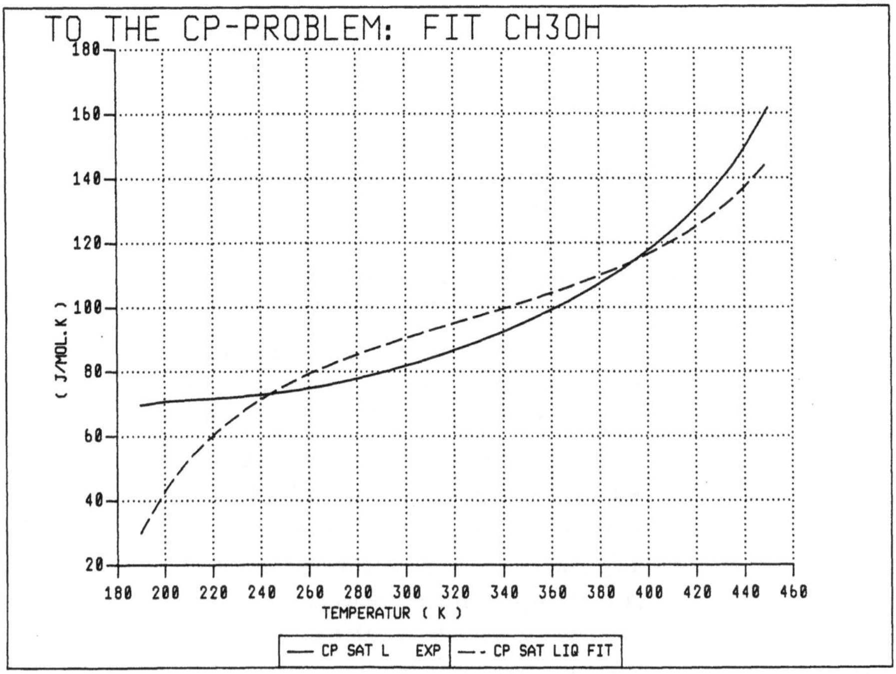

Strategy for Fitting EoS and GC-EoS Parameters

The typical problems encountered with the fit of parameters are given in table 7.

In order to fit the parameters for equations of state or for group-contribution equations of state the strategy described in table 8 was very successful.

Fitting EoS and GC–EoS Parameters

I. Situation

1. **Most eos or gc–eos parameters have no physical significance.**
   They are only fit parameters (e.g. all PFGC parameters).
2. **In a multiparameter eos most of the parameters are strongly correlated.**
3. **The weaker the correlation of the parameters is, the more characteristic for a group or compound the parameters are.**
4. **The advantages of an eos are exploited to the best, when all thermodynamically consistent properties can be predicted with the chosen eos.**
5. **Many examples have demonstrated, that a good fit of the vapour pressure results in an inadequate representation of other physical properties which involve derivatives**
   e. g. Hv, Cp.

Beilstein Workshop May 88		Table 7

Fitting EoS and GC–EoS Parameters

II. Strategy used :
Multiproperty–multicomponent and if necessary multiphase regression

1. In order to minimize the correlations of the parameters it is desirable to fit P and Cp, and perhaps also Hv and density, simultaneously.
2. The validity of an eos is determined by the temperature range of the fit and not the pressure. Best fit for components is in the temperature range $T_m < T < T_c$.
3. To get characteristic group parameters it is necessary to choose different components and their mixtures from an homologous series.
4. Fit infinite dilution data for binaries.
5. Fit algorithm is a weighted least–squares method. Because of weeknesses in all models a maximum–likelihood /22/ regression seems unnecessary.

Beilstein Workshop May 88	Fitting EoS and GC–EoS Parameters	Table 8

Conclusions and Perspectives

If the industrial requirements for a global applicable prediction method for complex mixtures have to be fullfilled, all investigated methods are insufficient.

Therefore at present only a complex and carefully selected test and fit strategy such as the one described above for a limited area of applications has to be applied.

Other group-contribution equations of state have been developed /23, 24, 25, 26/; they seem to be promising but have not yet tested for the industrial application problems discussed above.

Group-contribution methods are still of much interest for industrial applications. However, much work remains to be done to remove the weaknesses of the methods studied above.

Appendix 1

Table 9 Basic Data Requirements for the PFGC Group Contribution Equation of State

1. Component dependant data	
number of components in the mixture	nc
number of groups in the mixture	ng
matrix giving the frequency of groups of kind k in molecule i	$\tilde{\eta}_{ik}$ $i = 1, nc$ $k = 1, ng$
2. Group dependant data (empirical fit parameters)	
mole volume of group k in $m^3/kmol$	b_k^*
degree of freedom parameter of group k	s_k^*
specific energy parameters of group k in K, K^2, K^3	$(e_{k1}^*, e_{k2}^*, e_{k3}^*)$
binary interaction parameter of groups k and n or binary default value	$U_{k,n}^*$ $U_{k,n}^{default} = 1$

Appendix 2

Table 10 Basic Equations

1. Combination and mixing rules

(1) $b_i = \sum_k^{ng} \tilde{\eta}_{ik}\, b_k^*$ (2) $b = \sum_i^{nc} x_i\, b_i$

(3) $s_i = \sum_k^{ng} \tilde{\eta}_{ik}\, s_k^*$ (4) $s = \sum_i^{nc} x_i\, s_i$

(5) $h_k^* = \sum_{j=1}^{3} e_{kj}^*\; T^{1-j}$

(6) $d_{kn}^* = U_{kn}^* \cdot \dfrac{h_k^* + h_n^*}{2}$

2. Other principle equations, temperature derivatives and definitions

(7) $W_{kn}^* = \exp\,(-d_{kn}^*/T)$

(8) $\partial_T\, W_{kn}^* = \dfrac{W_{kn}^*\, U_{kn}^*}{2} \cdot \sum_{j=1}^{3} j \cdot (e_{kj}^* + e_{nj}^*) \cdot T^{-(j+1)}$

(9) $\partial_{T^2}^2\, W_{kn}^* = \dfrac{(\partial_T\, W_{kn}^*)^2}{W_{kn}^*} - \dfrac{W_{kn}^*\, U_{kn}^*}{2} \cdot \sum_{j=1}^{3} j(j+1)\; (e_{kj}^* + e_{nj}^*) \cdot T^{-(j+2)}$

(10) $\tilde{\beta}_{ik} = \tilde{\eta}_{ik}\, b_k^*$ (11) $q_k^* = \sum_i^{nc} \tilde{\beta}_{ik}\, x_i$

(11) $a_k^* = b - \sum_n^{ng} q_n^*\; W_{nk}^*$ (12) $\tilde{\alpha}_{ik} = \sum_n^{ng} \tilde{\beta}_{in}\, (1 - W_{nk}^*)$

3. Universal method constant

$c = 192.21528$

The PFGC Group Contribution Equation of State

$$(13)\quad \frac{P}{RT} = \frac{1-s}{v} - \frac{s}{b}\ln\left(1 - \frac{b}{v}\right) + \frac{c}{v}\sum_{k}^{ng} q_k^* \frac{a_k^*}{v-a_k^*}$$

Table 11 Expressions for the Physical Properties

Compressibility factors

$$Z = \frac{P\cdot v}{RT} = (1-s) - s\cdot\frac{v}{b}\cdot\ln\left(1-\frac{b}{v}\right) + c\cdot\sum_{k}^{ng} q_k^*\cdot\frac{a_k^*}{v-a_k^*}$$

Fugacity coefficients

$$\ln\phi_i = s_i\left[1 + \left(\frac{v}{b} - 1\right)\cdot\ln\left(1 - \frac{b}{v}\right)\right] - s\,\frac{b_i}{b}\left[1 + \frac{v}{b}\ln\left(1 - \frac{b}{v}\right)\right]\ldots$$

$$\ldots\quad + c\cdot\sum_{k}^{ng}\left[\tilde{\beta}_{ik}\cdot\ln\left(\frac{v}{v-a_k^*}\right) + q_k^*\cdot\frac{\tilde{\alpha}_{ik}}{v-a_k^*}\right] - \ln Z$$

Enthalpy departures

$$\Delta H = RT\left[(Z-1) - c\cdot T\cdot\sum_{k}^{ng} q_k^*\frac{\partial_T a_k^*}{v-a_k^*}\right]$$

Entropy departures

$$\Delta S = R\left\{s\left[\left(1 - \frac{v}{b}\right)\cdot\ln\left(1 - \frac{b}{v}\right) - 1\right]\ldots\right.$$

$$\left.\ldots + c\left[-b\cdot\ln v + \sum_{k}^{ng} q_k^*\left(\ln(v-a_k^*) - T\cdot\frac{\partial_T a_k^*}{v-a_k^*}\right)\right] + \ln Z\right\}$$

Table 11

Departures of specific heats

$$\Delta c_v = - c \cdot R \cdot \sum_k^{ng} g_k^* \left[\frac{T}{v-a_k^*} \left(2 \cdot \partial_T a_k^* + T \cdot \partial_{T^2}^2 a_k^*\right) - T \cdot \left(\frac{\partial_T a_k^*}{v-a_k^*}\right)^2 \right] - R$$

$$\left(\frac{\partial P}{\partial T}\right)_{v,N} = \frac{P}{T} + c \cdot RT \sum_k^{ng} g_k^* \frac{\partial_T a_k^*}{(v-a_k^*)^2}$$

$$\left(\frac{\partial P}{\partial V}\right)_{T,N} = RT \left\{ \frac{s-1}{v^2} - \frac{s}{v(v-b)} + c \cdot \left[\frac{b}{v^2} - \sum_k^{ng} \frac{g_k^*}{(v-a_k^*)^2} \right] \right\}$$

$$\Delta c_p = \Delta c_v - T \left(\frac{\partial P}{\partial T}\right)_{v,N}^2 / \left(\frac{\partial P}{\partial V}\right)_{T,N}$$

References

1.a. Prausnitz J.M., R.N. Lichtenthaler, E.G. de Azevedo: "Molecular Thermodynamics of Fluid-Phase Equilibria", Prentice-Hall, Englewood Cliffs, N.J. (1986)
b. Walas S.M.: "Phase Equilibria in Chemical Engineering", Butterworth, Boston (1985)
c. Van Ness H.C., M.M. Abbott, "Classical Thermodynamics of Nonelelectrolyte Solutions", McGraw-Hill Co., New York (1982)
2.a. Reid R.C., Prausnitz J.M., B.E. Poling: "The Properties of Gases and Liquids", McGraw-Hill Co., New York (1987)
b. Danner R.P., T.E. Daubert: "AIChE DIPPR Chemical Properties Prediction Manual", AIChE, New York (1983), revised and extended version (1988)
3. Pierotti G.J., C.H. Deal, E.L. Derr: Ind. Eng. Chem. 51, 95 (1959)
4.a. Derr E.L., C.H. Deal: Inst. Chem. Eng. Symp. Ser. London 3(32), 40 (1969)
b. Kojima K., K.Tochigi: "Prediction of Vaor-Liquid Equilibria by the ASOG Method", Elsevier, Amsterdam (1979)
5.a. Fredenslund A., R.L. Jones, J.M. Prausnitz: AIChE J. 21, 1086 (1975)
b. Fredenslund A., J. Gmehling, P. Rasmussen: "Vapor-Liquid Equilibria using UNIFAC", Elsevier, Amsterdam (1977)
c. Gmehling J., P. Rasmussen, A. Fredenslund: Ind. Eng. Chem. Process Des. Dev. 21, 118 (1982)
d. Macedo E.A., U. Weidlich, J. Gmehling, P. Rasmussen: Ind. Eng. Chem. Process Des. Dev. 26, 676 (1983)
e. Tiegs D., P. Rasmussen, J. Gmehling, A. Fredenslund: Ind. Eng. Chem. Res. 26, 159 (1987)
f. Fredenslund A., P. Rasmussen: Fluid Phase Eqilibria 24, 115 (1985)
g. Larsen B.L.: PhD Thesis, Instituttet for Kemiteknik, Danmarks Tekniske Hojskole, Lyngby/DK (1986)
6.a. Cunningham J.R.: MSc. Thesis, Bringham Young University, Provo, UT (1974)
b. Cunningham J.R., G.M. Wilson: Paper presented at GPA Meeting, Denver/CO (1974)
7.a. Moshfeghian A., A. Shariat, J.H. Erbar: Paper presented at AIChE Natl. Meet., Houston/TX (1979)
b. Moshfeghian A., A. Shariat, J.H. Erbar: ACS Symposium Series 133, 333 (1980)
8.a. Majeed A.I.: PhD Thesis, Oclahoma State University, Stillwater/OK (1983)
b. Majeed A.I., J. Wagner: ACS Symposium Series 300, 452 (1986)
c. Wagner J., R.C. Erbar, A.I. Majeed: Proceedings of the 62th Annual GPA Convention, San Francisco/CA, March 14-16, 65 (1983)

9. King C.J.: "Separation Processes", McGraw-Hill Co., New York (1980)
10. Kohl A.L., F.C. Riesenfeld: "Gas Purification", Gulf Publishing Co. Houston/TX (1985)
11. Maddox R.N.: "Gas Conditioning and Processing", Vol. IV, Campbell Petroleum Series, Norman/OK (1982)
12. Donohue M.D., J.M. Prausnitz: AIChE J. 24, 848 (1978)
13. Streich M., H. Kistenmacher: Hydrocarbon Processing, 237, May 1979
14.a. Prausnitz J.M., F.H. Shair: AIChE J 7, 682 (1961)
b. Chao K.C., G.D. Seader: AIChE J. 7, 598 (1961)
15. Prausnitz J.M. : EFCE Publication Series No. 11, 231 (1980)
16.a. Mathias P.M., J.P. O'Connell : Advances in Chemistry Series 182, 97 (1979)
b. O'Connell J.P.: EFCE Publication Series No. 11, 445 (1980)
c. Mathias P.M., J.P. O'Connell : Chem. Eng. Sci. 36, 1123 (1981)
17.a. Soave G.: Chem. Eng. Sci. 27, 1197 (1972)
b. Graboski M.S., T.E. Daubert: Ind. Eng. Chem. Process Des. Dev. 17, 443 and 448 (1978)
18.a. Peng D.-Y., D.B. Robinson: Ind. Eng. Chem. Fundam. 15, 59 (1976)
b. Peng D.-Y., D.B. Robinson: AIChE J. 23, 137 (1977)
19. Schmidt G., H. Wenzel: Chem. Eng. Sci. 35, 1503 (1980)
20. Wilson G.M.: J.Am.Chem.Soc. 86, 127 (1964)
21. Goodwin R.D.: J.Phys.Chem.Ref.Data 16, 799 (1987)
22.a Abbott M.M., H.C. Van Ness: AIChE J. 21, 62 (1975)
b. Anderson T.F., D.S.Abrams, E.A. Grens, Prausnitz J.M.: Paper presented at 69th Annual AIEchE Meeting, Chicago/IL (1976)
c. Anderson T.F., D.S. Abrams, E.A. Grens II: AIChE J. 24, 20 (1978)
d. Kemeny S., J. Manczinger, S. Skjold-Jorgensen, K. Toth: AIChE J. 28, 20 (1978)
e. Skjold-Jorgensen S.: Fluid Phase Equilibria 14, 273 (1983)
23.a. Skjold-Jorgensen S.: Fluid Phase Equilibria 16, 317 (1984)
b. Skjold-Jorgensen S.: Ind. Eng. Chem. Res. 27, 110 (1988)
24. Schwartzentruber J., L. Ponce-Ramirez, H. Renon: Ind. Eng. Chem. Process Des. Dev 25, 804 (1986)
25.a. Gupte P.A., P. Pasmussen, A. Fredenslund: Fluid Phase Equilibria 29, 485 (1986)
b. Gupte P.A., P. Pasmussen, A. Fredenslund: Ind. Eng. Chem. Fundam. 25, 636 (1986)
26. Gani R., N. Tzouvaras, P. Pasmussen, A. Fredenslund: SEP-8810, Instituttet for Kemiteknik, Danmarks Tekniske Hojskole, Lyngby/DK (1988)

Prediction of Mixture Properties Using UNIFAC

Jürgen Gmehling

Universität Dortmund, Lehrstuhl für Technische Chemie B,
Fachbereich Chemietechnik, Postfach 50 05 00, 4600 Dortmund,
Federal Republic of Germany

ABSTRACT

In the last 20 years great progress has been achieved in the calculation and prediction of the real behavior of multicomponent systems using g^E-models or EOS and binary information alone.

When the required binary data are missing, the group contribution method UNIFAC can be applied to complete the available experimental information; however, this method also shows some weaknesses.

In this paper results for the UNIFAC method will be presented together with the improvements which are obtained when a modified form of the UNIFAC method and a larger data base are used.

INTRODUCTION

For the rational design of chemical processes the knowledge of the real behavior of the mixtures is required in order for example to predict the required K-factors for the different processes, to calculate the conversion of chemical reactions, etc.. g^E-models based on the local composition concept or equations of state can be used to describe the real behavior. Both methods require only binary experimental information to describe the real behavior of multicomponent systems. If experimental data are not available, the missing information can be predicted by the UNIFAC method.

C. Jochum M. G. Hicks J. Sunkel (Eds.)
Physical Property Prediction
in Organic Chemistry

UNIFAC

The group contribution method UNIFAC treats the liquid mixture as a mixture containing various structural groups. This is shown in Fig. 1 for the system n-propanol – n-hexane. These two compounds can be divided into the groups CH_3-, CH_2- and OH-.

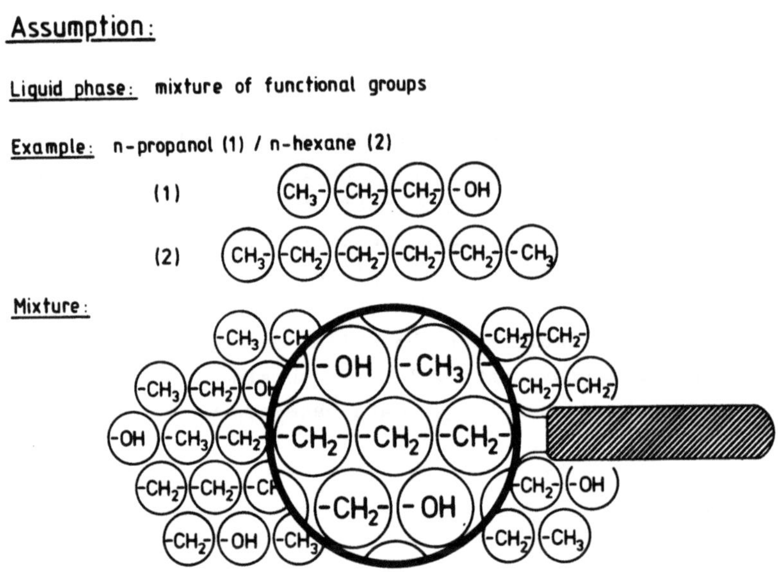

Fig. 1 Solution of groups concept

On the basis of the solution of groups concept, the liquid mixture consists only of these three different structural groups. If the interaction between these groups can be described, the thermodynamic properties, e.g. activity coefficients, heats of mixing etc. can be calculated. The great advantage of this procedure is that the number of different structural groups present in the liquid does not change when going to another alcohol/alkane or alcohol/alcohol system. This means that the real behavior of a large number of mixtures can be predicted using only a relatively small number of interaction parameters.

In UNIFAC (Fredenslund et al. (1975) ,Fredenslund et al. (1977)) the activity coefficient is calculated from a combinatorial part, which takes into account the size and the form of the molecules, and a residual part which considers the interactions between the different groups. The required group interaction parameters have been obtained for a large number of group combinations (Gmehling et al. (1982), Macedo et al. (1983), Tiegs et al. (1987)) using nearly all the published information on VLE.

At present the UNIFAC method incorporates 44 main groups, whereby ca. 45% of the group interaction parameters have been fitted with the help of consistent VLE data stored in the Dortmund Data Bank (Gmehling (1985)).

Because of the reliable results produced for VLE as well as their wide area of application, UNIFAC has become an important tool in chemical industry.

However, some weaknesses of the UNIFAC method are also known: for example the combinatorial part leads to negative deviations from Raoult's law which are too large when we consider molecules which differ very much in size. Furthermore only qualitative agreement is obtained for h^E- and γ^∞-, and the method is not directly applicable to supercritical compounds.

FURTHER DEVELOPMENT OF THE UNIFAC METHOD

With a view to the further development of the UNIFAC method we first extended the data base. We thus started to build up a data bank for LLE, and later also stored heats of mixing data, activity coefficients at infinite dilution, gas solubilities and excess heat capacities. These extensions were carried out in collaboration with the groups of Prof. Fredenslund in Denmark, Prof. Alessi and Prof. Kikic in Italy, Prof. Medina in Portugal and Prof. Knapp in Berlin.

The present status of the Dortmund Data Bank is shown in Fig. 2. The data bank now contains the required pure component properties for ca. 2000 compounds. Furthermore there is a file with the mixture data. Azeotropic data will also be added in the future.

The major part of these data has been published in an evaluated and unified form in the DECHEMA Chemistry Data Series (Gmehling et al. (1977), Sørensen and Arlt (1979), Christensen et al. (1984), Tiegs et al. (1986)).

Using the entire information in the Dortmund Data Bank, a modified form of the UNIFAC model has been developed.

A small empirical modification of the combinatorial part provides much better results for the activity coefficient at infinite dilution for molecules of all sizes. Furthermore, temperature-dependent group interaction parameters are used. In addition the values for the relative van der Waals volumes and surface areas were slightly modified and a larger data base was used to fit the required parameters. Using the standard state of a hypothetical liquid as well as gas solubility data the same model has also been applied to mixtures containing supercritical components.

Pure Component Properties — **File : Stoff (DA)**

contains appr. 2000 components

List of References — **Files : XXXLIT (DA)**

One of each type of data , XXX=VLE,LLE,HE,GAM,GLE,CPE
appr.6000 references

List of Journals (appr.470) — **File : JOUR (DA)**

Data on Mixtures — **PDA–File : XXX**

	since	No. of Isotherms or Isobars
Vapor–Liquid Equilibria (VLE)	1973	12400
Liquid–Liquid Equilibria (LLE)	1977	3000
Heats of Mixing (HE)	1980	5500
γ^{∞} (GAM)	1984	19000 values
Gas Solubilities (GLE)	1985	3400
c_P^E(CPE)	1986	400

Fig. 2 Present status of the Dortmund Data Bank

The modification of the combinatorial part leads to much better results for activity coefficients at infinite dilution.

Table 1 gives an impression of the data base, the temperature range and the compounds involved when fitting alkane/ketone parameters. While in the case of VLE the molecules only vary from acetone to heptanone and pentane to decane, molecules very different in size are taken into account by using γ^{∞}-information . Overall the data base contains alkanes from butane to hexatriacontane and ketones from acetone to nonadecanone and covers the temperature range from –40 to 120°C. The data used were checked as far as possible for thermodynamic consistency and plausibility.

Up to now this new version of the UNIFAC correlation has been used for predicting the behavior of systems including cyclic and aliphatic alkanes, alkenes, different types of aromatics, alcohols, ketones, esters , methanol, water and the gases carbon dioxide, methane, nitrogen, oxygen , ethane and ethylene. The parameters for most of the subcritical compounds have already been published (Weidlich and Gmehling (1987)).

type of data	number of data	temperature range (°C)	number of C-atoms
γ^{∞}	272 values	0–120	(1) 5–36 (2) 3–19
VLE	54 data sets	−35– 95	(1) 5–10 (2) 3–7
h^E	90 data sets	−40– 45	(1) 4–16 (2) 3–11

Table 1 Data base used for fitting the mod. UNIFAC parameters for alkane ketone systems

RESULTS OF MODIFIED UNIFAC

When we compare the results, the modified version of UNIFAC shows great improvements over the original UNIFAC method. Table 2 shows the mean deviations of both methods for binary and ternary systems. It can be seen that apart from the improvements for h^E and γ^{∞}, the VLE results are also slightly improved.

The possibility to predict reliably the real behavior of the liquid phase opens a wide field of applications (see Fig. 3).

The next figures will give an impression of the exactness of the predictions for h^E, azeotropic data, activity coefficients at infinite dilution, retention data , LLE and gas solubilities.

Fig. 4 shows h^E – x diagrams for 16 different alkane/ketone systems. In this case the diagrams not only contain the experimental values and those calculated from mod. UNIFAC , which are represented by dots and solid lines , but also the data obtained from UNIFAC, which are given by the dashed line . The curves show that great improvements are obtained when compared with UNIFAC. This means that

binary system	number of exp. values	deviation		
			orig.UNIFAC	mod.UNIFAC*
VLE	7690	$\lvert\Delta y\rvert$	0,0105	0,0081
h^E	12360	$\lvert\Delta h^E/h^E_{max}\rvert$	42,1%	12,6%
γ^∞	1773	$\lvert\Delta\gamma\rvert_{rel}$	21,1%	5,7%
ternary system				
VLE	3457	$\lvert\Delta y\rvert$	0,0151	0,0134
h^E	1478	$\lvert\Delta h^E/h^E_{max}\rvert$	50,7%	11,2%

Table 2   Results of the mod. and orig. UNIFAC method

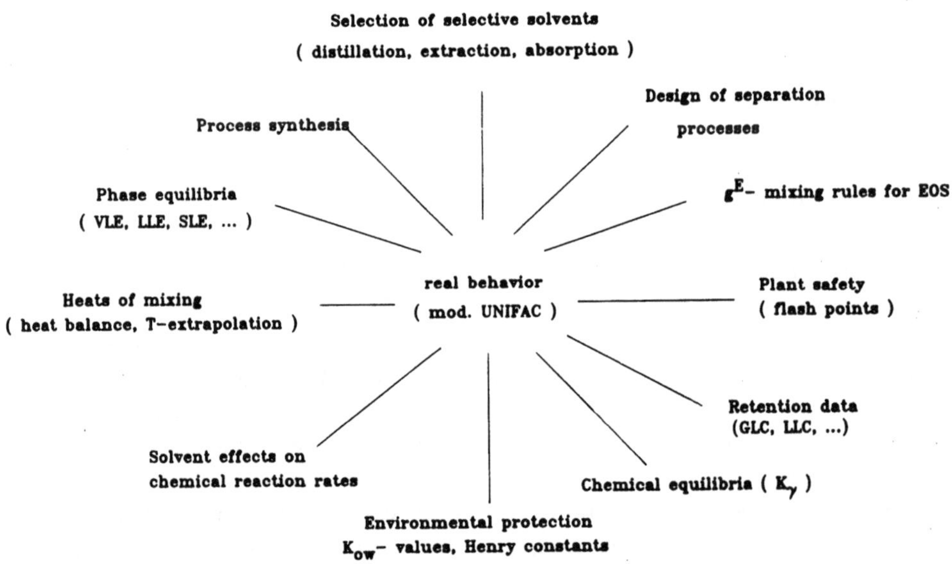

Fig. 3   Possible applications of the modified UNIFAC method

the temperature dependence of the phase equilibrium is much better described by the mod. UNIFAC model, and that small temperature extrapolations should be possible using this model, so that for example azeotropic data should be predicted more reliably for a larger temperature range.

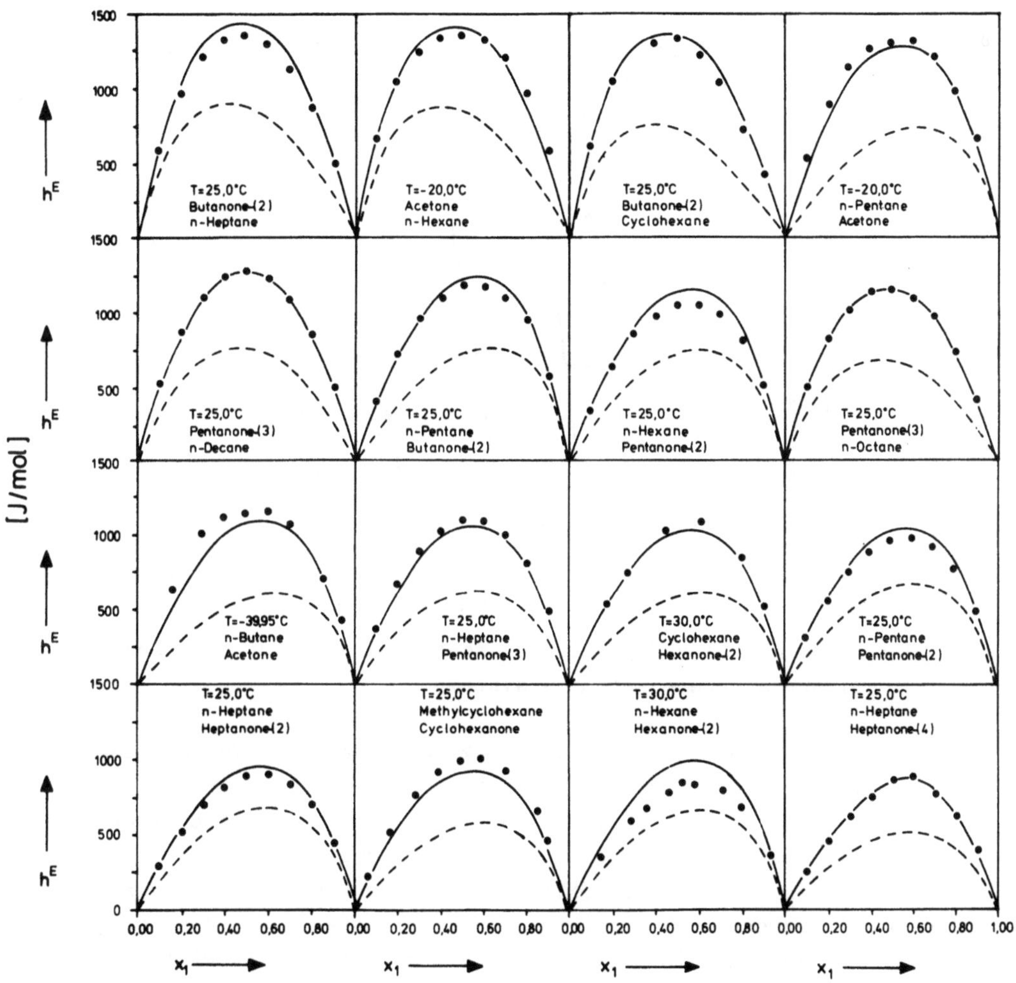

Fig. 4 h^E-x diagram for 16 different alkane - ketone systems
- - - - - UNIFAC ———— mod. UNIFAC

Fig. 5 will confirm this statement for the azeotropic data of binary alkane-ketone systems. Mod. UNIFAC predicts the correct temperature dependence for all binary systems shown on the diagram, whereby the data were taken from original references. Greater deviations are obtained using the data given in the book by Horsley (1973) because of conversion errors.

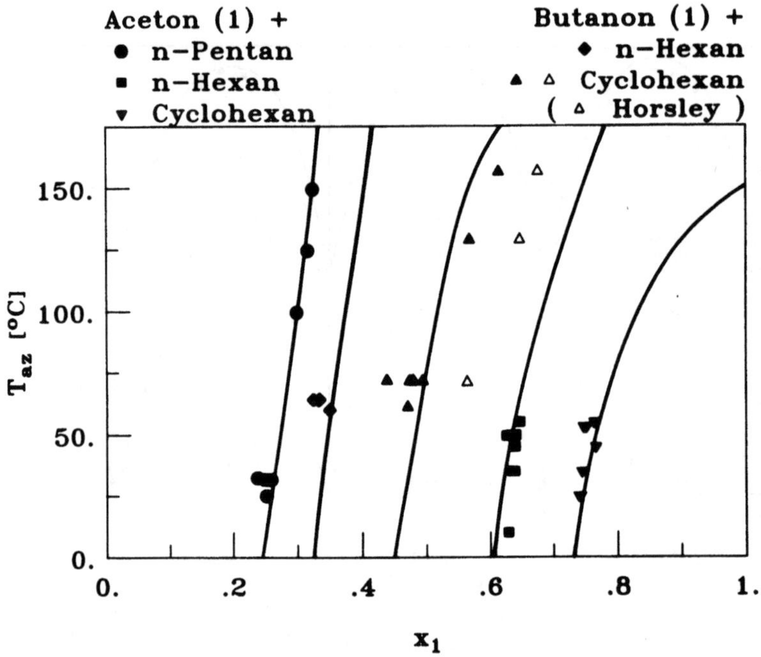

Fig. 5 Experimental and predicted azeotropic data for different binary alkane - ketone - systems

Fig. 6 shows γ^{∞} results for alkane - ketone systems. In the diagrams the predicted values for mod. UNIFAC and UNIFAC are plotted against the experimental values. In the case of a perfect model and error-free experimental data all values should lie on the 45 degree-line. Although this is not true for UNIFAC, it can be seen that the results have been improved by using the modified instead of the original version. Especially for systems with components very different in size, the values obtained using orig. UNIFAC are always too small. As shown in the error distribution diagram, deviations smaller than 5% are obtained for ca. 53% of all alkane-ketone sytems predicted by means of the modified form. In the case of original UNIFAC the major part of the predicted activity coefficients at infinite dilution shows a deviation between 20 and 25%. The improvement is mainly due to the modification of the combinatorial part and the use of γ^{∞} and h^E information.

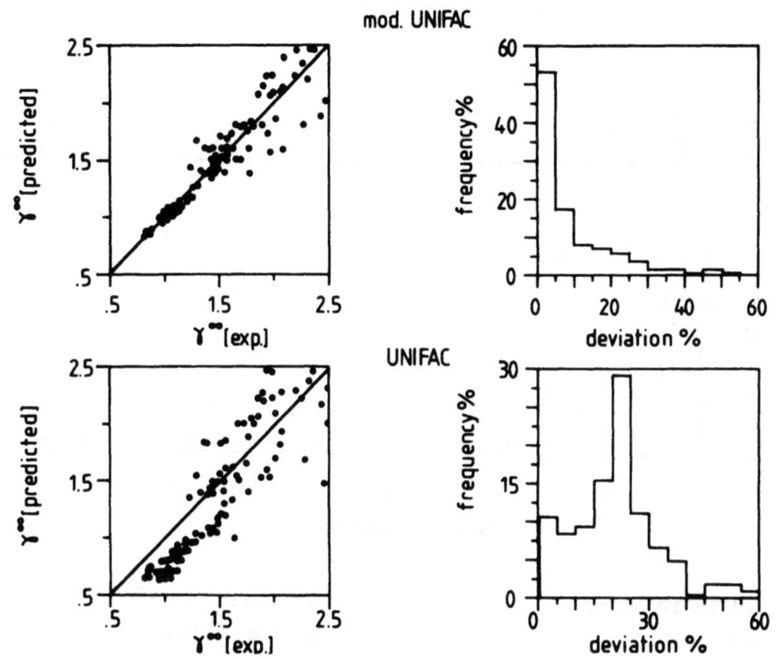

Fig. 6 Comparison of the γ^∞ results using orig. and mod. UNIFAC

A good description of γ^∞ values is of great importance for the design of separation processes, since the separation effort mainly depends on the real behavior in the area of high dilution. Furthermore a reliable knowledge of these values is of great importance for the selection of solvents for extractive distillation or extraction; at the same time these values can be used to predict retention data for gas-liquid- or liquid-liquid-chromatography.

Fig. 7 shows the experimental and the calculated gas chromatograms for the stationary phase squalane and different ketones or alkanes as solutes. It is obvious that, in contrast to mod. UNIFAC, orig. UNIFAC is not able to predict activity coefficients at infinite dilution with the desired accuracy: all solute peaks are shifted to higher retention times. This of course results from the fact that the calculated activity coefficients are too small compared with the experimental ones. Furthermore orig. UNIFAC predicts a smaller retention time for pentanone-2 than for n-hexane. This is in contradiction to the experimental results. Again mod. UNIFAC shows very good agreement with the experimental findings for all the solutes considered.

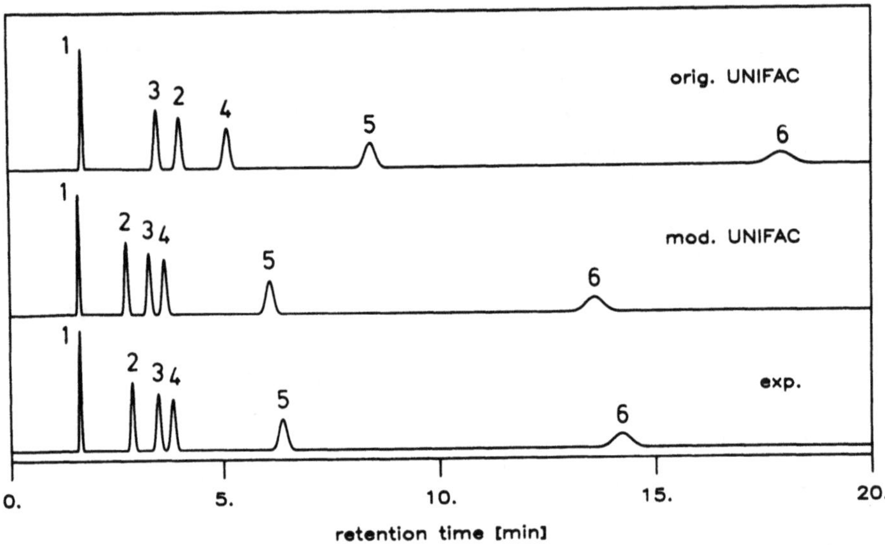

1 butanone-2
2 n-hexane
3 pentanone-2
4 pentanone-3
5 n-heptane
6 n-octane

Fig. 7 Experimental and predicted retention data for the stationary phase squalane at 89.8°C

Up to now no LLE data have been used to fit the required model parameters. Although it was not the aim of UNIFAC to predict LLE, excellent results are often obtained by the modified UNIFAC form, as shown in Fig. 8, in which heats of mixing results for the binary system butanone - water are shown as well as the liquid -liquid equilibrium results. Upper and lower consolute temperatures are predicted with the modified UNIFAC method. This is in agreement with the experimental findings of different authors, when they extend their range of measurement to the supercooled liquid. The reliable estimation is especially surprising because no LLE data were used to fit the required parameters.

With the possibility of predicting liquid-liquid equilibria partition coefficients for the system octanol-water can also be estimated. These values play an important role for the estimation of the bioaccumulation of organic pollutants. For the prediction of the partion coefficients an empirical method developed by Leo and Hansch (1979)

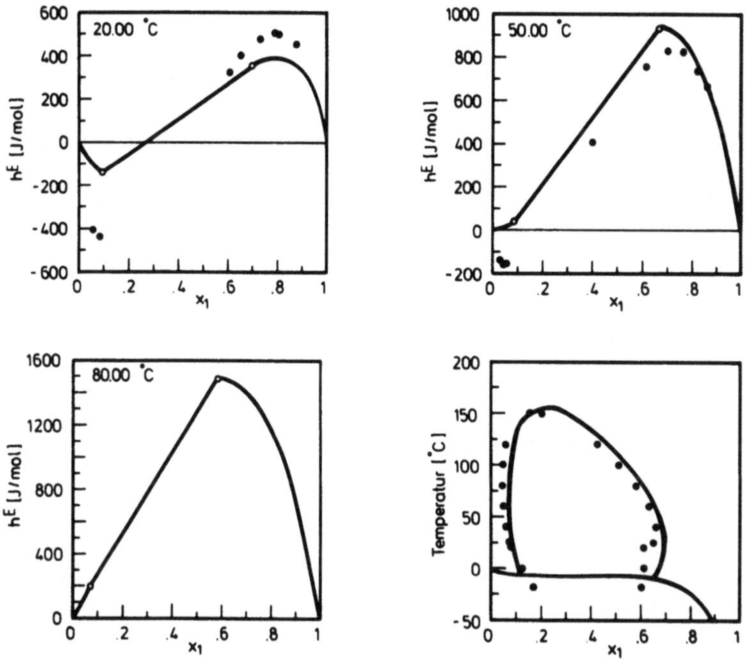

Fig. 8 Experimental and predicted h^E- and LLE data for the binary system butanone - water
.... experimental ———— mod. UNIFAC

is usually used. Fig. 9 shows a comparison of the results obtained by the UNIFAC method and the method of Leo and Hansch. As can be seen, similar results are obtained, although for the development of the method of Leo and Hansch experimental partition coefficients were used, in contrast to UNIFAC. The number of data points in the diagram on the right hand side is lower, since it is much easier to predict the data by the UNIFAC method than by the method of Leo and Hansch, where many of correction terms have to be taken into account. When we compare the different methods it must to be mentioned that the range of application for the prediction of K_{ow}-values in the case of UNIFAC is of course limited because up to now the parameters for several structural groups are missing which are required for example for pesticides, dyes, etc. But the range of applicability of UNIFAC will be extended by also applying K_{ow}-values to fit the required group interaction parameters. It will furthermore be necessary to refit a few of the existing parameters.

Fig.10 shows the predicted results for the gas solubility of oxygen in the ketones acetone, butanone, pentanone-2 and hexanone-2 at a partial pressure of 1 atm for different temperatures.

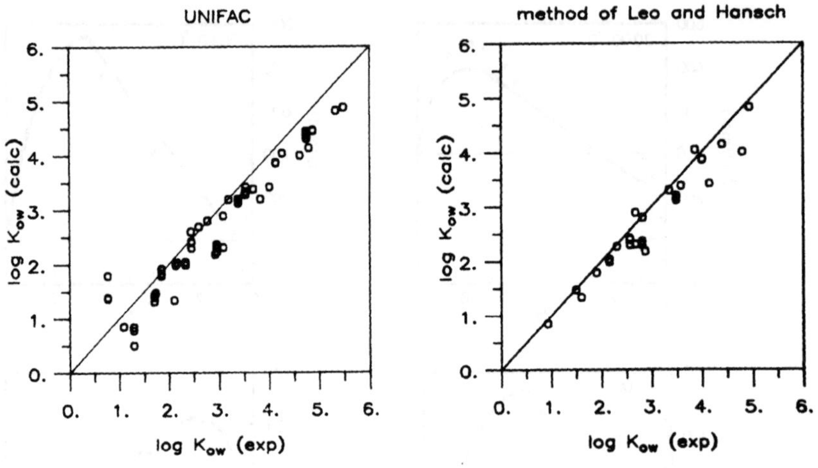

Fig. 9 Experimental and predicted K_{ow}-values for different aromatics

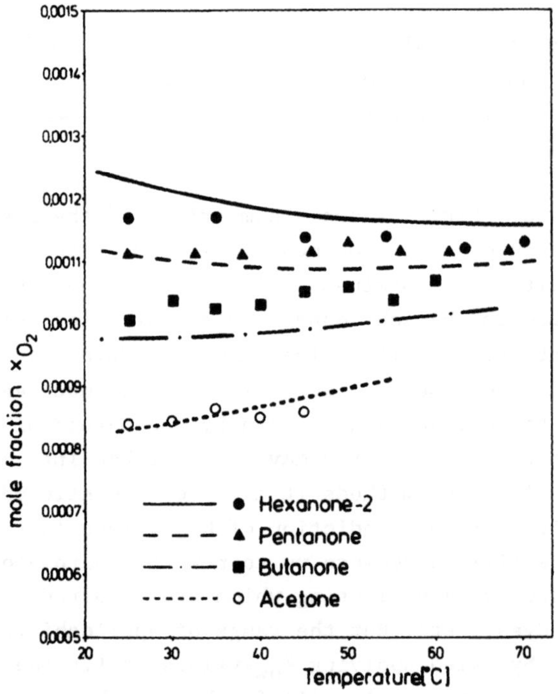

Fig. 10 Experimental and predicted gas solubilities of oxygen in different ketones using mod. UNIFAC

Although the solubilities are quite low, approximately 0.1 mol%, the predicted results shown by the different lines are in good agreement with the experimental findings, which are denoted by the different symbols.

CONCLUSION

Because of the possibility it offers of predicting the real behavior of liquid mixtures reliably, UNIFAC has become an important tool for the chemical engineer.

Mod. UNIFAC allows an improved simultaneous correlation of VLE and h^E data. This has been made possible by varying the combinatorial contribution, by introducing temperature-dependent parameters and by the use of a larger data base for parameter fitting. The prediction of these values has been greatly improved by including γ^∞ and h^E-data. The use of a hypothetical standard fugacity permits us to extend the method to treat supercritical components.

Up to now the number of fitted group interaction parameters for the mod. UNIFAC model is still limited; however, the Dortmund Data Bank is the ideal tool to extend the parameter matrix.

When this work is done, modified UNIFAC will replace the original form which is now used worldwide in chemical industry.

ACKNOWLEDGEMENT

The author is grateful to Prof. Dr. U. Onken, J. Menke, D. Tiegs and Dr. U. Weidlich for useful discussions; he thanks "Arbeitsgemeinschaft Industrieller Forschungsvereinigungen" for financial support.

REFERENCES

Christensen C., J. Gmehling, P. Rasmussen, U. Weidlich, 1984.
Heats of Mixing Data Collection,
DECHEMA Chemistry Data Series, Vol. III, 2 parts, Frankfurt.

Fredenslund A., J. Gmehling, P. Rasmussen, 1977.
Vapor-Liquid Equilibria Using UNIFAC
Elsevier, Amsterdam.

Fredenslund, A., Jones R.L., Prausnitz, J.M., 1975
AIChE J. 21: 1086

Gmehling J., U. Onken, W. Arlt, P. Grenzheuser, B. Kolbe, U. Weidlich,
1977-1984. Vapor-Liquid Equilibrium Data Collection,
DECHEMA Chemistry Data Series, Vol. I, 13 parts, Frankfurt.

Gmehling, J., P. Rasmussen, A. Fredenslund, 1982.
Vapor-Liquid Equilibria by UNIFAC Group Contribution
Revision and Extension II, Ind. Eng. Chem., Process Des. Dev., 21:118.

Gmehling J.
CODATA Bulletin 58, November 1985
Dortmund Data Bank - Basis for the Development of Prediction methods

Horsley, L.H., 1973
Azeotropic Data III, American Chemical Society, Washington

Hansch C., Leo A. 1979
Substituent Constants for Correlation Analysis in Chemistry and
Biology, Wiley, New York

Macedo, E., U. Weidlich, J. Gmehling, P. Rasmussen, 1983.
UNIFAC, Revision and Extension 3,
Ind. Eng. Chem., Process Des. Dev., 22:676.

Sørensen J.M., W. Arlt, 1979.
Liquid-Liquid Equilibrium Data Collection,
DECHEMA Chemistry Data Series, Vol. V, 3 parts, Frankfurt.

Tiegs D., J. Gmehling, A. Medina, M. Soares, J. Bastos, P. Alessi, I. Kikic, 1986.
Activity Coefficients at Infinite Dilution,
DECHEMA Chemistry Data Series, Vol. IX, 2 parts, Frankfurt.

Tiegs, D., J. Gmehling, P. Rasmussen, A. Fredenslund, 1987.
Vapor-Liquid Equilibria by UNIFAC Group Contribution:
Revision and Extension. 4, Ind. Eng. Chem. Research, 26:159.

Weidlich, U., J. Gmehling, 1987.
A Modified UNIFAC Model
Part I Prediction of VLE, h^E and γ^∞, Ind. Eng. Chem. Res., 26:1372.

Computer Analysis of Thermochemical Data of Organic Compounds

S. R. A. Cove and J. B. Pedley

School of Chemistry and Molecular Sciences, University of Sussex, Falmer – Brighton BN1 9QJ, United Kingdom

Thermochemical data has been collected, processed and assessed at the University of Sussex over a period of years. Particular emphasis has been placed on standard enthalpies of formation of organic compounds, culminating in 1986 in a book containing values for approximately 3000 compounds of the elements C, H, O, N, S and halogens. Since then, data on entropies and heat capacities have been acquired and all the data set up as computer files which may be searched for compounds containing or not containing specific structural features. Software has also been developed for calculating values of thermochemical properties using files of parameters defined to fit those experimental values which are considered to be most reliable.

Detailed analysis of the differences between experimental and calculated values has identified those types of compound for which the theoretical model is likely to give accurate predicted values. Where discrepancies are significantly large and of a similar size for most compounds in a structurally related set, these have been attributed to additional theoretical factors such as steric interactions, ring strain and conjugative effects which are not allowed for in the model. Doubtful experimental data have been revealed by discrepancies which do not correlate with those for the majority of compounds in the set, and these are indicated clearly in the data files. In these cases, the theoretically calculated values may be more reliable than those derived from original sources.

Table 1. shows a section of one of the experimental data files. The information stored for each compound include the formula, the Chemical

C. Jochum M. G. Hicks J. Sunkel (Eds.)
Physical Property Prediction
in Organic Chemistry

Abstracts Registry Number, the name (and possibly an alternative name) and a modified form of Wiswesser Line Notation (WLN), (see Tables 7. and 8.). Data consist of standard enthalpies of formation and vaporisation (or sublimation), followed by the symmetry number (excluding the internal symmetry of the methyl group), entropies and heat capacities.

Tables 2. demonstrates the computer retrieval of compounds by formula and structure. Note that even for the industrially important octenes there is only a limited amount of data, much of which is of doubtful accuracy (items labelled '#'). To be able to predict accurate values of thermochemical data or to identify suspect ones, it is essential to be able to identify trends in data for compounds having structural features in common. Therefore, the compounds in the data base are arranged according to their functional groups and ring systems. Allowed codes are shown in Table 3. Compounds containing more than one of the structural attributes are filed under the code appearing latest in the table, then by the next latest etc., as illustrated on the left hand side of Table 4. Compounds included under a set of specific attributes are further subdivided, where possible, into more detailed structural types as illustrated in Table 4. The program for search and retrieval can be used to identify compounds containing or not containing the attributes selected from Table 3. This allows retrieval of compounds scattered throughout the whole data base as illustrated in Table 5. These would be very difficult to locate by any other method of searching.

To predict values for thermochemical properties, the procedure outlined in Table 6. is used. Groups present in the structure of a compound and the bonded and non-bonded interactions between the groups are identified by computer analysis of the modified form of Wiswesser Line Notation (WLN) described in Tables 7. and 8. Tables 9. to 11. give examples of the application of the theory to data for some selected compounds.

Differences between experimental and calculated values indicate either that the theoretical model is inadequate or that the experimental data may be suspect. The latter can often be identified if data for an extended family of compounds is available as for example in linear 1-enes, at the beginning of Table 9. Data for 1-Hexene and 1-Octene appear to be slightly discordant and theoretical values may be regarded as more reliable than the experimental. In later sections of the table, the differences between experimental and calculated values labelled '!', indicate the possibility of nonbonding interactions not allowed for by the theory. These interactions probably arise between hydrogen atoms on groups separated by two, three or four other groups i.e. 1:4, 1:5 and 1:6 interactions respectively, as shown in the footnotes to Table 9. The 1:4 interactions for the liquid and gaseous states of the cis forms of the linear 2-enes, 3-enes etc: are averages for all compounds of this type, and estimated values are derived by examining trends in values for similar structures. Predicted values for standard enthalpies of formation of alkenes containing the basic structures represented in the footnotes must therefore be corrected by the amounts shown.

Table 10. gives the differences between experimental and calculated values of standard enthalpies of formation of some saturated carboxylic acids. For data on the crystalline state, the experimental values are compared with the calculated values for the liquid state and the differences between them correspond ideally to the enthalpies of fusion. For linear monocarboxylic acids this difference appears to increase linearly with the number of CH2 groups in the chain, except for C11H22O2. There also appears to be an alternating pattern of values for the gaseous states after C13H26O2. For linear dicarboxylic acids the differences for the crystalline state show a distinct alternating pattern, whereas those for the gaseous state appear to be erratic.

Table 11 contains data for three-membered rings; the large discrepancies in the table may be interpreted as arising from ring strain. The values for a given type of ring are very consistent except for those labelled '#', where the data may be doubtful, and those labelled '!' which are explained in the footnotes to the table.

Provisional analysis of data on entopies and heat capacities indicates that reasonable agreement between calculated and experimental values can be obtained without including values for bonding and nonbonding interactions (terms b(i,j) and c[i(j,k...)] in Table 6). However, further data has yet to be added to the system and a more detailed analysis will be carried out when that exercise is complete.

Table 1. Specimen Section of File A.PRP

```
C1H4O(1)            67561

Methanol

Q1

                    -239.16  0.29 HL   -201.50  0.29 HG     37.83  0.25 HV

 1              127.19 SL   239.70 SG                  81.13 CL    43.89 CG


C16H34O(1)          36653824

1-Hexadecanol

Q16

 -686.47  1.17 HC                      -517.02  2.38 HG    169.45  2.09 HV

 1                          829.69 SG   422.    CC    524.  *CL   384.47 CG


C2H6O(2)            115106

Dimethyl ether (Oxybismethane)

1O1

                    -203.34  1.13 HL   -184.05  0.50 HG     19.30  1.00 HV

 2                          267.06 SG                 102.30*CL    65.81 CG


C4H10O2(5)          25154534

1,1-Dimethoxyethane (Acetaldehyde dimethyl acetal)

1OY1&O1

                    -420.58  1.30 HL   -389.74  0.84 HG
```

HC, HL, HG, HV /kJ mol-1 - standard enthalpies of formation of crystalline, liquid and gaseous states and enthalpy of vaporization respectively at 298.15 K.

SC, SL, SG, CC, CL, CG /J mol-1 K-1 - standard entropies and heat capacities of crystalline, liquid and gaseous states respectively at 298.15 K

* - values estimated from data at temperatures other than 298.15 K

Table 2. Experimental Thermochemical Data for Octenes

! - systematic discrepancies
- possibly incorrect experimental data
* - values estimated from data at temperatures other than 298.15K

Name	WLN	DfH(c)	DfH(l)	DfH(g)	DvapH	DsubH
				/kJ mol-1		
C8H16(1)	7U1					
1-Octene			-121.8#	-81.3#	40.4	
			(1.2)	(1.2)	(0.2)	
C8H16(2)	3U1X1&1&1 &&CIS					
(Z)-2,2-Dimethyl-3-hexene			-126.4!	-89.3!	37.2	
			(2.8)	(2.8)	(0.2)	
C8H16(3)	3U1X1&1&1 &&TRANS					
(E)-2,2-Dimethyl-3-hexene			-144.9	-107.7	37.3	
			(1.6)	(1.6)	(0.2)	
C8H16(4)	2Y2&Y1&U1					
2-Methyl-3-ethyl-1-pentene			-137.9	-100.3	37.6	
			(1.3)	(1.3)	(0.2)	
C8H16(5)	1X1&1&1Y1&U1					
2,4,4-Trimethyl-1-pentene			-145.9!	-110.5!	35.8	
			(1.3)	(1.4)	(0.1)	
C8H16(6)	1X1&1&1UY1&1					
2,4,4-Trimethyl-2-pentene			-142.4!	-104.9!	37.5	
			(2.0)	(2.0)	(0.1)	
C8H16(18)	6U2 &&CIS					
(Z)-2-Octene			-135.7#			
			(4.1)			
C8H16(19)	6U2 &&TRANS					
(E)-2-Octene			-135.7#			
			(4.1)			
C8H16(20)	1Y1&1U1Y1&1 &&CIS					
(Z)-2,5-Dimethyl-3-hexene			-151.0#			
			(3.6)			
C8H16(21)	1Y1&1U1Y1&1 &&TRANS					
(E)-2,5-Dimethyl-3-hexene			-159.2#			
			(2.9)			

Name	WLN	Symmetry	S(c)	S(l)	S(g)	C(c)	C(l)	C(g)
					/J mol-1 K-1			
C8H16(1)	7U1							
1-Octene		1		360.5	462.5		241.2	178.1
C8H16(5)	1X1&1&1Y1&U1							
2,4,4-Trimethyl-1-pentene		1		306.3			235.4*	
C8H16(6)	1X1&1&1UY1&1							
2,4,4-Trimethyl-2-pentene		1		311.7			240.2	

Table 3. Codes for Functional Groups and Ring Systems

AN - alkanes					
EN - double bond.	i.e. =		HS - thiol	i.e. -SH	
YN - triple bond.	i.e. $		ES - sulphide	i.e. -S-	
HY - hydroxy	i.e. -OH		CS - thione	i.e. >C=S	
ET - ether	i.e. -O-		SX - thioic acid	i.e. O -C-SH	
CB - carbonyl	i.e. >C=O		PS - poly sulphide	i.e. -S-S-	
PX - peroxy	i.e. -O-O-		SI - sulphinyl	i.e. >S=O	
AC - carboxylic acid	i.e. O -C-OH		SY - sulphone	i.e. >SO2	
AM - amine	i.e. >N-		ST - sulphate and sulphite	i.e. >SO3 >SO4	
IM - imine	i.e. =N-		HL - halogen	i.e. -X	
NL - nitrile	i.e. -C$N		DL - dihalogen	i.e. -CX2	
AO - azo	i.e. -N=N-		TL - trihalogen	i.e. -CX3	
AZ - azide	i.e. -N=N=N		XL - tetrahalogen	i.e. CX4	
HZ - hydrazine	i.e. >N-N<		BZ - benzene		
NT - nitrite	i.e. -N=O		Rn - a ring of n atoms		
NR - nitro	i.e. -NO2		SP - a spiro compound		
NA - nitrate	i.e. -NO3		FU - a fused ring compound		

Table 4. Section of the Index File for Structural Attributes

```
HY                                        B       1
          Primary alcohols
          Secondary and tertiary alcohols
          Polyhydric alcohols
HY-EN                                     B     297
          =CH-CH2-OH
HY-YN                                     B     307
              |
          #C-C-OH
              |
ET                                        B     324
          -CH2-O-CH2-
          -O-CH2..CH2-O-
          >CH-O-
           |
          -C-O-
           |
          -O-CH2-O-
              |
          -O-CH-O-
              |
          -O-C-O-
              |
ET-EN                                     B     754
                      |
          =CH-O-; =C-O-
ET-HY                                     B     792
          -O-CH2-CH2-OH
          Miscellaneous
```

Codes refer to the name or substructure corresponding to the structural feature common to a given family of compounds.

Table 5. Experimental Thermochemical Data for Compounds Containing Both Imino and Amino Groups (See Table 4 for codes IM,AM,AC etc.)

Name	WLN	DfH(c)	DfH(l)	DfH(g) /kJ mol-1	DvapH	DsubH
*******************	IM-AM			********************		
C1H5N3(1)	ZYZUM					
Guanidine		-56.0 (1.0)				
*******************	IM-AM-AC			********************		
C4H9O2N3(1)	QV1N1&YUMZ					
Creatine (N-(Aminoiminomethyl)-		-537.2 (0.8)				
C6H14O2N4(1)	QVYZ3MYZUM					
d-Arginine		-623.5 (1.3)				
*******************	NL-IM-AM			********************		
C2H4N4(1)	NCMYZUM					
Guanidine-1-carbonitrile		24.9 (0.5)				
*******************	NR-IM-AM			********************		
C1H4O2N4(1)	WNMYZUM					
Nitroguanidine		-92.4 (4.1)				
*******************	NR-IM-AM-CB			********************		
C2H5O3N5(2)	WNNUYZMMVH					
1-Formamido-2-nitro-guanidine		-146.9 (2.5)				
C2H5O3N5(3)	NWMYUM&MVZ					
1-Nitro-3-guanidinourea		-313.3 (1.1)				
*******************	NR-HZ-IM-AM			********************		
C1H5O2N5(1)	ZMYUM&MNW					
Nitroaminoguanidine		22.1 (0.7)				
*******************	NR-HZ-IM-AM-CB			********************		
C3H7O3N5(1)	WNNUYZMMV1					
1-Acetamido-2-nitro-guanidine		-193.6 (5.8)				
*******************	CS-IM-AM			********************		
C3H7N3S(1)	2UNMYUSZ					
2-Ethylidene hydrazine-carbothioamide		64.0 (8.3)				
*******************	HL-IM-AM			********************		
C1N3F5(1)	FNFYNFFUNF					
Pentafluoroguanidine				95.7 (3.6)		

Table 6. Estimation Procedure

Values for thermochemical properties are calculated by the program using the following equations.

DfH(l) = group terms

DfH(g) = group terms

DvapH = group terms

S(l) = group terms - Ssig

S(g) = group terms - Ssig + Strans

Cp(l) = group terms

Cp(g) = group terms + 4R

where

Ssig = 19.145log10(symmetry)

Strans = 28.7177log10(mol wt) + 108.0233

R = gas constant

The group terms are of the form:-

$$\sum_{i} a(i) + \sum_{\substack{i,j \\ \text{pairs}}} b(ij) + \sum_{i} c[i(jk)]$$

where a, b and c represent repectively contributions from groups, bonded pairs of groups and clusters of groups joined to a central group.

For standard enthalpies of formation, c terms are included for alkanes only. For all other types of compound these terms are defined as zero. Other thermochemical properties are defined by the a terms only.

Table 7. Wiswesser Line Notation Symbols

Numeral	Number of carbon atoms in a saturated chain or segment
Y	Carbon atom attached to three atoms other than hydrogen
X	Carbon atom attached to four atoms other than hydrogen
C	Carbon atom attached to no more than two other atoms
O	Oxygen atom with no attached hydrogen
Q	Hydroxyl
V	Carbonyl
N	Nitrogen atom with no attached hydrogen
M	Imino or imido
Z	Amino of amido
S	Sulphur atom with or without hydrogen
F	Fluorine atom
G	Chlorine atom
E	Bromine atom
I	Iodine atom
W	Dioxo group in -NO2, SO2 etc.
U	Double bond between two atoms
UU	Triple bond between two atoms
H	Hydrogen atom not implied as part of another symbol
a,b..	Lower case letter used as "break points" in cyclic systems

Conventional WLN for cyclic compounds is very complex and is not used in the program to describe the ring fragments. Instead each ring is considered to be broken at a selected single bond and the WLN of the corresponing acyclic compound. The "break points" in the rings are indicated by lower case letters.

Table 8. Examples of Wiswesser Line Notation

WLN	Structural Formula
2Y2&3	CH3.CH2.CH(CH2.CH3).CH2.CH2.CH3
1X2&1&Y1&1	CH3.C(CH2.CH3)(CH3).CH(CH3)(CH3)
1UU2U2	CH$C.CH=CH.CH3
2UCU2	CH3.CH=C=CH.CH3
2O2	CH3.CH2.O.CH2.CH3
Q2Q	HO.CH2.CH2.OH
3V2O1	CH3.CH2.CH2.CO.CH2.CH2.O.CH3
2VQ	CH3.CH2.CO.OH
2E	CH3.CH2.Br
3N2&MZ	CH3.CH2.CH2.N(CH2.CH3).NH.NH2
3Y2NW&1SW	CH3.CH2.CH2.CH(CH2.CH2.NO2).CH2.SO2
4H	CH3.CH2.CH2.CH3
SH2VH	SH.CH2.CH2.CHO

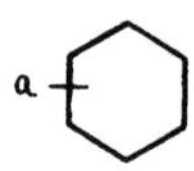

a6a

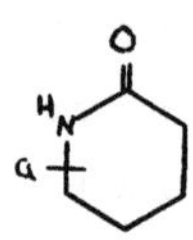

aMV4a

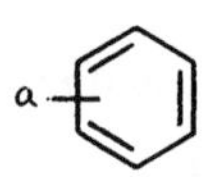

a1U2U2U1a

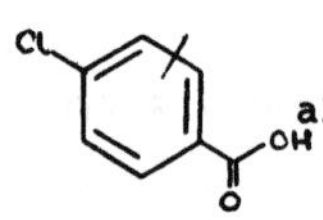

a1UYG1U1YVQU1a

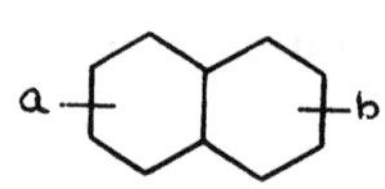

a2Y2bY2a2b
two 6 membered rings

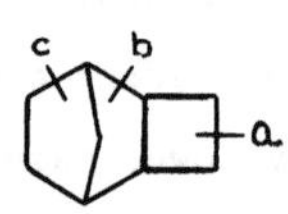

a1YbY1aY1Ybc2c
one 4 and two 5 membered rings

Table 9. Differences Between Experimental and Calculated Values of

Standard Enthalpies of Formation of Alkenes

\# - possibly incorrect experimental data
c - theoretically calculated value

Name	WLN	DfH(c)	DfH(l)	DfH(g) /kJ mol-1	DvapH	DsubH
1. =CH2						
C2H4(1) Ethylene (Ethene)	1U1		0.0c	0.0	0.0c	
2. -CH=CH2						
C3H6(1) Propene	2U1		0.0	0.0	0.0 (2.3)	
(a) >CH2 -CH=CH2						
C4H8(1) 1-Butene	3U1		1.0	0.9	-0.1	
C5H10(1) 1-Pentene	4U1		0.6	0.5	-0.1	
C6H12(1) 1-Hexene	5U1		-0.9#	-1.0#	-0.1	
C7H14(1) 1-Heptene	6U1		1.1	1.0	-0.1	
C8H16(1) 1-Octene	7U1		3.0#	2.8#	-0.3	
C10H20(1) 1-Decene	9U1		2.5	2.4	-0.1	
C12H24(2) 1-Dodecene	11U1		1.6	2.0	0.4	
C16H32(1) 1-Hexadecene	15U1		2.1	2.2	0.1	
C7H14(2) 5-Methyl-1-hexene	1Y1&3U1		5.9#	6.3#	0.4	
C6H12(8) 4-Methyl-1-pentene	1Y1&2U1		0.1	-0.1	-0.2	

Name	WLN	DfH(c)	DfH(l)	DfH(g) /kJ mol-1	DvapH	DsubH
C7H14(6) 4,4-Dimethyl-1-pentene	1X1&1&2U1		5.6	3.4	-0.2	
C5H8(4) 1,4-Pentadiene	1U3U1			2.6		
C6H10(1) 1,5-Hexadiene	1U4U1		1.9	2.0		
(b) >CH- -CH=CH2						
C5H10(5) 3-Methyl-1-butene	1Y1&1U1		2.9	2.8	-0.2	
C6H12(7) 3-Methyl-1-pentene	2Y1&1U1		2.0	1.6	-0.3	
(c) >C< -CH=CH2						
C6H12(16) 3,3-Dimethyl-1-butene	1X1&1&1U1		2.9	3.8	0.2	
3. >C=CH2						
C4H8(4) 2-Methylpropene (Isobutene)	1UY1&1		0.0	0.0	0.0	
(a) >CH2 >C=CH2						
C5H10(4) 2-Methyl-1-butene	2Y1&U1		2.1	2.6	0.7	
C6H12(6) 2-Methyl-1-pentene	3Y1&U1		-1.0#	-0.8#	0.1	
C6H12(14) 2-Ethyl-1-butene	2Y2&U1		1.9	2.5	0.6	
C7H14(5) 2,4-Dimethyl-1-pentene	1Y1&1Y1&U1		4.6	4.2	-0.4	
C8H16(5) 2,4,4-Trimethyl-1-pentene	1X1&1&1Y1&U1		11.7	11.3	-0.2	

Name	WLN	DfH(c)	DfH(l)	DfH(g) /kJ mol-1	DvapH	DsubH
(b) >CH- >C=CH2						
C6H12(15) 2,3-Dimethyl-1-butene (Tetramethylethylene)	1Y1&Y1&U1		2.7	4.8	0.6	
C7H14(10) 3-Methyl-2-ethyl-1-butene	1Y1&Y2&U1		7.5	8.5	1.0	
C8H16(4) 2-Methyl-3-ethyl-1-pentene	2Y2&Y1&U1		9.5	8.6	-0.9	
(c) >C< >C=CH2						
C7H14(11) 2,3,3-Trimethyl-1-butene	1X1&1&Y1&U1		14.2	15.5	1.2	
4. -C=C-						
(a) -CH= =CH-						
C4H8(2) (Z)-2-Butene	2U2 &&CIS		3.3	5.4	2.1	
C4H8(3) (E)-2-Butene	2U2 &&TRANS		-0.3	1.1	1.4	
C5H10(2) (Z)-2-Pentene	3U2 &&CIS		5.1	5.7	1.3	
C5H10(3) (E)-2-Pentene	3U2 &&TRANS		0.7	1.4	1.2	
C6H12(2) (Z)-2-Hexene	4U2 &&CIS		0.7#	1.8#	1.1	
C6H12(3) (E)-2-Hexene	4U2 &&TRANS		-0.9	0.2	1.2	
C7H14(24) (Z)-2-Heptene	5U2 &&CIS		5.2			
C7H14(25) (E)-2-Heptene	5U2 &&TRANS		0.8			
C8H16(18) (Z)-2-Octene	6U2 &&CIS		0.4# (2.5)			
C8H16(19) (E)-2-Octene	6U2 &&TRANS		0.4 (2.5)			

Name	WLN	DfH(c)	DfH(l)	DfH(g) /kJ mol-1	DvapH	DsubH
C6H12(4) (Z)-3-Hexene	3U3 &&CIS		5.6	6.5	0.9	
C6H12(5) (E)-3-Hexene	3U3 &&TRANS		-1.5	-0.3	1.2	
C7H14(26) (Z)-3-Heptene	4U3 &&CIS		6.0			
C7H14(27) (E)-3-Heptene	4U3 &&TRANS		1.0			

Name	WLN	DfH(c)	DfH(l)	DfH(g)	DvapH	DsubH
C6H12(12) (Z)-4-Methyl-2-pentene	2U1Y1&1 &&CIS		4.4	5.3	0.9	
C6H12(13) (E)-4-Methyl-2-pentene	2U1Y1&1 &&TRANS		-0.1	1.3	1.4	
C7H14(8) (Z)-4,4-Dimethyl-2-pentene	2U1X1&1&1 &&CIS		22.1	24.0	1.6	
C7H14(9) (E)-4,4-Dimethyl-2-pentene	2U1X1&1&1 &&TRANS		5.7	7.8	1.9	
C8H16(2) (Z)-2,2-Dimethyl-3-hexene	3U1X1&1&1 &&CIS		26.7	28.1	1.2	
C8H16(3) (E)-2,2-Dimethyl-3-hexene	3U1X1&1&1 &&TRANS		8.3	9.8	1.3	
C8H16(20) (Z)-2,5-Dimethyl-3-hexene	1Y1&1U1Y1&1 &&CIS		-1.3#			
C8H16(21) (E)-2,5-Dimethyl-3-hexene	1Y1&1U1Y1&1 &&TRANS		-9.5#			
C10H20(2) (Z)-2,2,5,5-Tetramethyl-3-hexene	1X1&1&1U1X1&1&1 &&CIS		58.2			
C10H20(3) (E)-2,2,5,5-Tetramethyl-3-hexene	1X1&1&1U1X1&1&1 &&TRANS		14.3	15.2	0.5	

(b) -CH= =C<

Name	WLN	DfH(c)	DfH(l)	DfH(g)	DvapH	DsubH
C5H10(6) 2-Methyl-2-butene	2UY1&1		5.9	7.7	2.2	
C6H12(9) 2-Methyl-2-pentene	3UY1&1		1.7#	3.3#	1.6	
C6H12(10) (Z)-3-Methyl-2-pentene	2Y1&U2 &&CIS		5.8	7.9	2.1	
C6H12(11) (E)-3-Methyl-2-pentene	2Y1&U2 &&TRANS		5.7	7.1	1.3	

Name	WLN	DfH(c)	DfH(l)	DfH(g) /kJ mol-1	DvapH	DsubH
C7H14(3)	3UY2&1 &&CIS					
(Z)-3-Methyl-3-hexene			10.1#	11.6#	1.5	
C7H14(4)	3UY2&1 &&TRANS					
(E)-3-Methyl-3-hexene			13.3#	14.2#	0.9	

C7H14(7)	1Y1&U1Y1&1					
2,4-Dimethyl-2-pentene			9.8	11.0	1.2	
C8H16(6)	1X1&1&1UY1&1					
2,4,4-Trimethyl-2-pentene			26.5	28.6	1.9	

(c) >C= =C<

~~~~~~

C6H12(17)	1Y1&UY1&1					
2,3-Dimethyl-2-butene (Tetramethylethylene)			14.6	18.2	3.0	
~~~~~~

Footnotes

Corrections for some alkene structures.

The top number is for the liquid state and the other for the gaseous state

(a) - average value

(e) - estimated value

```
2(a)                          C           (b)      /            (c)  \ /
       --C      1.5(a)       / \     5.6       --C      2.5(a)     --C      2.9
          \     2.0(a)    --C   C==  3.4          \     2.2(a)        \     3.8
           C==               / \                   C==                 C==

3(a) --C        2.0(a)        C      4.6          C     11.7
        \       2.6(a)     \ / \     4.2       \ / \    11.3
         C==              --C   C==            --C  C==
        /                       /              /   /

 (b)   |                     |                     C--
     --C        2.7        --C        7.5     \   /     9.5
        \       4.8           \       8.5     C--C      8.6
         C==                   C==               \
        /                     /                   C==
                           --C                   /

 (c)  \ /
     --C        14.2
        \       15.5
         C==
        /

4(a)                   |                  \ /                   \ /     \ /
   \    / 5.0(a)     --C          4.4    --C           24.4(a) --C      C-- 58.2
    C==C  5.5(a)        \      /  5.3       \      /   26.1(a)    \    /    60.0(e)
                         C==C                C==C                  C==C

                      |                  \ /                  \ /
   \      0.0(a)    --C          -0.1   --C          7.0(a)  --C             14.3
    C==C  1.0(a)       \          1.3      \         8.8(a)     \            15.2
        \               C==C                C==C                 C==C
                            \                   \                    \
                                                                      C--

 (b)                  |                   \ /
   \    / 5.8(a)    --C           9.8    --C           26.5
    C==C  7.6(a)       \       / 11.0       \      /   28.6
        \               C==C                 C==C
                            \                    \

 (c) \    /  14.6
      C==C   18.2
     /    \

     Estimated values
      \ /                        \ /
     --C          35.0(e)      --C            21.0(e)
        \     /   38.0(e)         \           24.0(e)
         C==C                      C==C
        /                         /    \
```

Table 10. Differences Between Experimental and Calculated Values of Standard Enthalpies of Formation of Acids

! - systematic discrepancies
- possibly incorrect experimental data

1. Linear carboxylic acids

alternating pattern of data for gaseous state after C13H26O2(1)

Name	WLN	DfH(c)	DfH(l)	DfH(g) /kJ mol-1	DvapH	DsubH
C1H2O2(1) Formic acid	VHQ		0.0	0.0	0.0	
C2H4O2(1) Acetic acid	QV1		0.0	0.0	0.0	
C3H6O2(3) Propanoic acid	QV2		0.0	0.0	0.0	
C4H8O2(1) Butanoic acid (Butyric acid)	QV3		2.6	0.6	-2.0	
C5H10O2(1) Pentanoic acid	QV4		2.8	5.4	0.9	
C6H12O2(1) Hexanoic acid	QV5		4.1	6.2	2.3	
C7H14O2(1) Heptanoic acid	QV6		3.5	2.8	-0.9	
C8H16O2(1) Octanoic acid	QV7		3.4	5.5	1.9	
C9H18O2(1) Nonanoic acid	QV8		5.5	3.3	-2.2	
C10H20O2(1) Decanoic acid	QV9	-22.8	6.6	6.5		29.3
C11H22O2(1) Undecanoic acid	QV10	-19.2#	6.5#	7.6#		26.9
C12H24O2(1) Dodecanoic acid	QV11	-32.1	4.6	1.1		33.2
C13H26O2(1) Tridecanoic acid	QV12	-38.4	4.7	3.7		42.1
C14H28O2(1) Tetradecanoic acid	QV13	-39.5	5.1	-9.1		30.5
C15H30O2(1) Pentadecanoic acid	QV14	-42.0	8.0	6.5		48.6

Name	WLN	DfH(c)	DfH(l)	DfH(g) /kJ mol-1	DvapH	DsubH
C16H32O2(1) Hexadecanoic acid	QV15	-46.0	7.3	-10.8		35.3
C17H34O2(1) Heptadecanoic acid	QV16	-53.2	5.6			
C18H36O2(1) Octadecanoic acid	QV17	-50.7#	12.3#	-13.2#		37.5
C19H38O2(1) Nonadecanoic acid	QV18	-61.3	6.3	3.5		64.8
C20H40O2(1) Eicosanoic acid (Icosanoic acid)	QV19	-63.5	8.5	-2.8		60.7
C5H10O2(3) 3-Methylbutanoic acid	QV1Y1&1		7.4 (5.0)	-4.0 (2.8)	-10.9	

2. Branched carboxylic acids

Name	WLN	DfH(c)	DfH(l)	DfH(g)	DvapH	DsubH
C4H8O2(6) 2-Methylpropanoic acid	QVY1&1				-1.3	
C5H10O2(2) 2-Methylbutanoic acid	QVY2&1		5.6# (5.0)			
C8H16O2(12) 2-Ethylhexanoic acid	QVY4&2		2.2	2.7	1.6	
C5H10O2(4) 2,2-Dimethylpropanoic acid	QVX1&1&1		0.0 (3.0)	0.0	0.0	

3. Dicarboxylic acids

alternating pattern of data for crystalline state

Name	WLN	DfH(c)	DfH(l)	DfH(g)	DvapH	DsubH
C2H2O4(1) Oxalic acid (Ethanedioic acid)	VQVQ	3.4		0.1		6.7
C3H4O4(1) Propanedioic acid (Malonic acid)	QV1VQ	8.8				
C4H6O4(2) Butanedioic acid (Succinic acid)	QV2VQ	-14.9		4.5		19.5
C5H8O4(2) Pentanedioic acid (Glutaric acid)	QV3VQ	-8.6				
C6H10O4(6) Hexanedioic acid (Adipic acid)	QV4VQ	-17.3		4.1		21.3
C7H12O4(2) Heptanedioic acid	QV5VQ	-6.6				

Name	WLN	DfH(c)	DfH(l)	DfH(g) /kJ mol-1	DvapH	DsubH
C8H14O4(5) Octanedioic acid	QV6VQ	-9.4		15.9		25.3
C9H16O4(1) Nonanedioic acid	QV7VQ	0.0				
C10H18O4(2) Decanedioic acid	QV8VQ	-2.5		30.5		33.0
C11H20O4(1) Undecanedioic acid	QV9VQ	6.4				
C12H22O4(2) Dodecanedioic acid	QV10VQ	1.6		17.2		15.6
C13H24O4(1) Tridecanedioic acid	QV11VQ	9.1				

! anomalies possibly caused by non-bonded terms particularly in tetraethyl compound

C5H8O4(1) Methylbutanedioic acid (Methylsuccinic acid)	QVY1&1VQ	-9.0
C6H10O4(5) 2-Ethylbutanedioic acid (Ethyl succinic acid)	QVY2&1VQ	-14.2
C6H10O4(2) meso-2,3-Dimethyl-butanedioic acid	QVY1&Y1&VQ &&CIS	-4.6
C6H10O4(3) racemic-2,3-Dimethyl-butanedioic acid	QVY1&Y1&VQ &&TRANS	-10.9
C6H10O4(4) (-)-2,3-Dimethylbutanedioic acid	QVY1&Y1&VQ &&-	-9.6
C8H14O4(3) meso-2,3-Diethylbutanedioic acid	QVY2&Y2&VQ &&CIS	5.2!
C8H14O4(4) racemic-2,3-Diethylbutanedioic acid	QVY2&Y2&VQ &&TRANS	-1.9
C6H10O4(1) 2,2-Dimethylbutanedioic acid	QVX1&1&1VQ	-8.3
C8H14O4(2) 2,2-Diethylbutanedioic acid (2,2-Diethylsuccinic acid)	QVX2&2&1VQ	-1.7
C7H12O4(1) Trimethylbutanedioic acid (Trimethylsuccinic acid)	VQY1&X1&1&VQ	2.3
C10H18O4(1) Triethylbutanedioic acid (Triethylsuccinic acid)	QVY2&X2&2&VQ	14.0!
C8H14O4(1) Tetramethylbutanedioic acid (Tetramethylsuccinic acid)	QVX1&1&X1&1&VQ	20.9!
C12H22O4(1) Tetraethylbutanedioic acid (Tetraethylsuccinic acid)	QVX2&2&X2&2&VQ	39.9!

Footnotes

$C_2H_4O_2$(1) - $C_{20}H_{40}O_2$(1)

See Figure 1.

Data for (c) in reasonable agreement except for $C_{11}H_{22}O_2$(1) and $C_{18}H_{36}O_2$(1).

Alternating pattern of data for gaseous state after $C_{13}H_{26}O_2$(1).

$C_3H_4O_4$(1) - $C_{13}H_{24}O_4$(1)

See Figure 2.

Regular alternation in values for crystalline state.

Data for $C_{10}H_{18}O_4$(2) (g) appears anomalous.

$C_5H_8O_4$(1) - $C_{12}H_{22}O_4$(1)

Consistent set of values except for $C_8H_{14}O_4$(3), $C_{10}H_{18}O_4$(1), $C_8H_{14}O_4$(1) and $C_{12}H_{22}O_4$(1).

Anomalies possibly caused by steric interactions particularly in tetraethyl compound.

Table 11. Differences Between Experimental and Calculated Values of

Standard Enthalpies of Formation of Three Membered Rings

! - systematic discrepancies
- possibly incorrect experimental data

Name	WLN	DfH(c)	DfH(l)	DfH(g) /kJ mol-1	DvapH	DsubH
C3H6(2) Cyclopropane	a3a		112.4	115.7	3.3	
C4H8(6) Methylcyclopropane	a2Ya1		111.6			
C5H10(7) Ethylcyclopropane	a2Ya2		110.8			
C5H8(8) Ethenylcyclopropane (Vinylcyclopropane)	1U1Y1a&1a		132.6# (3.1)			
C5H10(13) 1,1-Dimethylcyclopropane	a1X1&1&1a		112.5	117.5	4.8	
C5H10(10) cis-1,2-Dimethylcyclopropane	a1Y1&Ya1		116.2			
C5H10(11) trans-1,2-Dimethyl- cyclopropane	a1Y1&Ya1		111.7			
C7H14(22) cis-1,2-Diethylcyclopropane	aY2&1Ya2		114.1			
C7H14(23) trans-1,2-Diethyl- cyclopropane	aY2&1Ya2		110.7			
C6H12(20) 1,1,2-Trimethylcyclopropane	aY1&Y1&Ya1		78.9#			
C7H14(12) 1,1-Dimethyl-2-ethyl- cyclopropane	a1X1&1&Ya2		114.0			
C8H16(7) 1,1-Dimethyl-2-propyl- cyclopropane	a1X1&1&Ya3		113.9			
C11H22(1) 1,1-Dimethyl-2-hexyl- cyclopropane	a1X1&1&Ya6		114.2			
C7H14(21) 1,1,2,2-Tetramethyl- cyclopropane	aX1&1&X1&1&1a		94.7#			
C3H7N(1) Cyclopropylamine	a1YZ1a		113.7	118.4	4.6	
C4H5N(2) Cyclopropanecarbonitrile	a1YCN&1a		110.4	117.2	6.9	
C7H2N4(1) 1,1,2,2-Cyclopropane- tetracarbonitrile	a1XCNCNXaCNCN	204.7! (9.8)				

Name	WLN	DfH(c)	DfH(l)	DfH(g) /kJ mol-1	DvapH	DsubH
C5H8CL2(1)	aY1&Y1&XaGG					
cis-1,1-Dichloro-2,3-dimethylcyclopropane			166.6!			
C5H8CL2(2)	aY1&Y1&XaGG					
trans-1,1-Dichloro-2,3-dimethylcyclopropane			163.2!			
C6H10(7)	a2YaYb2b					
Bicyclopropyl			219.8! (2.1)	229.4! (2.3)	9.7!	
C4H6(6)	a1YU1&1a					
Methylenecyclopropane				175.2!		
C3H4(3)	a1U2a					
Cyclopropene				226.5		
C4H6(7)	a2UY1&a					
1-Methylcyclopropene				229.9		
C2H4O(2)	a1O1a					
Oxirane (Ethylene oxide)			100.1	112.3	12.3	
C3H6O(4)	aY1&O1a					
Methyloxirane (Propene oxide)			95.3	105.8	10.6	
C4H8O(5)	a1OYa2					
Ethyloxirane (Butene oxide)			75.2#			
C5H10O2(17)	aO1Ya1O2					
(Ethoxymethyl)oxirane			101.0			
C6H12O2(17)	aO1Ya1O3					
(Propoxymethyl)oxirane			102.3			
C8H16O2(11)	aO1Ya1O5					
((Pentyloxy)methyl) oxirane (Glycidyl pentyl ether)			106.3			
C7H14O2(12)	aO1Ya1OX1&1&1					
(tert-Butoxymethyl)oxirane			101.2			
C8H16O2(10)	aO1Ya1OX1&1&2					
((1,1-Dimethylpropoxy)-methyl)oxirane			104.5			
C10H20O2(3)	aO1Ya1OX1&1&4					
((1,1-Dimethylpentyloxy)-methyl)oxirane			106.5			
C3H6O2(1)	a1OYa1Q					
2,3-Epoxy-1-propanol (Oxiranemethanol)			101.9			
C3H5O1CL(1)	a1YOa1G					
(Chloromethyl)oxirane			110.9	121.0 (3.1)	10.1 (3.5)	
C5H7O6N2F(1)	a1OYa1O1XFNWNW					
((2-Fluoro-2,2-dinitro-ethoxy)methyl)oxirane			202.4!			

Name	WLN	DfH(c)	DfH(l)	DfH(g) /kJ mol-1	DvapH	DsubH
C2H5N(1) Aziridine	a1M1a		99.8	114.9	15.1	
C2H3N2CL(1) 3-Methyl-3-chlorodiazirene	aNUNXa1&G			106.5# (35.8)		
C2H4S(1) Thiirane	a1S1a		73.6	80.0	6.4	
C3H6S(1) Methylthiirane	a1SYa1		66.4	74.6	8.2	
C4H8S(1) 2,2-Dimethylthiirane	a1SX1&1&a		63.0	70.8	7.8	
C4H8S(2) cis-2,3-Dimethylthiirane	aY1&SY1&a		63.8	71.0	7.2	
C4H8S(3) trans-2,3-Dimethylthiirane	aY1&SY1&a		58.6	63.3	4.7	
C5H10S(3) Trimethylthiirane	aY1&SX1&1&a		59.9	69.0	9.1	
C6H12S(3) Tetramethylthiirane	aSX1&1&Xa1&1	69.7				

Footnotes

C7H2N4(1)

Correction due to interactions between -CN groups.

(c.f. more extensive set of corrections in nitrile table).

C5H8CL2(1) and C5H8CL2(2)

Corrections due to interactions between -Cl atoms.

(c.f. more extensive set of corrections in halide table).

C4H6(6)

Corrections due to extra strain arising from exocyclic double bond.

C6H10(7)

Correction double that of cyclopropane.

C5H7O6N2F(1)

Correction due to interactions in the group -C(NO2)2F.
(c.f. similar interactions in the halide table).

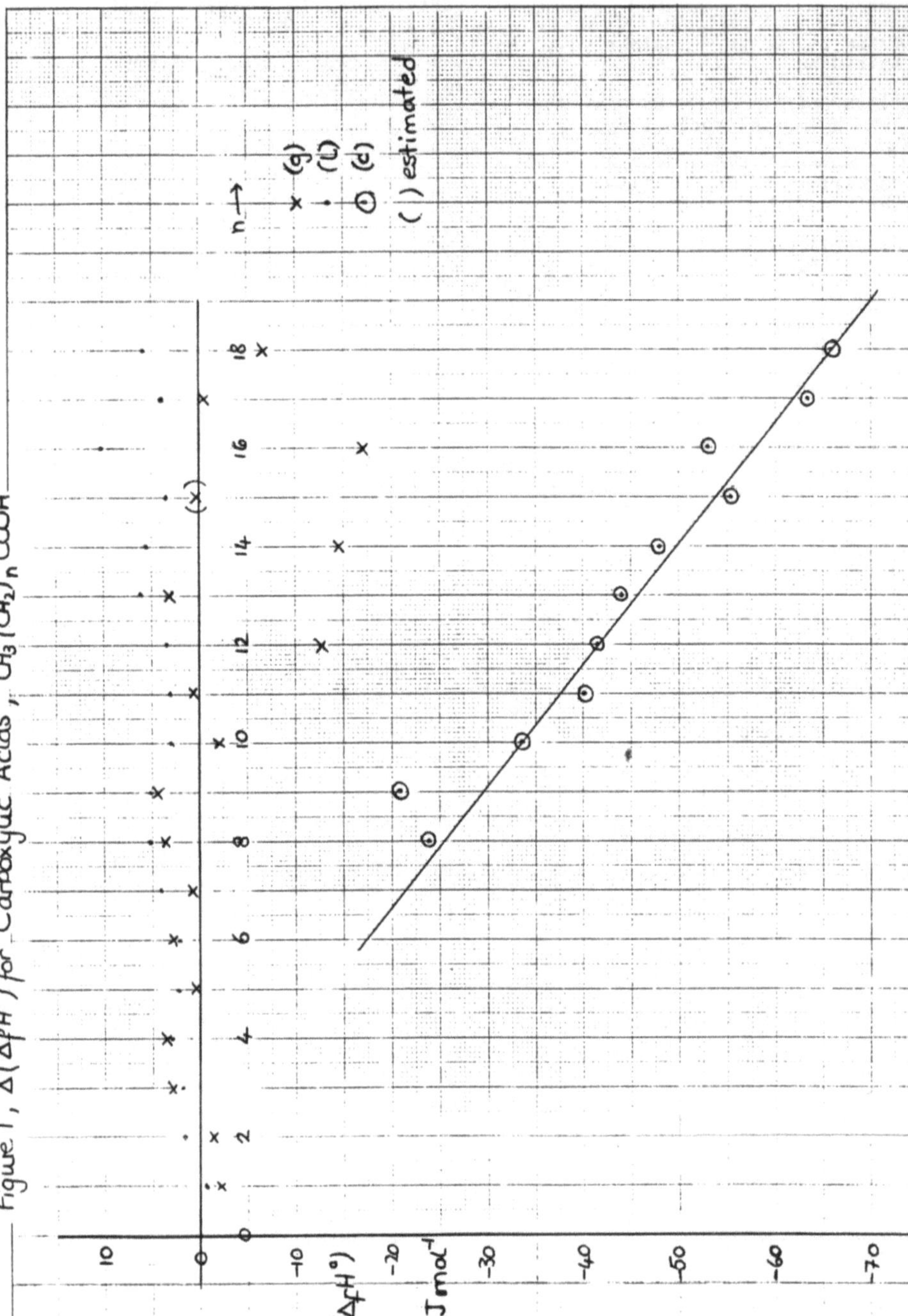

Figure 1, $\Delta(\Delta_f H°)$ for Carboxylic Acids, $CH_3(CH_2)_nCOOH$

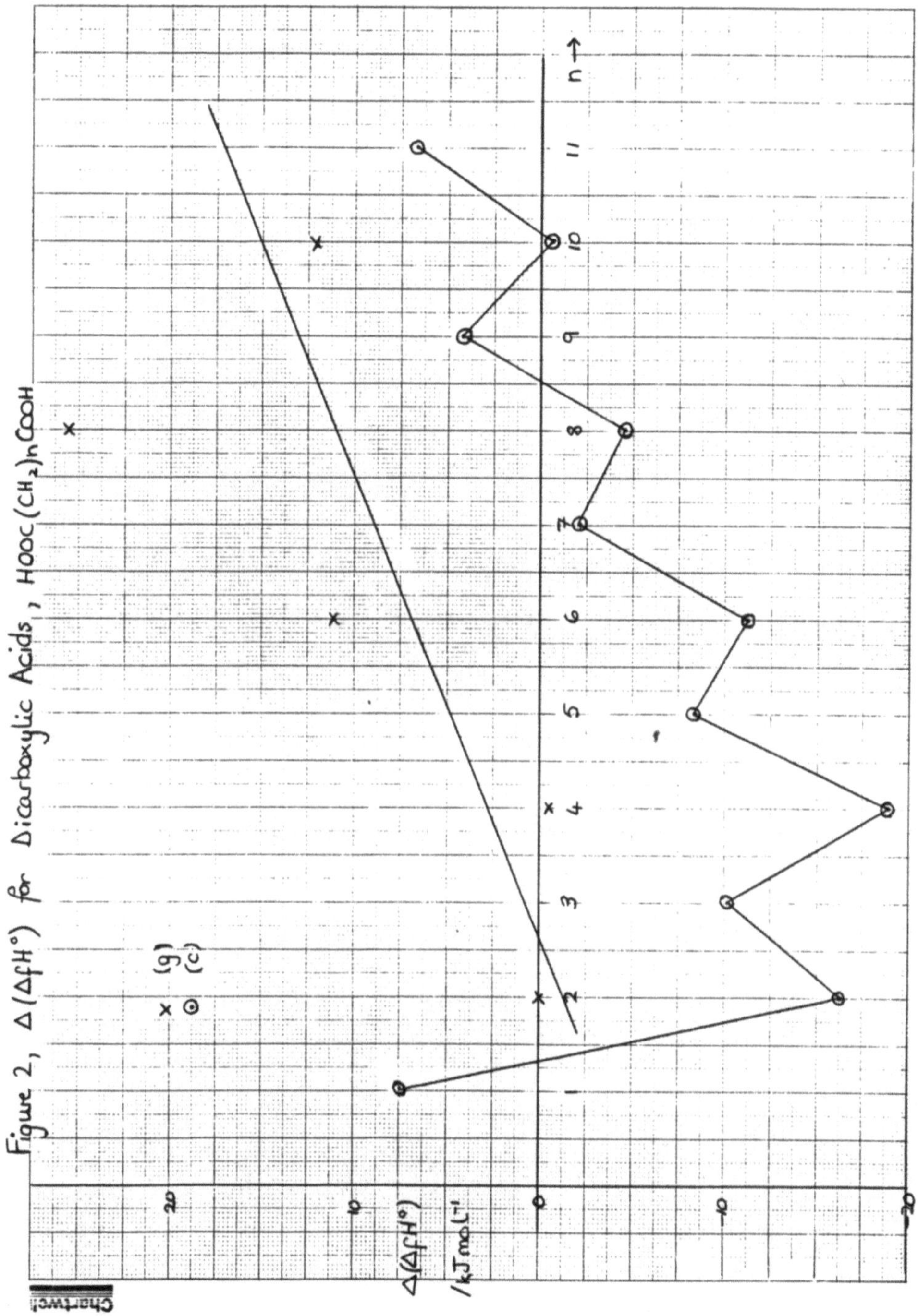

Figure 2, Δ(ΔfH°) for Dicarboxylic Acids, $HOOC(CH_2)_nCOOH$

Prediction of Physicochemical Properties Using a Semi-Empirical Group Contribution Approach

J. Howard Rytting

Pharmaceutical Chemistry Department, The University of Kansas,
Lawrence, KS 66046, USA

I. INTRODUCTION

Considerable effort has been directed in the past toward understanding the relationship between molecular structure and the distribution-transport characteristics of chemicals and drugs. Although distribution is an equilibrium process and transport is a rate process, transport is subject to distribution for all passive diffusion processes and for most active transport phenomena in their presaturation phase. For example, a substantial portion of drug absorption involves partitioning into the rate-controlling membrane. The lipophilic-hydrophilic balance of ophthalmic drugs, for instance, largely determines their transcorneal absorption rate. Similarly, the rate of GI absorption of passively absorbed drugs usually depends on their ability to partition into the rate-determining barrier membrane. This condition is also true for dermal agents.

Such basic considerations can be used to design and develop more effective drugs and form the rationale for the use of structure-activity relationships. Over the years, several a priori methods for predicting the thermodynamics of organic molecules in solution have been proposed. However, no single approach has been completely satisfactory. The thermodynamic parameters that define solution behavior provide a basis for predicting solubility and the distribution properties of molecules. A method that allows a priori estimation of solution behavior should have application in understanding the distribution and transport of drug molecules in the body.

Ideally, it is preferable to be able to predict how a particular solute will behave in a given solvent simply from the physical properties of the pure components (e.g., molar volume, solubility parameter, dipole moment, and polarizability). However, rigorous methods based on this approach are limited almost entirely to mixtures of nonpolar species. The extension of statistical thermodynamics to binary systems has yet to

C. Jochum M. G. Hicks J. Sunkel (Eds.)
Physical Property Prediction
in Organic Chemistry

provide good estimates of nonideal behavior, and often the derived equations are based on mathematically convenient approximations that have little physical significance. Consequently, a semiempirical group contribution approach appears to be an acceptable alternative.

II. METHODS OF DETERMINING GROUP CONTRIBUTIONS

In the group contribution approach, a molecule is considered to be composed of groups acting independently of the rest of the molecule and having certain associated thermodynamic properties. Therefore, the activity coefficient, free energy or partition coefficient can be found from the sum of the values for the different groups comprising the molecule. In contrast to the more theoretical approaches, this method is derived solely from an analysis of empirical data. However, the results frequently assume a form similar to those mathematically derived from statistical mechanics. In cases where two polar grouping are close to each other on a solute molecule, some modification of the additive concept is necessary. However, it should be possible to develop methods for their calculation. This concept was originally introduced by Langmuir[1] as the "principle of independent surface action". It was refined and verified by Butler and coworkers in papers on the thermodynamics of hydration[2,3].

The group contribution approach has been applied to many different situations. It forms the basis for the extensive correlations made using structure activity relationships. It has been applied to the transfer of whole molecules from one phase to another phase[4].

A. ION-PAIR EXTRACTION STUDIES.

In a study from our laboratories, its application to the transfer of various hydrocarbon groups from water to nonpolar solvents was demonstrated[5]. This study involved the determination of ion pair extraction constants for 26 alkyl sulfates as a function of temperature. These data were used to calculate free energy, enthalpy and entropy changes for the transfer of 18 different organic groups. Group contribution constants were then calculated for these various groups.

The trends found in the thermodynamic quantities calculated here were interpreted in terms of interfacial interactions, water structuring, and specific solvation in the aqueous phase. It was further demonstrated that for saturated hydrocarbon groups, excellent correlation was found for the free energy of transfer from aqueous to nonpolar phases with the surface areas of the groups. The surface areas were found to be of greater significance than the volume of the solute in determining the solution behavior of many organic molecules.

B. HENRY'S CONSTANTS

In another study from our laboratories, the group contribution approach was extended to polar groups including the hydroxyl and amino groups. In this study the contributions of the hydroxyl, amino and methylene groups to the free energy of transfer from water to the organic phase, from water to the vapor phase, and from heptane to the vapor phase were determined from the Henry's constants of aliphatic series of hydrocarbons, alcohols and amines in dilute aqueous and heptane solutions[6].

In Figure 1, a plot of the log of the Henry's constant values found in the study referred to above versus carbon number for the alcohols and amines studied using a gas chromatographic head space technique in water as well as for alkanes in water is shown. The free energy of the methylene group is calculated from the slope of the line:

$$(\Delta G) = -RT \ln (\text{slope}) \qquad \text{(Eq. 1)}$$

The result was -148 cal/mole for the transfer of a methylene group from water to the vapor phase. In addition, from Fig. 1, the hydroxyl and amine group contributions may be obtained by subtracting the alkane in water values from the alcohol in water and amine in water values for each corresponding carbon number.

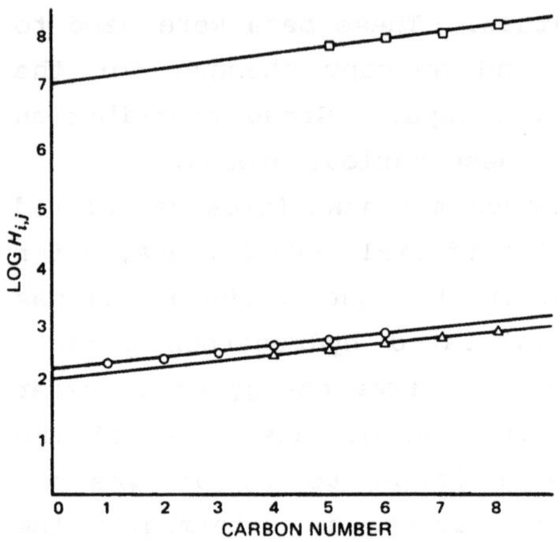

Figure 1. Log Henry constants versus carbon number for homologous series of alkanes (□), alcohols (o), and amines (△) in water. (From Rytting, et al. Ref. 6).

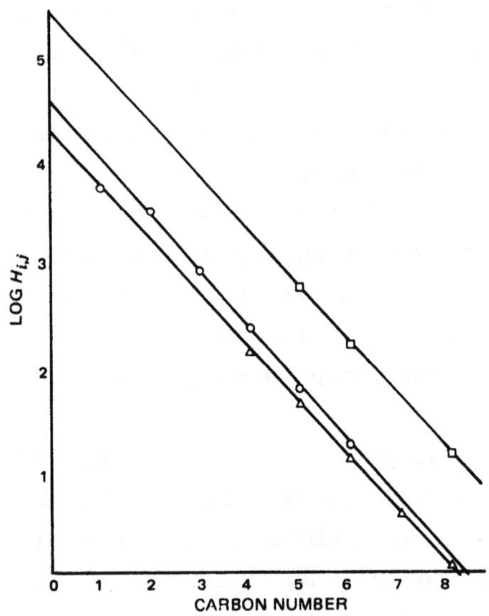

Figure 2. Log Henry constants versus carbon number for homologous series of alkanes (□), alcohols (o), and amines (△) in heptane. (From Rytting, et al. Ref. 6).

The results are shown in Table I.

Figure 2 shows a plot of the Henry's constant values versus the carbon number for the alcohols, amines, and alkanes in heptane. The free energy of the methylene group was approximately 730 cal/mole for the transfer of a methylene group from organic to vapor phases. The hydroxyl and amino group contributions also may be obtained from Fig. 2 by subtracting the alkane in heptane values for each corresponding carbon number from the values for the alcohol or amine as described earlier. For these data, the values were 1320 cal/mole for the hydroxyl group and 1490 cal/mole for the amino group from organic to vapor phase.

Combining these values one can calculate the group contribution for transfer of a methylene, hydroxyl or amino functional group from water to the organic phase. This approach has an advantage over direct measurement of partition coefficients in that the mutual solubility of the two solvents do not affect the results. In some cases where the partition coefficient is very high, it may also be experimentally easier to measure Henry's constants than partition coefficients.

Group contribution approaches have been quite useful in predicting partitioning behavior. They have also been more successful in predicting free energy changes than in predicting enthalpy and entropy changes. Properties such as solubility, melting point, and thermodynamic activity of drugs in their pure liquid or solid state are less predictable especially for relatively polar molecules. We have been particularly interested in the application of these approaches to solubility and dissolution phenomena.

III. THE PREDICTION OF SOLUBILITY

Attempts to understand and predict the solubilities and solution behavior of drugs and chemical substances in general often lead to studies of the thermodynamic properties of mixtures. The interpretation of solution nonideality usually has followed two dissimilar lines: the "physical" approach originated by van Laar[7] and the "chemical" approach proposed by Dolezalek[8]. The physical approach may be described by a random distribution of molecules throughout the entire solution, while the chemical approach may be characterized by a specific geometric orientation of one molecule with respect to an adjacent molecule.

The physical approach generally relies on the premise that the solubility of solute in a given solvent is related only to the bulk properties of the pure components. This approach is also the basis of predictive methods derived from regular solution theory in which solubility is predicted from the solubility parameters of the pure components[9]. Although this approach was initially intended for systems which involve only London dispersion forces (nonpolar molecules), its use has been expanded to include quite polar solutions by using various extended solubility parameter approaches. Although such attempts generally lack a firm theoretical foundation, they do demonstrate that some accounting of specific interactions is often very important. Conversely, even in systems known to contain specific interactions, the need to properly account for nonspecific interactions should be recognized.

One approach that appears to adequately account for many situations is based on the Nearly Ideal Binary Solvent (NIBS) model developed by Bertrand and co-workers[10,11] for systems containing only nonspecific interactions. Using a somewhat modified approach of this model, we found that predictions of benzil solubility in mixtures of carbon tetrachloride and isooctane based on this model agreed with experimental solubilities quite well if the equations were based on the Flory-Huggins approach rather than on regular solution theory[12]. For example, such a comparison is shown in Figure 3. The success of the NIBS approach for this system is significant since the mole fraction solubility of benzil changes by a factor of 14 as one goes from pure isooctane to pure carbon tetrachloride. Our

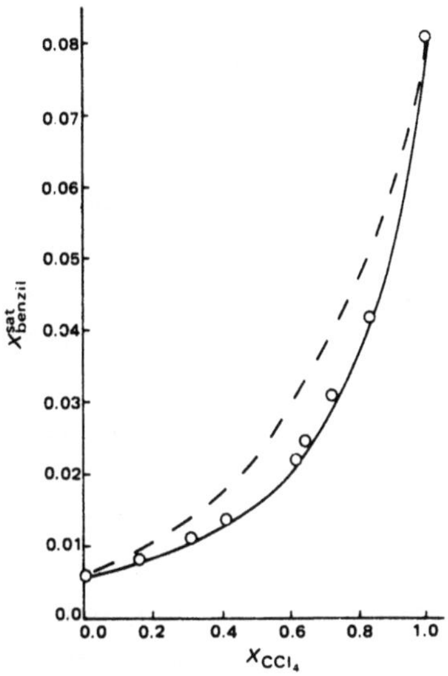

Figure 3. Comparison between experimental solubilities (0) and the NIBS predictions for benzil in binary mixtures of carbon tetrachloride and isooctane using regular solution theory (- - -) and the Flory-Huggins model (----). (From Acree and Rytting, Ref. 12).

studies[12,13,14] indicate expressions derived from the NIBS model adequately account for the solubility and solution behavior of a number of systems as long as nonspecific interactions are the only important considerations. In situations where specific interactions are important it is necessary to extend the model to account for these. We have derived expressions to extend the NIBS model to include association between a solute and one of the solvent components for ternary systems[15]. Several expressions were tested for their ability to describe anthracene solubilities in binary solvent mixtures containing benzene. The best description of the experimental solubilities requires the formation of a 1:1 anthracene-benzene complex with an equilibrium constant of $K_{AC} = 0.11\ M^{-1}$. In comparison, a stoichiometric complexation model which attributes all solubility enhancement to the formation of anthracene-benzene complexes requires a larger equilibrium constant ($K^{c}_{AC} = 0.23\ M^{-1}$) to describe the solubility behavior of anthracene in the benzene-n-heptane solvent system. The fact that the two equilibrium constants differ by a factor of two demonstrates the importance of including nonspecific interactions in equilibrium constant calculations, particularly in the case of weak association complexes. Furthermore, the determination of solute-solvent equilibrium constants from solubility data depends on the theoretical model used and the manner in which nonspecific interactions are incorporated into the model.

IV. CHOICE OF REFERENCE AND STANDARD STATE

The choice of the reference or standard state in making comparisons among molecules can have a profound effect on the development of various group contribution approaches and methods. One can choose any set of conditions desired as a standard state. However, the convenience of certain conditions has led to a few choices being usually adopted. Traditionally the use of the pure substance as a standard state is often used. An advantage of this choice is that the reference state is independent of solvent which makes it feasible to make comparisons among different solutions of the same solute. A limitation to this choice is that a different state is used for each solute making comparisons among solutes difficult.

The vapor state is another possible choice as a reference state. Since intermolecular interactions in the vapor state are small and usually negligible at low vapor pressures, the relative vapor pressures of different compounds above their respective pure states are a function primarily of the differences in the energetics of their condensed phases. Unfortunately, vapor pressures are low and difficult to measure for most drug substances at room temperature.

Rytting, et al.[16] proposed that a hydrocarbon solution at infinite dilution is a convenient reference state. The standard state is then defined as a hypothetical 1 molar solution of the drug in an alkane solvent. This choice leads to similar activities in nonpolar solvents and can be extended to group contributions to the distribution properties of a molecule between various phases. For example the methylene group contribution is found to be relatively constant over a large range of nonpolar solvents, indicating that at infinite dilution the free energy contribution of a methylene group is nearly independent of solvent[16]. Hydrocarbon solvents are also useful because they interact with the solute primarily through weak dispersion and induction forces so that the solubility of a drug in this type of solvent reflects primarily interactions in the pure solid phase. These interactions between the solute and hydrocarbon solvent are relatively insensitive to minor molecular structural changes in the solute and thus most differences observed in the hydrocarbon solubility for a series of structurally related drugs can be attributed primarily to differences in interactions present in the solid state. One exception to this generalization occurs when one is dealing with relatively small highly polar solutes that can interact relatively strongly with the hydrocarbon solvent through dipole-induced dipole (Debye) interactions.

An example of this situation has been reported from our laboratories[17,18] in a comparison of the physical and solution properties of gamma-butyrolactone and ethyl acetate. Although gamma-butyrolactone ($C_4H_6O_2$) closely resembles ethyl acetate ($C_4H_8O_2$) in its empirical chemical composition, its physical properties are very different from those of the noncyclic ester. For example, gamma-butyrolactone boils at 206° C whereas ethyl acetate boils at 77.1° C. The vapor pressures of both esters in isooctane solutions as a function of concentration were measured by gas chromatographic analyses of the head space over these solutions. Henry's constants for each ester were found to be 4.0 torr for gamma-butyrolactone and 46.7 torr for ethyl acetate.

The results suggest that gamma-butyrolactone interacts significantly with isooctane molecules. The interaction appears to involve dipole-induced dipole interactions because of the marked differences in the dipole moments of gamma-butyrolactone and ethyl acetate along with the consideration of the molecular structures of these two esters. This interpretation is supported by the fact that the free energy of solvation of gamma-butyrolactone in isooctane at infinite dilution is 1.5 kcal/mole less than that found for ethyl acetate. If dipole-induced dipole interactions are so important in these situations, these effects should be even more pronounced in other solvents generally considered to have low polarity but greater polarizability. Such studies in solvents such as toluene, benzene and cinnamaldehyde confirm these suggestions[18].

V. SOLUBILITY AND DISSOLUTION PREDICTIONS

The ability to modify aqueous solubility and design agents having desired solubility and dissolution characteristics depends directly on how well one can predict solubility from molecular structure. Often there is a temptation to rely on simple rules of thumb such as "like dissolves like". However, it is often the case that such rules are inadequate in quantitative prediction of solubility especially in the case of solids.

For liquid organic compounds, Hansch has shown that in some special cases there is a relationship between the log of the inverse of solubility and the log of the octanol/water partition coefficient. Such correlations are reasonably good for changes in chain length and nonpolar groupings. However, they are not nearly as effective for polar modifications in the chemical structure. In this regard we have been interested in developing approaches to separating and predicting the components of thermodynamic activity responsible for differences in solubility and dissolution properties.

For example if one takes a simple example of the aqueous solubility of a moderately polar liquid such as 1-butanol. The overall free energy of solution for the alcohol can be considered as the sum of the contributions reflecting the breaking of bonds in the pure alcohol, the solvation of the hydroxyl group in 1-butanol by water molecules and the hydrophobic interactions between water and the hydrocarbon chain of the alcohol. It was found that the standard free energy of transfer of 1-butanol from its pure state to an alkane solvent is 2.0 kcal/mole[19,20] which reflects primarily the free energy associated with breaking the hydrogen bonds in pure 1-butanol. The energy involved in solvating 1-butanol can be obtained from a ratio of Henry's constants in an alkane solvent and water, which we found to be about 0.3 kcal/mole[6]. The hydroxyl group contribution of -4.5 kcal/mole[6] for this process can be combined with the previous value to calculate a hydrocarbon chain contribution of 4.8 kcal/mole. The sum of these values yields a free energy change of solution for butanol of 2.3 kcal/mole which can be converted to a mole fraction solubility of 0.02[20]. By separating the aqueous solution process into these terms one can see that although the dominant term is that due to hydrophobic bonding, the relatively low aqueous solubility of 1-butanol is partially due to the extensive hydrogen bonding and self-association of 1-butanol in the pure alcohol since the terms for aqueous solution nearly cancel. This type of detailed information cannot be obtained from solubility data alone and therefore provide a rationale for thorough studies of the solution process which includes a consideration of the energetics associated with each step in the dissolution process. The prediction of the solubility of solids is complicated by the additional complex forces associated with the crystal lattice.

A typical approach is to separate the aqueous solution energetics of solids into processes reflecting the removal of the solute from the crystal and the transfer into water considers the solid going to a supercooled melt and the solute then dissolving in water[20]. The first step in this process represents the ideal solubility of the substances as calculated from the melting point and heat of fusion. The supercooled melt is usually a hypothetical state which is not attained experimentally. The second step is analogous to a partition coefficient of the drug between water and its own melt. This type of an approach is thermodynamically correct in that the individual free energies for each process add up to the total. However, the individual steps are not very revealing. The free energy associated with the transfer of a drug from the solid to its supercooled melt does not clearly differentiate between specific interactions in the crystalline state and specific interaction which can occur in the melt particularly for polar substances that can interact through hydrogen bonding or other strong specific interactions. Furthermore, the melt cannot act as a unchanging reference solvent when comparing various solutes.

Alternatively, we have considered a scheme that considers dissolution as a series of steps involving the dissolution of the solute in a hydrocarbon solvent followed by partitioning into water (Figure 4). Using this approach, the solubilities of 16 substituted melamines in water and isooctane have been measured as a function of temperature using ultraviolet spectroscopy[21]. Fourteen of the compounds required synthesis and the solubilities of 6 of the compounds were also measured in 1-octanol. The

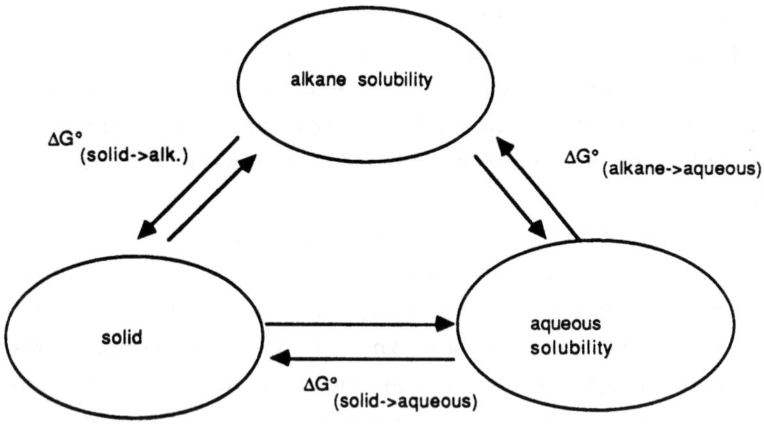

HESS' LAW

$$\Delta G^{o}_{(solid\text{->}aqueous)} = \Delta G^{o}_{(solid\text{->}alkane)} + \Delta G^{o}_{(alkane\text{->}aqueous)}$$

term A **term B**

EQUILIBRIUM

$$\Delta G^{\circ}_{(solid\text{->}alkane)} = -2..303\ RT \log\ (X_{alk.\ solubility})$$

$$\Delta G^{\circ}_{(alkane\text{->}aqueous)} = -2.303\ RT \log\ (1/PC)$$

Figure 4. Schematic illustration of the free energy model proposed to predict solubility of a solid in an aqueous solvent. (From Braxton, Ref. 21).

enthalpy changes of solution were determined from the temperature dependence of the solubilities. The entropy changes were determined from the free energy changes and the enthalpy changes of solution. Differential scanning calorimetry was used to determine the enthalpy and entropy changes of fusion for each of the compounds.

For these compounds it was found that the signs of the entropy changes for solution in isooctane were positive while in water they were negative. This suggests that the compounds dissolve via different mechanisms in the two solvents. The positive entropy change in isooctane reflect minimal solvent-solute interactions in the hydrocarbon system. However, the entropy change of solution in water is negative for all compounds indicating extensive structuring of water around the dissolved solute or other extensive solute-solvent interactions.

The aqueous enthalpy change of solution is small while in isooctane it is large and positive for all compounds. For 4 of the compounds the value is close to the experimentally determined enthalpy of fusion suggesting that they form nearly ideal solutions in isooctane. As illustrated in figure 5, the solution process in isooctane is enthalpy dominated and the solution of the solids in water is entropy controlled for nearly all of the compounds studied. For the compounds that deviate from this generalization, the process of forming a solution in isooctane appears to result in significant hydrogen bond breaking as reflected in an increase in the total enthalpy change.

Thermodynamic values were also obtained in 1-octanol for a few select compounds. As expected, compounds not capable of hydrogen bonding in the solid phase or in solution have similar solubilities in 1-octanol and in isooctane. Those compound capable of hydrogen bonding have solubilities in 1-octanol which are significantly greater than in isooctane and this is probably due to interaction with the hydroxyl group in 1-octanol.

The melting point is often considered a measure of crystal lattice energies but there appears to be a relatively weak free

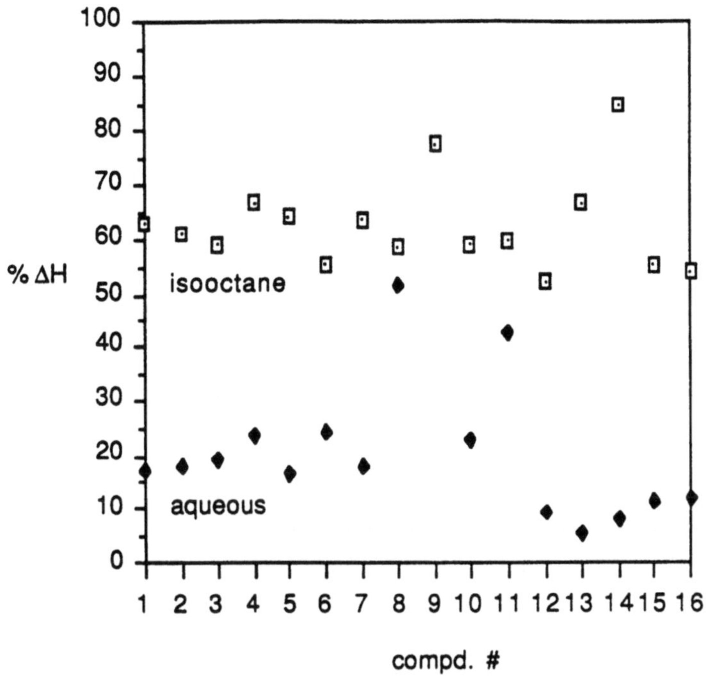

Figure 5. Schematic illustration of the percent enthalpy change contribution to the total free energy of solution for each compound studied in isooctane and 0.1 M borate buffer (pH = 8.0). (From Braxton, Ref. 21).

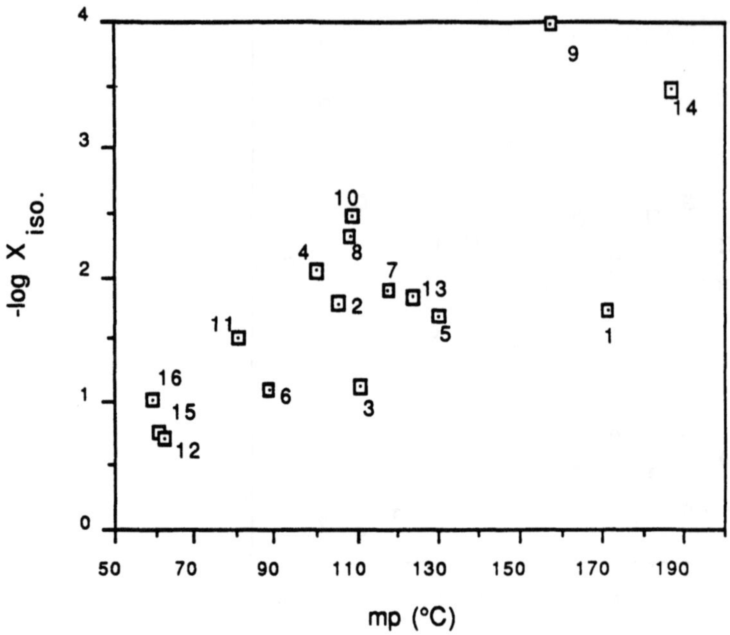

Figure 6. Plot of the logarithm of the solubility in isooctane at 25° C versus the melting point. (From Braxton, Ref. 21)

energy relationship between melting point and isooctane solubility as shown in Figure 6. This poor correlation is particularly apparent for compounds that are capable of hydrogen bonding in the solid state. However, correlations for solubility prediction using both a melting point model[22] and the free energy model described here were similar. The regression coefficients in both cases were high ($r = 0.975$ and 0.987 respectively). However, there exists a statistically significant intercept and a larger than unity slope using the melting point model. Multiple linear regression procedures show that the coefficient for the melting point is approximately two orders of magnitude smaller and statistically less significant than the coefficient determined for hydrocarbon solubilities. Thus, it appears that hydrocarbon solubility is more sensitive to changes in molecular structure than is the melting point and may yield better predictions of solubility. However, melting points are easier to measure. On the other hand, hydrocarbon solubilities may be more amenable to prediction using group contribution approaches.

VI. CONCLUSION

Historically, group contribution approaches have found a great deal of utility in predicting physicochemical properties of chemical species in various environments. They have been particularly helpful in the design of new drugs and other chemical entities using various types of structure activity relationships. There are a number of limitations and challenges involved in their development and use. However, as our ability to effectively use these approaches increases, they will become increasingly valuable in the design of drugs and prediction of various physicochemical properties.

Table I —Group Contribution to the Free Energy of Transfer, $\Delta(\Delta G)_x$, at 25° (From Rytting, et al., Ref. 6)

Group	Process	System Studied	$\Delta(\Delta G)$, cal/mole: Measured	$\Delta(\Delta G)$, cal/mole: Literature
Methylene	Water to vapor phase	Alcohols and amines in water	−148	—
Methylene	Heptane to vapor phase	Alcohol in heptane	759	—
		Amines in heptane	703	—
		Alkanes in heptane	729	—
Methylene	Water to heptane phase	Alcohols	−907	−917[a], −877[b], −872[c], −850[d]
		Amines	−851	−849[e], −880[e]
Hydroxyl	Water to vapor phase	Alcohol in water	6880	—
Hydroxyl	Heptane to vapor phase	Alcohol in heptane	1320	—
Hydroxyl	Water to heptane phase	—	5570	5280[f], 5650[g]
Amino	Water to vapor phase	Amines in water	6470	—
Amino	Heptane to vapor phase	Amines in heptane	1490	—
Amino	Water to heptane phase	—	4980	5140[f]

REFERENCES

1. I. Langmuir, Colloid Symp. Monogr., **3**, 48 (1925).

2. J.A.V. Butler, D. W. Thomson, and W. H. Maclennan, J. Chem. Soc., **1933**, 674.

3. J. A. V. Butler and P. Harrower, Trans. Faraday Soc., **33**, 171 (1937).

4. S.S. Davis, T. Higuchi and J. H. Rytting, in "Advances in Pharmaceutical Sciences," vol. 4, H. S. Bean, A. H. Beckett and J. E. Carless, Eds., Academic, London, England, 1974, pp. 73-261.

5. M. J. Harris, T. Higuchi and J. H. Rytting, J. Phys. Chem., **77**, 2694 (1973).

6. J. H. Rytting, L. P. Huston, and T. Higuchi, J. Pharm. Sci., **67**, 615 (1978).

7. J. J. van Laar, **Z. Phys. Chem.**, **72**, 723 (1910).

8. F. Dolezalek, **Z. Phys. Chem.**, **64**, 727 (1908).

9. J. H. Hildebrand and R. L. Scott, "The Solubility of Nonelectrolytes," Reinhold, New York, N.Y., 1950.

10. E. L. Taylor and G. L. Bertrand, **J. Soln. Chem.**, 3, 479 (1974).

11. W. E. Acree and G. L. Bertrand, **J. Phys. Chem.**, **83**, 2355 (1979).

12. W. E. Acree, Jr., and J. H. Rytting, **J. Pharm. Sci.**, **71**, 201 (1982).

13. W. E. Acree, Jr., and J. H. Rytting, **Intern. J. Pharmaceutics**, **10**, 231 (1982).

14. W. E. Acree, Jr., and J. H. Rytting, **J. Pharm. Sci.**, 72, 292, (1983).

15. W. E. Acree, Jr., D. R. McHan, and J. H. Rytting, **J. Pharm. Sci.**, **72**, 929 (1983).

16. J. H. Rytting, S. S. Davis, and T. Higuchi, **J. Pharm. Sci.**, **61**, 816 (1972).

17. D. J. W. Grant, T. Higuchi, Y. T. Hwang, and J. H. Rytting, **J. Soln. Chem.**, 13, 297 (1984).

18. J. H. Rytting, D. R. McHan, T. Higuchi, and D. J. W. Grant, J. Soln. Chem., **15**, 693 (1986).

19. B. D. Anderson, Ph. D. Dissertation, The University of Kansas, 1977.

20. B. D. Anderson in "Physical Chemical Properties of Drugs", S. H. Yalkowsky, A. A. Sinkula and S. C. Valvani, eds., Marcel Dekker, Inc., pp. 231-266 (1980).

21. Bryan Braxton, Ph.D. Dissertation, The University of Kansas, 1988.

22. S. H. Yalkowsky and S. C. Valvani, J. Pharm. Sci., **69**, 912 (1980).

Acknowledgements. Figures 1, 2, 3 and Table 1 have been reproduced with permission of the American Pharmaceutical Association.

Estimation of the Aqueous Solubility of Organic Compounds

S.H. Yalkowsky

College of Pharmacy, University of Arizona, Tucson, AZ 85721, USA

INTRODUCTION

In spite of its importance to a number of scientific disciplines, there are few reliable techniques for the estimation of the aqueous solubility of a crystalline organic compound. One of the major reasons for the lack of progress in this field is the fact that solubility is not strictly an additive, constituative property, and thus cannot be predicted by group contributions alone.

THEORY

Aqueous Solubility

It has been shown (1,2) that the aqueous solubility of a crystalline solute S^c_w can be described by two independent terms; one describing the effect of crystal structure (called the crystal-liquid solubility ratio, or simply the ideal solubility) and another describing the effect of solute polarity or mixing (called the activity coefficient). Mathematically this can be written as

$$\log S^c_w = \log X^c_i + \log Y_w \qquad 1$$

where X^c_i is the ideal mole fractional solubility of the crystal and Y_w is the aqueous activity coefficient of the solute. These terms will be discussed in more detail below.

Ideal Solubility

The ideal solubility is given by the van't Hoff equation which is often approximated as (1,2)

$$\log X^c_i = - \frac{\Delta S_f \,(T_m - T)}{2.303\ RT} \qquad 2$$

where ΔS_f is the entropy of fusion of the solute, T_m is the

C. Jochum M.G. Hicks J. Sunkel (Eds.)
Physical Property Prediction
in Organic Chemistry

Kelvin melting point and T is the temperature in Kelvin. At 25°C equation 2 becomes

$$\log X^c_i = - \frac{\Delta S_f \text{ (MP - 25)}}{1364} \qquad 3$$

where MP is the centigrade melting point of the solute.

For rigid organic molecules ΔS_f can be approximated by 13.5 e.u. (For flexible molecules the value is higher.) Using this approximation the ideal solubility of a rigid molecule becomes

$$\log X^c_i \sim - 0.01 \text{ (MP - 25)} \qquad 4$$

Equation 4 indicates that each one hundred degree increase in melting point will reduce solubility ten fold. (A three hundred degree increase in melting point will decrease solubility by a factor of one thousand.)

Aqueous Activity Coefficient

The aqueous activity coefficient can be related to the octanol-water partition coefficient K_{ow} by

$$K_{ow} \sim \frac{Y_o}{Y_w} \qquad 5$$

where Y_o is the activity coefficient of the solute in octanol. For most organic compounds Y_o is equal to the molarity of the pure solvent. (1,2) (Thus for most crystalline solutes the solubility in pure octanol is equal to the ideal solubility as given by equations 2, 3 and 4.) If this is so, it can be shown (1,2) that

$$\log Y_w \sim - \log K_{ow} + 0.54 \qquad 6$$

Combining equations 4 and 6 gives

$$\log S_w = -0.01\ MP - \log K_{ow} + 0.8 \qquad 7$$

Equation 7 is a generally useful equation for the estimation of the aqueous solubility of any rigid organic nonelectrolyte. It is also applicable to the unionized form of weak electrolytes.

Many techniques have been shown to be applicable to a series of closely related compounds. Unfortunately these techniques must be altered significantly each time a new series of compounds is used. Any true test of the validity of a method must involve a wide variety of chemical structures.

APPLICATIONS

Aqueous Solubility

In this report equation 7 is applied to a structurally diverse group of nonelectrolytes and weak electrolytes that are of interest to the pharmaceutical industry. The aqueous solubility data for the compounds of Table I are shown in Figure 1 as open and filled circles for nonelectrolytes and weak electrolytes respectively as a function of the values predicted by equation 7. The solid line indicates perfect prediction without any adjustable parameters. The dashed line corresponds to the regression equation

$$\log S_w = -0.012\ MP - 1.13 \log P + 1.62 \qquad 8$$

$$r = 0.955 \qquad s = 0.402$$

It can be seen from Figure 1 that equation 7 does an adequate job in predicting the solubilities of the 36 non-electrolytes (open circles) and weak electrolytes (filled circles) considered.

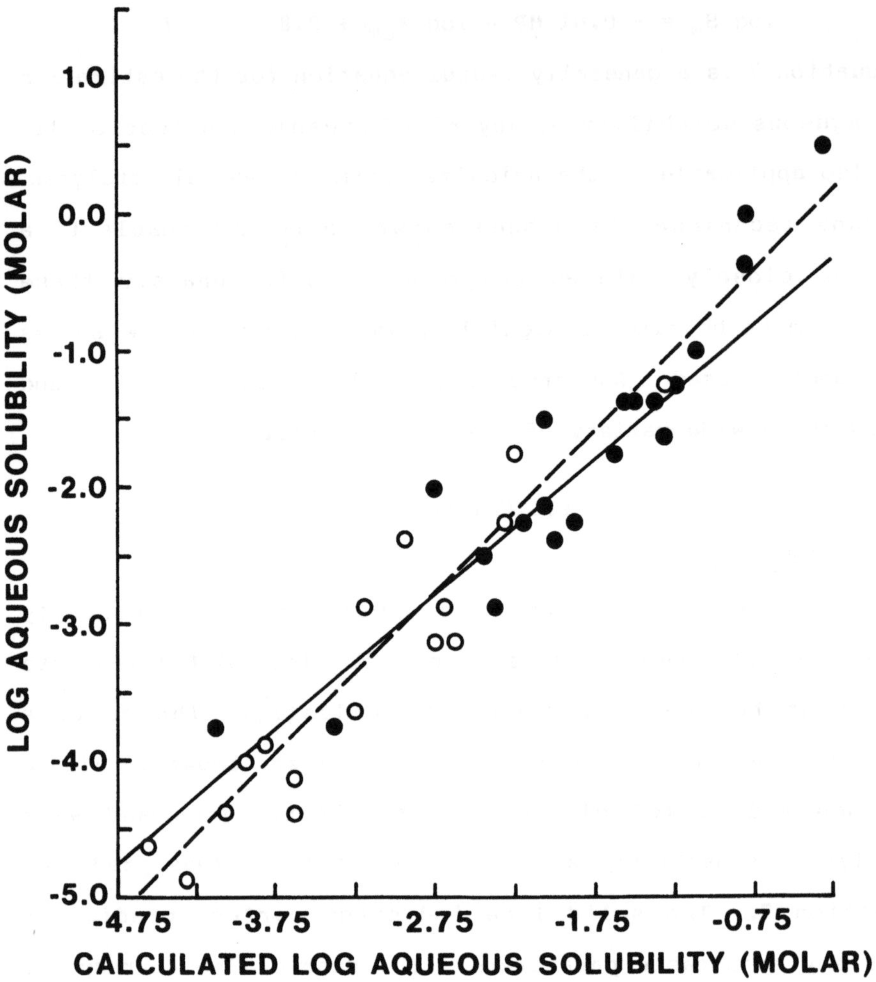

FIGURE 1 Observed and predicted aqueous solubilities of nonelectrolytes (O) and weak electrolytes (O). Key: (-) theoretical line described by Eq. 7; (---) regression line described by Eq. 8.

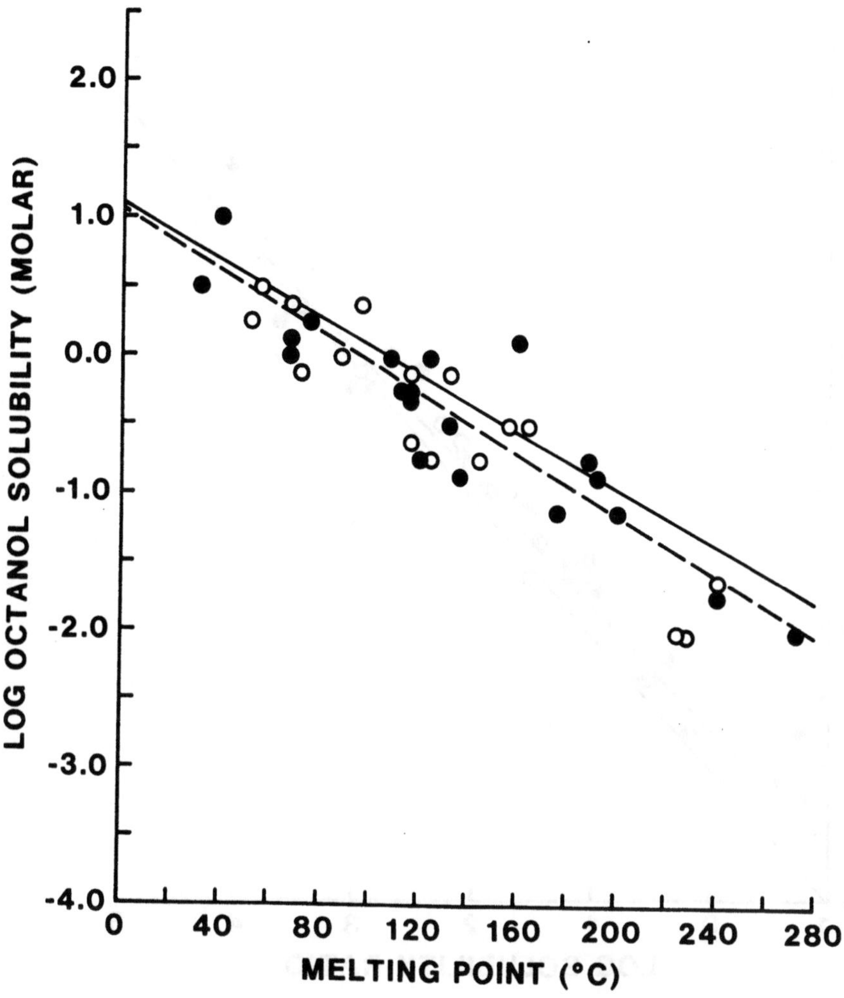

FIGURE 2 Octanol solubilities and melting points of nonelectrolytes (O) and weakelectrolytes(O). Key: (-) theoreticalline described by Eq. 4; (---) regression line.

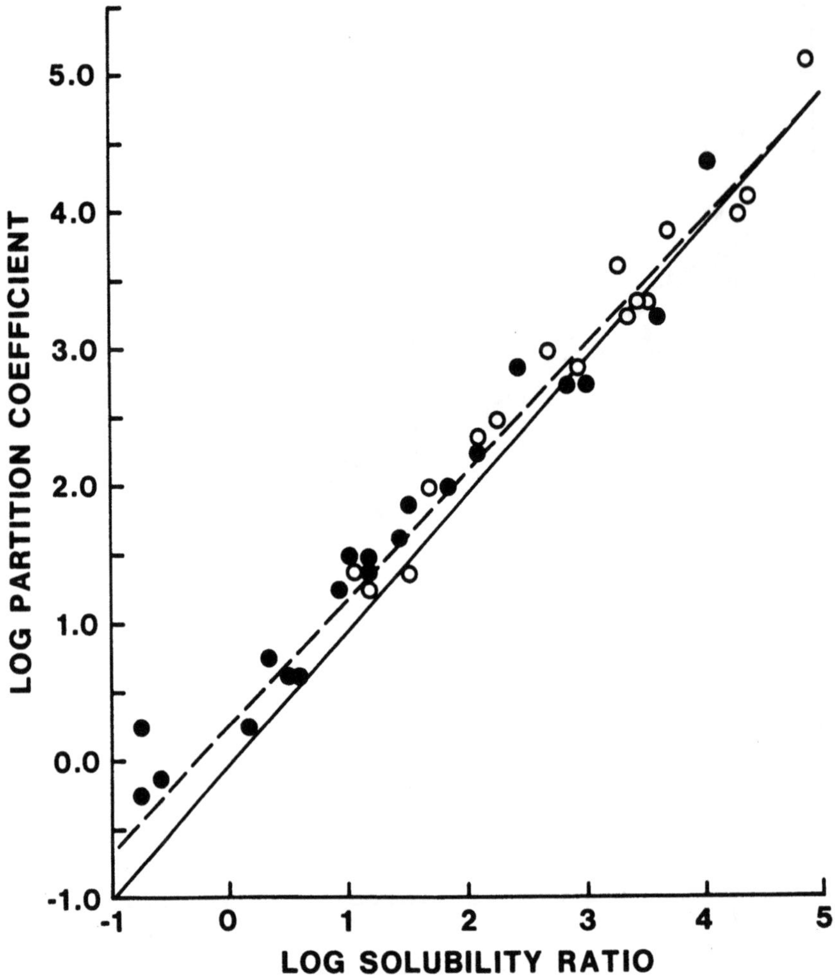

FIGURE 3 Partition coefficients and solubility ratios of nonelectrolytes (O) and weakelectrolytes(O). Key: (-) theoreticalline described by Eq. 5; (---) regression line.

Assumptions

The two major assumptions used to obtain equation 7 are (1) the solubility of the solute in octanol is equal to the ideal solubility (equation 4) and (2) the octanol-water partition coefficient is equal to the octanol-water solubility ratio (equation 6).

Octanol Solubility

The octanol solubility data is given in Table II and Figure 2. The solid line in Figure 2 represents the octanol solubility predicted by equation 4. The open and filled circles again represent the data for nonelectrolytes and weak electrolytes respectively, and the dashed line is the regression line. The closeness of the two lines validate the use of equation 4 to estimate the octanol solubility of organic solutes.

Partitioning

Table III and Figure 3 give a comparison of the experimentally determined partition coefficients of the 36 solutes and their octanol-water solubility ratios. As in the previous figures the solid line represents the predictions (in this case equivalence) and the dashed line is the regression line. Again the two lines are in good agreement with each other.

EXTENTION TO WEAK ELECTROLYTES

Equation 6 can be extended to partially ionized weak electrolytes. For a weak electrolyte the total solubility S_T is related to the solubility of the unionized form by (3,4)

$S_T = S_w + S_i$ where S_w is the solubility of the unionized solute predicted by equation 6 and S_i is the concentration of

TABLE I Estimation of Aqueous Solubility[a]

Solute	log PC (obs)	mp°,b	log S_w (obs)	log S_w (calc)	Difference
Acetylsalicylic acid	(1.21)	135	(-1.60)[c]	-1.51	0.094
p-Aminobenzoic acid	(0.58)	187	-1.35	-1.40	-0.048
Aminopyrine	(0.80)	108	(-0.36)[d]	-0.83	-0.470
Antipyrine	(0.26)	112	0.53	-0.33	-0.865
Barbital	(0.67)	190	(-1.41)[e]	-1.52	-0.106
Benzoic acid	(1.87)	122	-1.53	-2.04	-0.507
Butyl-p-aminobenzoate	2.72	58	-2.84	-2.25	0.593
Caffeine	-0.20	238	-0.98	-1.13	-0.154
Flurbiprofen	3.26	111	-3.74	-3.32	0.424
Ethyl-p-aminobenzoate	1.96	89	-2.17	-1.80	0.368
Fumaric acid	0.28	200	-1.29	-1.23	0.062
Ibuprofen	4.43	76	-3.76	-4.14	-0.382
Methyl-p-aminobenzoate	1.35	114	-1.70	-1.44	0.259
Phenobarbital	(1.48)	176	-2.26	-2.19	0.072
Phenacetin	(1.58)	135	-2.28	-1.88	0.399
Phenol	(1.48)	41	(-0.04)[f]	-0.84	-0.796
Prostaglandin E_2	2.82	67	-2.47	-2.44	0.017
Prostaglandin F_{2a}	2.72	30	-2.38	-1.97	0.409
Salicylic acid	(2.23)	158	-1.95	-2.76	-0.813
Theophylline	-0.09	272	-1.38	-1.58	-0.200
Acetanilide	(1.21)	114	-1.31	-1.30	0.007
Biphenyl	(3.98)	70	-4.34	-3.63	0.710
Butyl-p-hydroxybenzoate	3.57	69	-2.93	-3.21	-0.283
Cortisone	1.47	222	-3.12	-2.64	0.479
Desoxycorticosterone	2.88	142	(-3.45)[g]	-3.25	0.201
Diphenylethane	5.12	52	-4.63	-4.84	-0.210
Ethyl-p-hydroxybenzoate	2.47	116	-2.20	-2.58	-0.380
Fluorene	4.18	117	-4.91	-4.30	0.607
Methyl-p-hydroxybenzoate	1.96	131	-1.78	-2.22	-0.438
15-s-15-Methylprostaglandin F_{2a} methyl ester	3.21	55	-2.88	-2.71	0.173
Methyltestosterone	3.36	163	-3.99	-3.94	0.046
Prednisolone	(1.42)	240	-3.10	-2.77	0.335
Propyl-p-hydroxybenzoate	(3.04)	96	-2.35	-2.95	-0.600
Progesterone	3.87	131	(-4.15)[h]	-4.13	0.023
Testosterone	3.32	155	(-3.87)[h]	-3.82	0.048
Triazolam	2.42	224	-4.09	-3.61	0.485

a) Values in parentheses were obtained from the literature. b) Melting point data taken from M. Windholz, Ed., "The Merck Index," 9th ed., Merck & Co., Rahway, N.J. (1976) or the manufacturers' specifications. c) L.J. Edwards, Trans. Faraday Soc., 57, 1191 (1951). d) R. Charonnat, Compt. Rend., 185, 284 (1927). e) F.A. Long and W.F. McDevitt, Chem. Rev., 51, 119 (1952). f) L. Erichsen and E. Dobbert, Brennstoff Chem., 36, 338 (1955). g) H. Tomida, T. Yotsuyanagi, and K. Ikeda, Chem. Pharm. Bull. Tokyo, 26, 2832 (1978). h) K. Uekema, T. Fujinaga, F. Hirayama, M. Otagiri, and M. Tamasaki, Int. J. Pharm., 10, 1 (1982).

TABLE II Estimation of Octanol Solubility

Solute	mp^{o,a}	log X_o (obs)	log X_o (calc)	Difference
Acetylsalicylic acid	135	-1.58	-0.30	0.395
p-Aminobenzoic acid	187	-1.68	-0.82	-0.023
Aminopyrine	108	-0.89	-0.03	-0.029
Antipyrine	112	-1.06	-0.07	0.122
Barbital	190	-1.80	-0.85	0.071
Benzoic acid	122	-0.95	-0.17	-0.109
Butyl-p-aminobenzoate	58	-0.72	0.47	0.340
Caffeine	238	-2.61	-1.33	0.395
Flurbiprofen	111	-1.05	-0.06	0.140
Ethyl-p-aminobenzoate	89	-1.19	0.16	0.469
Fumaric acid	200	-2.00	-0.95	0.169
Ibuprofen	76	-0.55	0.29	0.017
Methyl-p-aminobenzoate	114	-1.42	-0.09	0.444
Phenobarbital	176	-1.97	-0.71	0.382
Phenacetin	135	-1.73	-0.30	0.543
Phenol	41	-0.06	0.64	-0.300
Prostaglandin E_2	67	-0.81	0.38	0.406
Prostaglandin F_{2a}	30	0.11	0.75	0.263
Salicylic acid	158	-0.73	-0.53	-0.679
Theophylline	272	-2.83	-1.67	0.276
Acetanilide	114	-1.00	-0.09	0.027
Biphenyl	70	-0.98	0.35	0.448
Butyl-p-hydroxybenzoate	69	-0.48	0.36	0.021
Cortisone	222	-2.85	-1.17	0.797
Desoxycorticosterone	142	-1.58	-0.37	0.340
Diphenylethane	52	-0.66	0.53	0.337
Ethyl-p-hydroxybenzoate	116	-0.83	-0.11	-0.150
Fluorene	117	-1.45	-0.12	0.445
Methyl-p-hydroxybenzoate	131	-0.96	-0.26	-0.182
15-s-15-Methylprostaglandin F_{2a} methyl ester	55	0.12	0.50	0.046
Methyltestosterone	163	-1.31	-0.58	-0.130
Prednisolone	240	-2.50	-1.35	0.270
Propyl-p-hydroxybenzoate	96	-0.55	0.09	-0.267
Progesterone	131	-1.58	-0.26	0.447
Testosterone	155	-1.35	-0.50	-0.013
Triazolam	224	-2.93	-1.19	0.856

a) Melting point data taken from M. Windholz, Ed., "The Merck Index," 9th ed., Merck & Co., Rahway, N.J. (1976) or the manufacturers' specifications.

TABLE III Estimation of Partition Coefficient[a]

Solute	log S_o (obs)	log S_w (obs)	log PC (obs)	log SR (calc)	Difference
Acetylsalicylic acid	-0.69	(-1.60)[b]	1.21	0.91	0.301
p-Aminobenzoic acid	-0.80	-1.35	(0.58)	0.56	0.024
Aminopyrine	-0.00	(-0.36)[c]	(0.80)	0.36	0.441
Antipyrine	-0.19	0.53	(0.26)	-0.73	0.987
Barbital	-0.92	(-1.41)[d]	(0.67)	0.49	0.177
Benzoic acid	-0.06	-1.53	(1.87)	1.47	0.399
Butyl-p-aminobenzoate	0.13	-2.84	2.72	2.97	-0.253
Caffeine	-1.72	-0.98	-0.20	-0.75	0.548
Flurbiprofen	-0.20	-3.74	3.26	3.54	-0.284
Ethyl-p-aminobenzoate	-0.31	-2.17	1.96	1.86	0.101
Fumaric acid	-1.12	-1.29	0.28	0.17	0.107
Ibuprofen	0.27	-3.76	4.43	4.03	0.399
Methyl-p-aminobenzoate	-0.53	-1.70	1.35	1.17	0.185
Phenobarbital	-1.09	-2.26	(1.48)	1.17	0.310
Phenacetin	-0.84	-2.28	(1.58)	1.44	0.145
Phenol	0.94	(-0.04)[e]	(1.48)	0.98	0.496
Prostaglandin E_2	-0.03	-2.46	2.82	2.43	0.389
Prostaglandin F_{2a}	0.49	-2.38	2.72	2.87	-0.146
Salicylic acid	0.15	-1.95	2.23	2.10	0.134
Theophylline	-1.95	-1.38	-0.09	-0.57	0.476
Acetanilide	-0.12	-1.31	(1.21)	1.19	0.021
Biphenyl	-0.10	-4.34	(3.98)	4.24	-0.262
Butyl-p-hydroxybenzoate	0.34	-2.93	3.57	3.27	0.304
Cortisone	-1.97	-3.12	1.47	1.15	0.318
Desoxycorticosterone	-0.71	(-3.45)[f]	2.88	2.74	0.139
Diphenylethane	0.19	-4.63	5.12	4.82	0.299
Ethyl-p-hydroxybenzoate	0.04	-2.20	2.47	2.24	0.229
Fluorene	-0.56	-4.91	4.18	4.34	-0.162
Methyl-p-hydroxybenzoate	-0.08	-1.78	1.96	1.70	0.256
15-s-15-Methylprostaglandin F_{2a} methyl ester	0.45	-2.88	3.21	3.34	-0.127
Methyltestosterone	-0.45	-3.99	3.36	3.54	-0.176
Prednisolone	-1.62	-3.10	1.42	1.49	-0.065
Propyl-p-hydroxybenzoate	0.36	-2.35	(3.04)	2.71	0.333
Progesterone	-0.71	(-4.15)[g]	3.87	3.45	0.424
Testosterone	-0.49	(-3.87)[g]	3.32	3.38	-0.061
Triazolam	-2.05	-4.09	2.42	2.05	0.371

a) Values in parentheses were obtained from the literature. b) L.J. Edwards, Trans. Faraday Soc., 57, 1191 (1951). c) R. Charonnat, Compt. Rend., 185, 284 (1927). d) F.A. Long and W.F. McDevitt, Chem. Rev., 51, 119 (1952). e) L. Erichsen and E. Dobbert, Brennstoff Chem., 36, 338 (1955). f) H. Tomida, T. Yotsuyanagi, and K. Ikeda, Chem. Pharm. Bull. Tokyo, 26, 2832 (1978). g) K. Uekema, T. Fujinaga, F. Hirayama, M. Otagiri, and M. Tamasaki, Int. J. Pharm., 10, 1 (1982).

ionized solute which is given by

$$S_T = S_w \left(\frac{K_a + [H^+]}{[H^+]}\right) \qquad 9$$

so that

$$S_T = S_w \left(1 + \left(\frac{K_a + [H^+]}{[H^+]}\right)\right) \qquad 10$$

and

$$\log S_T = \log S_w + \log \left(\frac{K_a + [H^+]}{[H^+]}\right) \qquad 11$$

Therefore

$$\log S_T = - 0.01\ MP - \log P + \log \left(\frac{K_a + [H^+]}{[H^+]}\right) \qquad 12$$

where MP and log P refer to the unionized form of the solute. Pinal (3,4) and Tsakanikas (5) have applied equation 12 to barbiturates and hydantoins and to triazine herbicides with excellent results. Their results provide further evidence of both the applicability and the validity of Yalkowsky's original equation (equation 7 for rigid molecules).

REFERENCES

1. S.H. Yalkowsky and S.C. Valvani, J. Pharm. Sci., 69, 912 (1980).

2. S.H. Yalkowsky, S.C. Valvani and T.J. Roseman, J. Pharm. Sci., 72, 866 (1983).

3. S.H. Yalkowsky, R. Pinal, and S. Banerjee, J. Pharm. Sci., 77, No. 1, 74 (1988).

4. R. Pinal and S.H. Yalkowsky, J. Pharm. Sci., 76, No. 1, 75 (1987).

5. P.D. Tsakanikas and S.H. Yalkowsky, Tox. Env. Chem., 17, 19 (1988).

Recommended g^E-Model Parameters by Simultaneous Fitting of Different Excess Properties

Jürgen R. Rarey-Nies, Dieter Tiltmann, and Jürgen Gmehling

Universität Dortmund, Lehrstuhl für Technische Chemie B, Fachbereich Chemietechnik, Postfach 500500, 4600 Dortmund, Federal Republic of Germany

ABSTRACT

Many problems of practical interest require the knowledge of the equilibrium properties of multicomponent mixtures. With the availability of modern computers complex semiempirical models for the description of the real behavior of these mixtures became popular. These models typically require two or three parameters per binary pair of components, which have to be fitted to experimental mixture data.

Although the description of the real behavior as a function of the liquid phase composition is mostly of good accuracy, the parameters used are only valid for a limited temperature region.

This paper describes the structure of an interactive program package for the simultaneous correlation of different excess properties.

Results for the system ethanol-water are presented, which stress the importance of temperature-dependent interaction parameters.

The program-package RECVAL is directly interfaced to the Dortmund Data Bank (DDB), makes excessive use of graphics for data representation and may be extended for the correlation and prediction of pure component properties.

INTRODUCTION

With the development of expressions for the excess Gibbs energy (g^E) based on the local composition concept (Wilson (1964) ; Renon and Prausnitz (1968); Abrams and Prausnitz (1975)) it became possible to predict the behavior of multicomponent mixtures from binary data alone. Being able to describe the real behavior of all components in the mixture it is now possible to simulate complex multistage separation processes or even whole chemical plants. Fig. 1 gives an overview

C. Jochum M. G. Hicks J. Sunkel (Eds.)
Physical Property Prediction
in Organic Chemistry

on some of the practical applications for models describing the real behavior of mixtures.

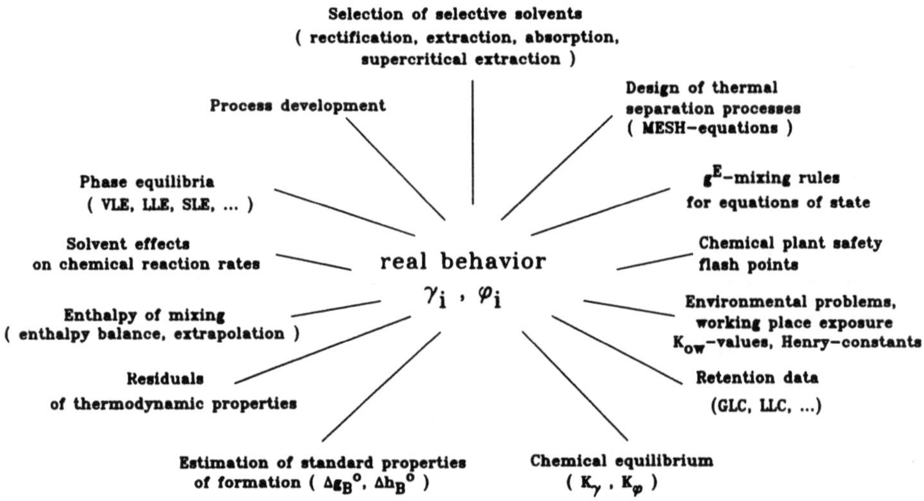

Fig. 1 Examples for practical applications, where the real behavior of the mixture is of great importance.

VAPOR-LIQUID EQUILIBRIA

In principal two different approaches may be used to describe vapor-liquid-equilibria (VLE). The real behavior of the components in both phases may be calculated by using an equation of state (EOS). This method leads to good results in case of mixtures of non-polar and simple molecules. For typical mixtures of practical interest, which contain alcohols, water, carboxylic acids etc., the use of a g^E-model for the description of the real liquid phase is preferred. Fig. 2 shows the basic equations for both approaches. A detailed discussion of this topic can be found in various textbooks, e.g. Prausnitz et al. (1987) or Gmehling and Kolbe (1988).

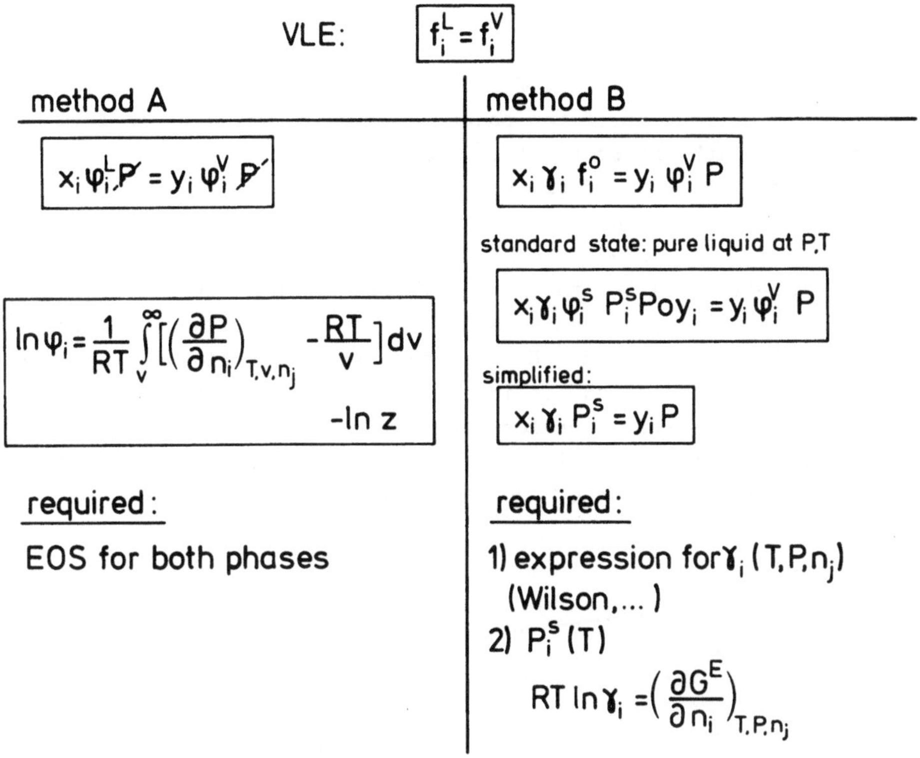

Fig. 2 Different methods for the solution of the vapor-liquid equilibrium problem

PARAMETER FITTING IN CASE OF g^E-MODELS

Up to now a number of models for the description of the excess Gibbs energy have been developed, the most commonly known are the Wilson-, NRTL- and UNIQUAC-equation. All these models require interaction parameters, which have to be obtained from the correlation of binary experimental phase equilibrium data. This is usually done by varying the parameters in order to minimize an objective function, which is a measure of the difference between calculated and experimental data. Fig. 3 contains the 3-dimensional representation of the sum of relative squared residuals in γ for the system ethanol-water measured by Mertl (1972). It can be seen from this figure that the parameters u_{12}-u_{22} and u_{21}-u_{11} of the UNIQUAC-equation are strongly intercorrelated. Using the fitted parameters different properties of the mixture can be calculated.

Fig. 4 shows the results of a correlation for five different models together with the fitted parameters, the experimental data and the deviations between the calculated and experimental values as well as a graphical representation of the phase equilibrium behavior in form of an x-y diagram (Gmehling et al. (1977-1988)).

Instead of fitting the parameters to experimental VLE-data, they can also be obtained from activity coefficients at infinite dilution (γ^{∞}). Fig. 5a and 5b show the activity coefficients calculated from experimental data for the system methanol - n-heptane and calculated curves using parameters calculated from activity coefficients at infinite dilution (a) and VLE-data (b). The prediction of the VLE from activity coefficients at infinite dilution results in a fair description of the real behavior over the whole composition range while the prediction of the infinite dilution behavior from VLE-data is not satisfactory in case of low methanol concentrations.

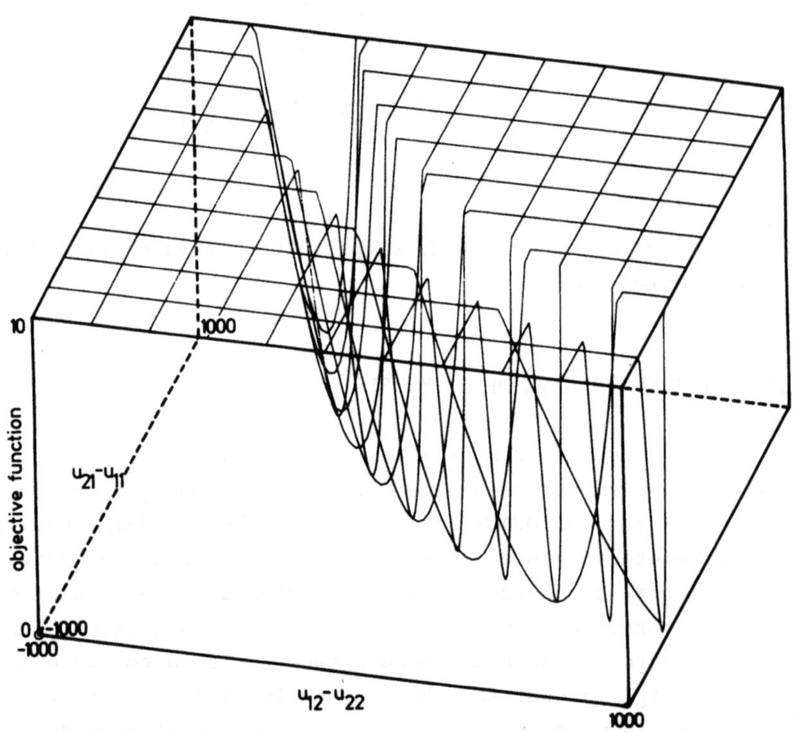

Fig. 3 Sum of relative squared deviations in γ for a typical VLE-data set as a function of the parameters of the UNIQUAC-equation. (objective function values were limited to a value of 10)

(1) ETHANOL	C2H6O
(2) WATER	H2O

+++++ ANTOINE CONSTANTS				REGION +++++	CONSISTENCY	
(1)	8.11220	1592.864	226.184	20- 93 C	METHOD 1	+
(2)	8.07131	1730.630	233.426	1- 100 C	METHOD 2	+

TEMPERATURE= 70.00 DEGREE C

LIT: MERTL I.
COLLECT.CZECH.CHEM.COMMUN. 37,366(1972).

CONSTANTS:	A12	A21	α_{12}	γ_1^∞	γ_2^∞	OBJECTIVE FUNCTION	
MARGULES	1.6346	.8563		5.13	2.35	.0304	G
VAN LAAR	1.7967	.9236		6.03	2.52	.0127	G
WILSON	471.1394	883.4301		7.24	2.62	.0120	G
NRTL	-121.3014	1338.4060	.2974	5.90	2.50	.0136	G
UNIQUAC	-30.1929	337.0028		6.04	2.52	.0126	G

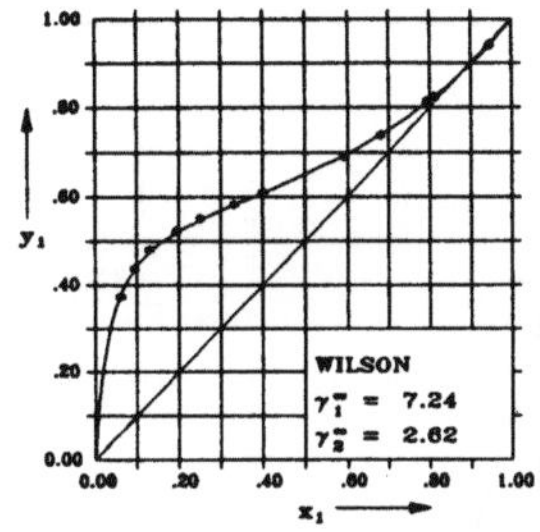

EXPERIMENTAL DATA			MARGULES		VAN LAAR		WILSON		NRTL		UNIQUAC	
P MM HG	X1	Y1	DIFF P	DIFF Y1	DIFF P	DIFF Y1	DIFF P	DIFF Y1	DIFF P	DIFF Y1	DIFF P	DIFF Y1
362.50	.0620	.3740	11.70	.0031	3.34	-.0097	-1.23	-.0149	4.21	-.0085	3.34	-.0096
399.00	.0950	.4390	9.52	-.0080	4.45	-.0118	4.23	-.0083	4.77	-.0118	4.53	-.0116
424.00	.1310	.4820	4.26	-.0166	3.52	-.0127	6.64	-.0042	3.25	-.0136	3.66	-.0124
450.90	.1940	.5240	-2.63	-.0273	2.35	-.0157	7.53	-.0057	1.47	-.0175	2.49	-.0155
468.00	.2520	.5520	-4.15	-.0280	3.16	-.0143	7.81	-.0063	2.14	-.0161	3.26	-.0142
485.50	.3340	.5830	-3.28	-.0246	3.96	-.0130	6.68	-.0096	3.11	-.0142	3.98	-.0131
497.60	.4010	.6110	-1.11	-.0156	4.70	-.0082	5.90	-.0081	4.09	-.0088	4.68	-.0083
525.90	.5930	.6910	3.83	-.0020	5.32	-.0070	4.17	-.0112	5.23	-.0063	5.24	-.0072
534.30	.6800	.7390	3.54	.0020	3.72	-.0051	2.18	-.0084	3.76	-.0043	3.64	-.0051
542.70	.7930	.8160	3.94	.0040	2.74	-.0007	.97	-.0015	2.90	-.0003	2.66	-.0007
543.10	.8100	.8260	3.52	.0008	2.16	-.0031	.40	-.0035	2.33	-.0028	2.08	-.0031
544.50	.9430	.9410	2.03	-.0022	.58	-.0007	-.44	.0006	.73	-.0009	.55	-.0007
544.50	.9470	.9450	2.02	-.0021	.64	-.0006	-.34	.0007	.78	-.0008	.60	-.0006
MEAN DEVIATION:			4.27	.0105	3.13	.0079	3.73	.0064	2.98	.0081	3.13	.0079
MAX. DEVIATION:			11.70	.0280	5.32	.0157	7.81	.0149	5.23	.0175	5.24	.0155

Fig. 4 Typical output of a parameter fitting program for experimental VLE-data

However, the behavior in the diluted region is of special importance for most thermal separation processes, environmental protection and many other practical applications, so that data from this concentration range should be included in the parameter fit.

The excess Gibbs energy is connected to the excess enthalpy by the Gibbs-Helmholtz equation :

$$h^E = \left(\frac{\partial \, g^E/T}{\partial \, 1/T} \right)_{P,x}$$

With the help of this equation it is possible to fit binary interaction parameters for g^E-models using heats of mixing (h^E) data.

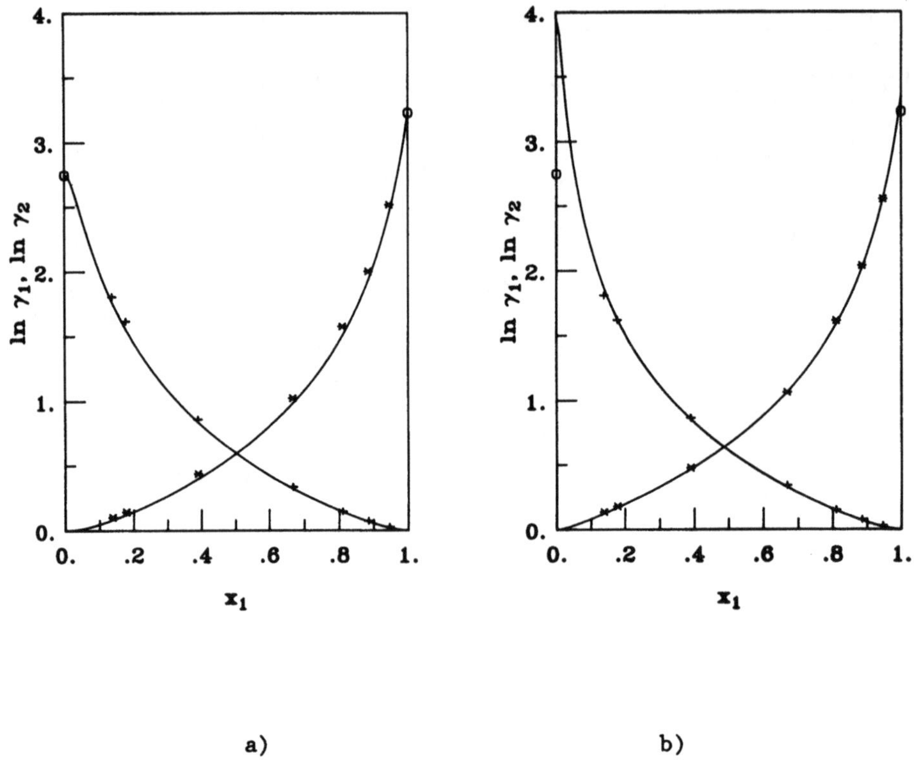

Fig. 5 Experimental activity coefficients and calculated values for the system methanol-n-heptane from the Wilson equation.
(exp. VLE-data taken from Benedict et al. (1945)
exp. γ^∞-data taken from Tochigi et al. (1976))

a) parameters calculated from activity coefficients at infinite dilution
b) parameters calculated from VLE-data

Fig. 6 shows data for a binary mixture of ethanol and water measured by Kolbe et al. (1983) together with the curve calculated from parameters fitted to h^E-data alone. For most mixtures no reasonable result can be obtained by fitting parameters only to h^E-data.

On the other hand, when these data are used in addition to VLE- and γ^∞-data, h^E-data are most valuable for the correlation of the temperature dependence of the real behavior of mixtures.

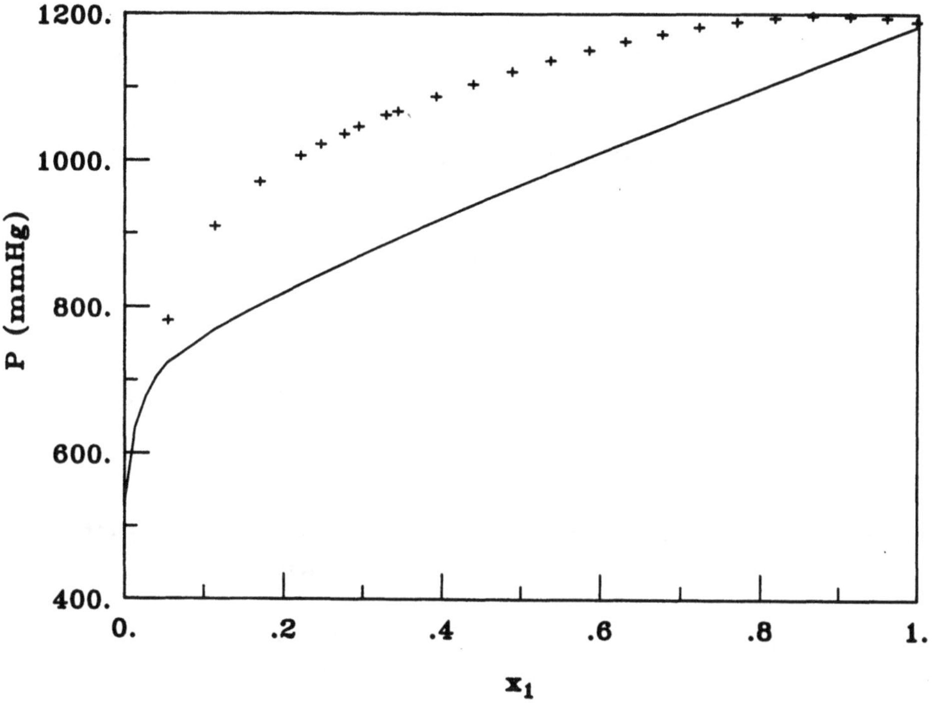

Fig. 6 Predicted x,P-data at 363.25 K using parameters calculated from experimental h^E-data
—— calculated using the Wilson equation with a quadratic temperature dependence of the parameters
\+ exp. data from Kolbe and Gmehling (1985)

THE DORTMUND DATA BANK AS A SOURCE FOR PURE COMPONENT AND MIXTURE DATA

In 1973 a data bank for thermodynamic data on pure component and mixture properties has been established at the University of Dortmund (FRG). This data bank now contains nearly all published experimental results on vapor-liquid equilibria as well as heats of mixing, activity coefficients at infinite dilution and liquid-liquid equilibria for non-electrolyte systems (Gmehling (1985)).

Beside the regular update of these data compilations work was started on data banks containing gas solubilities, excess heat capacities and azeotropic data. In addition to these data on mixtures information on pure component properties was stored.

Fig. 7 gives an overview on the current contents of the Dortmund Data Bank. A great part of the data was published in form of the DECHEMA Chemistry Data Series (Gmehling et al.(1977-1988), Sørensen and Arlt (1979-1988), Christensen et al. (1984-1988), Tiegs et al.

(1986-1988)). The data are stored in a form which makes it easy to interface them directly to various calculation and correlation programs. This possibility was extensively used for the further development of the UNIFAC (Fredenslund et al. (1977)) and mod. UNIFAC (Weidlich and Gmehling (1985)) method as well as for the test and improvement of different g^E-models and equations of state.

To use the data for the simultaneous correlation of different data types in order to produce recommended, temperature dependent parameters for different models, a program package had to be developed.

Pure Component Properties appr. 2200 components

List of References

One for each type of data , XXX=VLE,LLE,HE,GAM, GLE,CPE .AZD appr. 6000 references

---

Data on Mixtures	since	No. of Isotherms or Isobars
Vapor–Liquid Equilibria (VLE)	1973	12800
Liquid–Liquid Equilibria (LLE)	1977	3000
Heats of Mixing (HE)	1980	5500
γ^{∞} (GAM)	1984	22000 values
Gas Solubilities (GLE)	1985	4000
c_p^E (CPE)	1986	400
Azeotropic Data (AZD)	1988	—

Fig. 7 Current status of the Dortmund Data Bank

A PROGRAM PACKAGE FOR THE INTERACTIVE SIMULTANEOUS CORRELATION OF DIFFERENT MIXTURE DATA

Fig. 8 gives an overview on the structure of the program package RECVAL. The whole system is to a great extent self-explanatory and accessed by the user through a set of hierarchical menus.

The first step is always to generate data files containing mixture data for the binary system of interest. As data source either the Dortmund Data Bank (DDB) or data supplied by the user may be used.

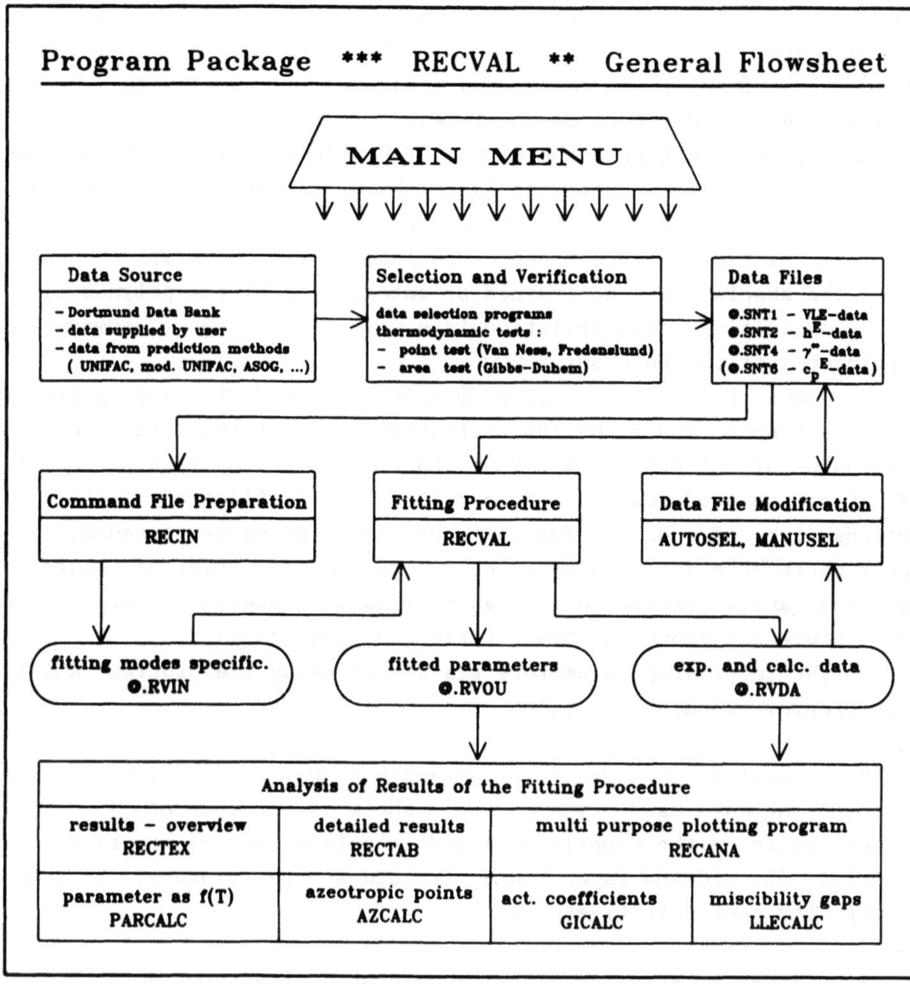

Fig. 8 Structure of the program package RECVAL

In case of complete (x,y,P,T) vapor-liquid equilibrium (VLE) data two different thermodynamic consistency tests are performed.

The so-called point-to-point-test (Van Ness et al., 1973) uses a flexible expression for the excess Gibbs energy for fitting experimental x,P,T-data. With the aid of the fitted parameters the vapor composition y for each experimental point (x,P,T) is calculated and compared with the experimental value. If the mean deviation between experimental and calculated y values is smaller than 0.01, the data set is considered to be consistent.

The second test uses the method of Redlich-Kister (1948) and Herington (1947, 1951). Here the logarithm of the ratio of the activity coefficients is calculated for each experimental point. These values are then fitted by a third degree polynomial and the areas above and below the abscissa are determined by integration of the polynomial.

The resulting area deviation is a measure of the thermodynamic consistency of the data set.

All data are stored in files @.SNT#, where @ is any arbitrary (valid) filename and # the identification code for the data type. With the help of the program RECIN an input file for the fitting routine RECVAL can be produced, which contains the specification of different options (type of equation for the pure component vapor pressure, real vapor phase behavior, selected data sets including type of objective function and weighting factors as well as the functional form for the temperature dependence of the interaction parameters).

In the next step parameters are fitted using the program RECVAL. This program produces two files :

@.RVOU contains the resulting parameters and information on pure component data used, type of vapor phase calculation etc..

@.RVDA contains the experimental and calculated values for each data point (liquid and vapor mole fractions, temperature, pressure, activity coefficients etc.).

These data can be analysed with the help of various programs:

RECTEX produces a formatted overview of the options selected for the fit and the results of the correlation. A file containing detailed results is generated by RECTAB.

The best method to analyse the fitting results is through graphical representation. Using the program RECANA, a wide variety of plots can be produced containing a selected part of the input data. Some possible choices are

- K-factors ($K_i = y_i/x_i$)
- separation factors ($\alpha_{12} = K_1/K_2$)
- temperature or pressure
- activity coefficients
- contribution of the data point to the objective function
- excess functions (g^E, h^E)

These data are plotted as a function of liquid or vapor composition, the output can be sent to the screen, laser printer, plotter or a file.

Several properties may be plotted as a function of temperature using special programs as indicated in Fig. 9 :

- interaction parameters (PARCALC)
- azeotropic points (AZCALC)
- activity coefficients at infinite dilution (GICALC)
- composition of two immiscible phases in equilibrium (LLECALC)

With the help of these programs one typically finds data points or even complete data sets, which differ significantly from the rest of the data. This may be due to improper measuring methods, impure substances or other systematic errors.

Phase equilibrium data found in literature often exhibit significant deviations from each other and only less than 40% of the data sets pass both consistency tests mentioned above.

Therefore the most difficult task in fitting recommended parameters for g^E-models is the selection of a reliable set of data.

After the first fit is made and the results are analysed with the help of different representations, some of the data may be discarded and the weighting factors of the single data sets are modified. This is done using the programs AUTOSEL or MANUSEL.

After a few cycles of fitting, analysing and selecting/weighting a reliable set of parameters should be found.

RESULTS

For demonstrating the use of the program package RECVAL, calculations were made for the system ethanol-water.

Vapor-liquid equilibria of ethanol-water mixtures were intensively studied. More than 130 data sets were published up to now and very precise data are available over a wide temperature range (Pemberton and Mash (1978), Kolbe and Gmehling (1985)).

Interaction parameters were fitted for the Wilson-equation using only VLE-data. Fig. 9 shows the experimental and calculated total pressures for several data sets as a function of temperature. The data sets are denoted by the data set number in the DDB. Table 1 gives additional information on the data.

It can be seen, that the non-temperature dependent parameters give a reasonable representation of the data.

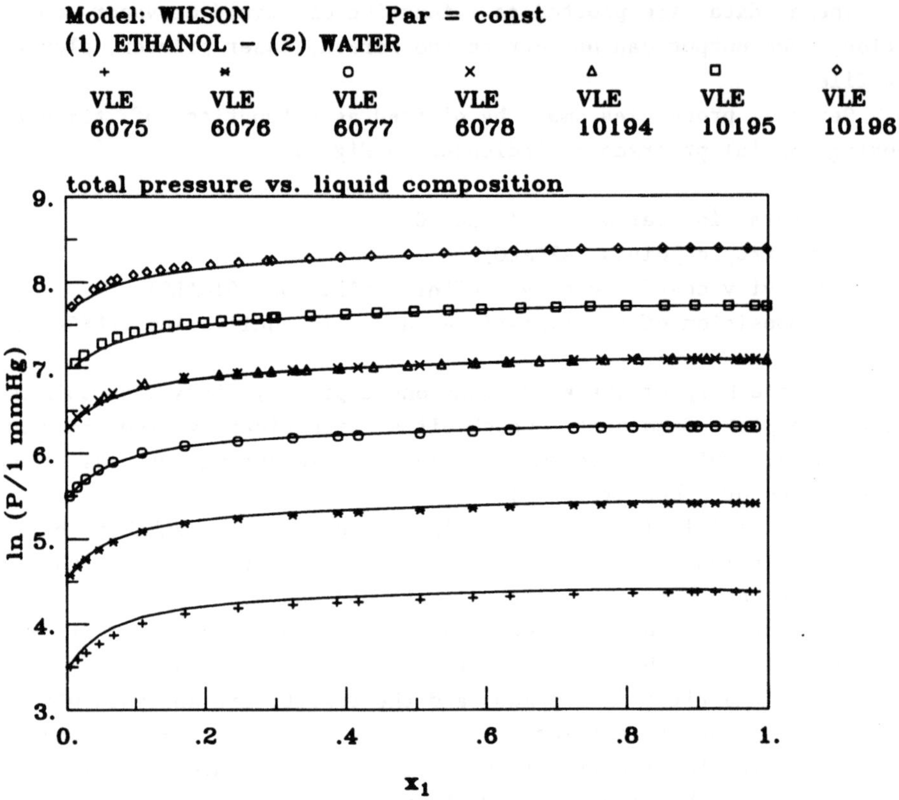

Fig. 9 x,P-diagram for several isothermal data sets and curves calculated from the Wilson-equation using temperature-independent parameters

One of the mixture properties which is of great importance for thermal separation processes is the azeotropic point. At this point the separation factor is equal to unity and no further separation is possible by rectification.

Fig. 10 shows the dependence of the azeotropic composition in the system ethanol-water as a function of temperature calculated using the Wilson equation. It can be seen that the experimental values (Kolbe (1983)) are not reproduced by the model parameters within the required accuracy in case of temperature independent parameters.

The reason for this failure lies in the inability of the models to describe the heats of mixing data, which are a measure of the temperature dependence of the real mixture behavior as shown for the Wilson-equation in Fig. 11.

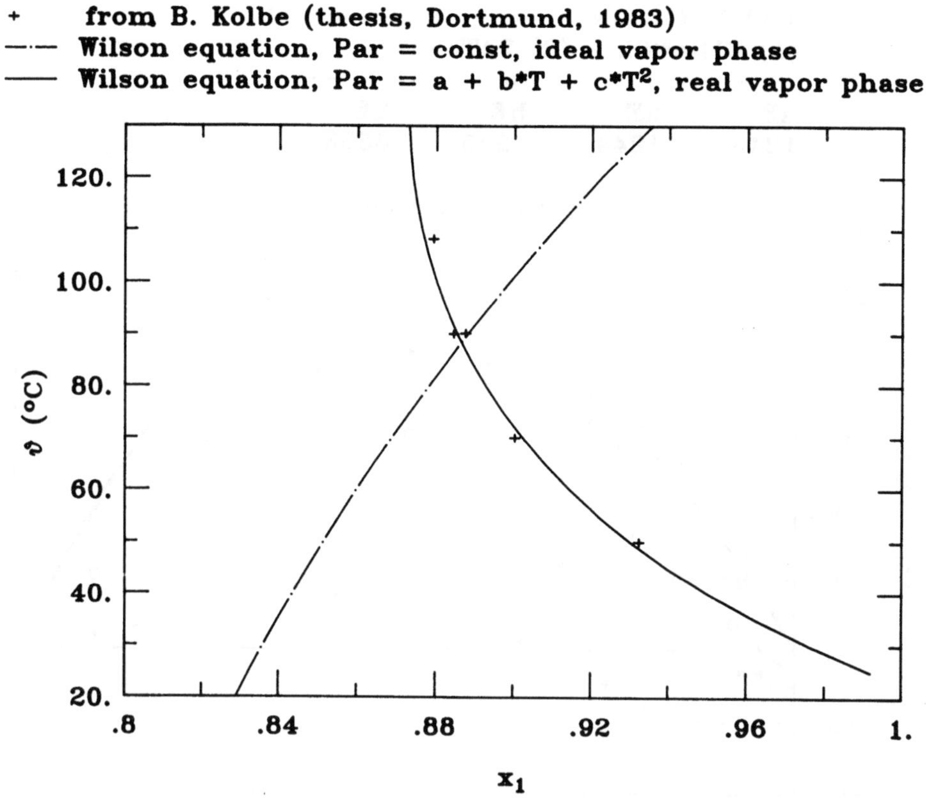

Fig. 10 Temperature dependence of the azeotropic point in the system ethanol-water

Table 1 References for some data sets of the Dortmund Data Bank

Author	data type	data set number
Christensen and Izatt (1984)	h^E	4558
Kolbe and Gmehling (1985)	VLE	10194, 10195, 10196
Larkin (1975)	h^E	1243, 1244, 1245
Pemberton and Mash (1978)	VLE	6075, 6076, 6077, 6078

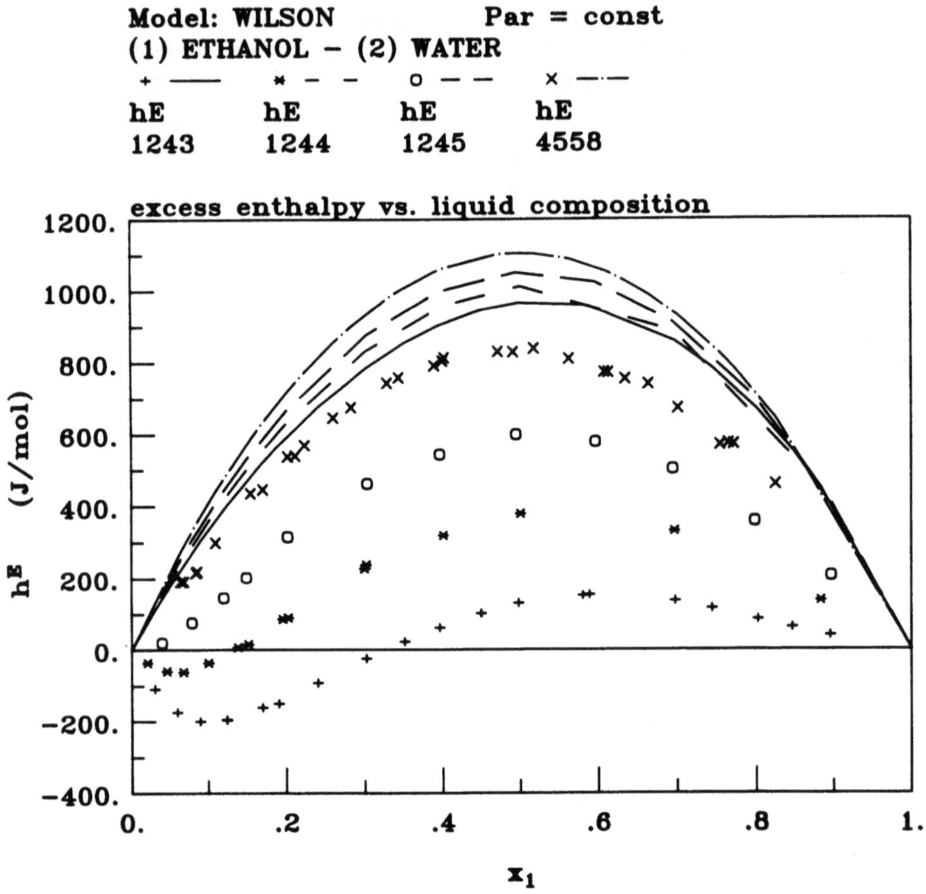

Fig. 11 h^E-data for several temperatures and curves calculated using the Wilson-equation with temperature-independent parameters

This problem can be solved by introducing temperature dependent parameters and thereby overriding the inadequate implicite temperature dependence of the model equations.

Fig. 12 shows model calculations performed by using temperature dependent parameters from a simultaneous correlation of VLE- and h^E-data. It can be seen that the description of the heats of mixing has been greatly improved.

Using these temperature-dependent parameters the azeotropic composition can be calculated over a large temperature range with high accuracy (Fig. 10).

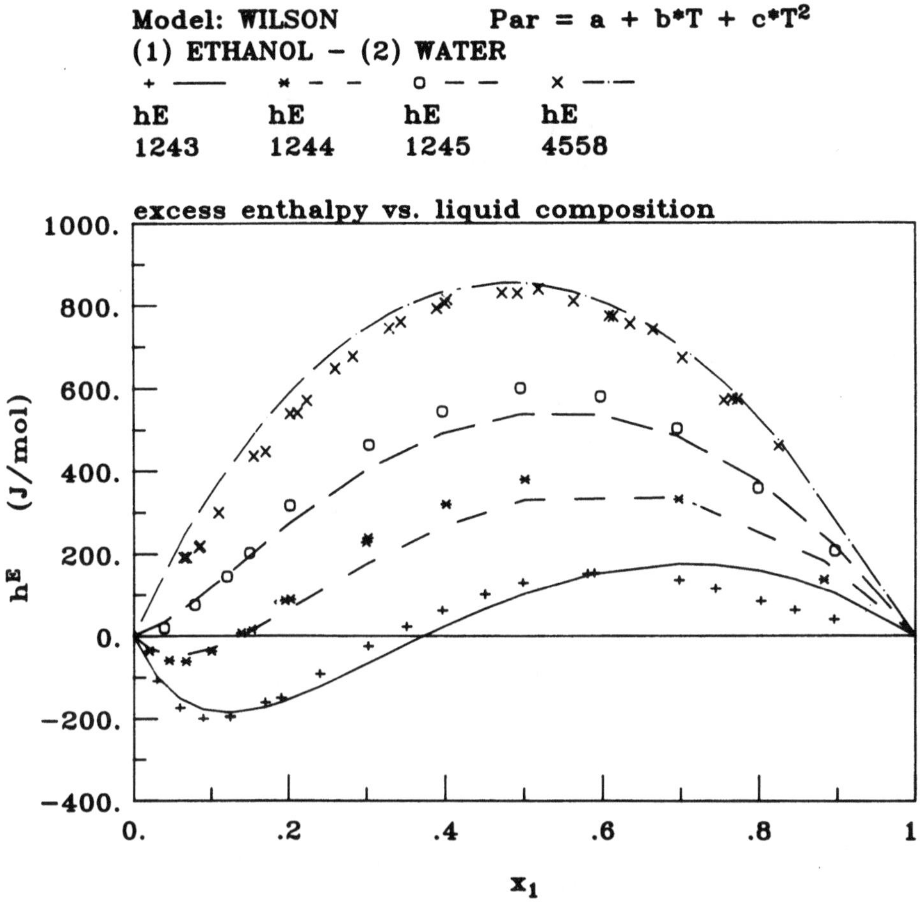

Fig. 12 h^E-data for several temperatures and curves calculated using the Wilson-equation with temperature-dependent parameters.

CONCLUSION

The description of the real behavior of fluid mixtures over a wide temperature range is possible using g^E-models with temperature dependent interaction parameters. To produce a reliable set of parameters, VLE-, γ^∞- and h^E-data should be correlated simultaneously.

To perform this correlation a program package was developed, which makes it possible to judge the reliability of the available experimental data and select a consistent data base for the parameter fit.

This program will be intensively used to examine the large amount of data stored in the DDB in order to generate a reduced and evaluated set of experimental information and produce a table containing recommended values for the binary mixtures. With the help of these parameters both reliability and accuracy of simulations in thermal separation processes and chemical process design will be greatly improved. A similar procedure should be developed in the future for other mixture data and various pure component properties.

ACKNOWLEDGEMENTS

The authors are grateful to Prof. U. Onken and to Mr. Z. Weigl from Warsaw, Poland for useful discussions and to "Max-Buchner Forschungsstiftung" for financial support.

LIST OF SYMBOLS

g^E	-	molar excess Gibbs energy
h^E	-	molar excess enthalpy
$u_{ij}-u_{ii}$	-	interaction parameters of the UNIQUAC equation
α_{12}	-	separation factor between component 1 and 2
γ	-	activity coefficient
γ^∞	-	activity coefficient at infinite dilution

REFERENCES

Abrams, D.S., J.M. Prausnitz: AIChE J. 21, 116 (1975)

Benedict, M., C.A. Johnson, E. Solomon, L.C. Rubin:
Trans. Am. Inst. Chem. Eng. 41, 371 (1945)

Christensen C., J. Gmehling, P. Rasmussen, U. Weidlich, T. Holderbaum:
Heats of Mixing Data Collection, Vol. III, 2 parts,
Supplement in preparation
DECHEMA Chemistry Data Series, Frankfurt 1984-1988.

Christensen, J.J., R.M. Izatt: Thermochim. Acta 73, 117 (1984)

Fredenslund, A., J. Gmehling, P. Rasmussen:
Vapor-Liquid Equilibria Using UNIFAC
Elsevier, Amsterdam (1977)

Gmehling J., U. Onken, W. Arlt, P. Grenzheuser, B. Kolbe, U. Weidlich, R.J. Rarey-Nies:
Vapor-Liquid-Equilibrium Data Collection, Vol. I, 14 parts,
Supplements (6 parts) in preparation
DECHEMA Chemistry Data Series, Frankfurt 1977 - 1988.

Gmehling, J., B. Kolbe: Thermodynamik
Georg Thieme Verlag, Stuttgart (1988)

Gmehling, J.: CODATA Bulletin No. 58, "Thermodynamic Databases" Pergamon Press, Oxford (1985)

Herington, E.F.G., Nature 160, 610 (1947)

Herington, E.F.G., J. Inst. Petrol., 37, 457 (1951)

Kolbe, B.: Thesis, Dortmund 1983

Kolbe, B., J. Gmehling: Fluid Phase Equilibria 23, 213 (1985)

Larkin, J.A.: J. Chem. Thermodyn. 7, 137 (1975)

Mertl, I.: Collect. Czech. Chem. Commun. 37, 366 (1972)

Pemberton, R.C., C.J. Mash: J. Chem. Thermodyn. 10, 867 (1978)

Prausnitz, J.M., R.N. Lichtenthaler, E.G. Azevedo Molecular Thermodynamics of Fluid-Phase Equilibria: Prentice-Hall, New Jersey (1987)

Redlich, O., A.T. Kister: Ind. Eng. Chem. 40, 345 (1948)

Renon, H., J.M. Prausnitz: AIChE J. 14, 135 (1968)

Sørensen J.M., W. Arlt: Liquid-Liquid Equilibrium Data Collection, Vol. V, 3 parts, DECHEMA Chemistry Data Series, Frankfurt 1979 - 1980

Tiegs D., J. Gmehling, J. Bastos, A.G. Medina, M. Soares, P. Alessi, I. Kicić: Activity Coefficients at Infinite Dilution Data Collection, Vol. IX, 2 parts, Supplement in preparation DECHEMA Chemistry Data Series, Frankfurt 1986.

Tochigi, K., K. Kojima: J. Chem. Eng. Jpn. 9, 267 (1976)

Van Ness, H.C., S.M. Byer, R.E. Gibbs: AIChE J. 19, 238 (1973)

Weidlich, U., J. Gmehling: Ind. Eng. Chem. Process Des. Dev. 26, 1372 (1987)

Wilson, G. M.: J. Am. Chem. Soc. 86, 127 (1964)

The Arizona Database: An Aqueous Solubility Database for Nonelectrolytes

R.-M. Dannenfelser and S.H. Yalkowsky
College of Pharmacy, University of Arizona, Tucson, AZ 85721, USA

INTRODUCTION

The aqueous solubility of organic compounds is important in a large number of areas of science. For this reason, solubility is a frequently measured property. Solubilities are reported in scientific publications representing nearly every scientific discipline. Yet in spite of the tremendous amount of solubility data that is available in the literature, it is difficult to find the solubility of a particular compound from a normal literature search. When values are available in the literature there are often major discrepancies among the values reported for a particular compound by different authors and it is difficult to select the best value from among all of the reported values. Hence, the ARIZONA dATAbASE was started.

The ARIZONA dATAbASE contains an extensive compilation of aqueous solubility data for unionized organic compounds. Each of its over ten thousand solubility data have been evaluated and checked several times to insure a high degree of accuracy. For ease in confirmation, each value is given in the units reported in the original reference. For the users convenience all solubilities are converted to moles per liter and all temperatures to centigrade.

The ARIZONA dATAbASE contains the most extensive compilation of aqueous solubility data for unionized organic compounds

C. Jochum M.G. Hicks J. Sunkel (Eds.)
Physical Property Prediction
in Organic Chemistry

TABLE 1 REFERENCE STRUCTURE

Field	Field Name	Type	Width
1	REF	CHAR	5
2	LOC_CIT	CHAR	5
3	F_INIT	CHAR	6
4	F_NAME	CHAR	30
5	S_INIT	CHAR	6
6	S_NAME	CHAR	30
7	T_INIT	CHAR	6
8	T_NAME	CHAR	30
9	L_INIT	CHAR	6
10	L_NAME	CHAR	30
11	TITLE	CHAR	240
12	BOOK_TITLE	CHAR	80
13	CODEN	CHAR	5
14	VOLUME	CHAR	5
15	BEG_PAGE	NUM	6
16	END_PAGE	NUM	6
17	YEAR	NUM	4
18	PUBLISHER	CHAR	60
19	CITY_STATE	CHAR	50
20	LANG	CHAR	4
** TOTAL **			615

TABLE 2 SOLUBILITY STRUCTURE

Field	Field Name	Type	Width
1	NAME	CHAR	100
2	SOLUBILITY	CHAR	12
3	UNIT_SOLUB	CHAR	3
4	TEMP	CHAR	5
5	UNIT_TEMP	CHAR	1
6	REFERENCE	CHAR	5
7	COMMENTS	CHAR	100
8	EVAL_TEMP	NUM	1
9	EVAL_PUR	NUM	1
10	EVAL_EQUIL	NUM	1
11	EVAL_ANAL	NUM	1
12	EVAL_ACC	NUM	1
13	EVAL_TOTAL	NUM	2
14	FILE	CHAR	10
15	SOLUB_G	CHAR	10
16	SOLUB_M	CHAR	10
17	TEMP_C	CHAR	6
** TOTAL **			270

available anywhere. Currently the database is available in hard copy or on machine readable tape.

DATA ENTRY

Data is entered into the ARIZONA dATAbASE with microcomputers using dBASE III in three components: reference data, compound data, and solubility data.

Reference Data

Each reference is given a four character code. The first character is alphabetic and the next three are numeric. This code appears in the solubility data section and identifies each data point with a reference. The reference data (Table 1) contains:

Reference code
Authors' last name and initials
Journal CODEN
Volume
Year
Inclusive paging
Complete title
Language.

Complete citations are also given for chapters and books.

Compound Data

Compound data (Table 2) is comprised of:

Name
Synonyms
Elements
Molecular formula
Molecular weight
Melting point
Boiling point
CAS registry number

These data are entered using the original article and various handbooks, such as The Merck Index.

TABLE 3 COMPOUND STRUCTURE

Field	Field Name	Type	Width	Dec
1	NAME	CHAR	100	
2	SYNONYM1	CHAR	100	
3	SYNONYM2	CHAR	100	
4	SYNONYM3	CHAR	100	
5	SYNONYM4	CHAR	100	
6	SYNONYM5	CHAR	100	
7	C	NUM	2	
8	H	NUM	3	
9	BR	NUM	2	
10	CL	NUM	2	
11	F	NUM	2	
12	I	NUM	2	
13	N	NUM	2	
14	O	NUM	2	
15	P	NUM	2	
16	S	NUM	2	
17	W	NUM	2	
18	OTHER_ELEM	CHAR	2	
19	NUM	NUM	2	
20	FORMULA	CHAR	25	
21	FORM_SORT	CHAR	30	
22	REG_NO	CHAR	10	
23	MP	CHAR	6	
24	BP	CHAR	6	
25	MW	NUM	7	2
** TOTAL **			712	

Solubility Data

Solubility data (Table 3) consists of the following:

- Solubility and units given in the original reference
- Temperature and units given in the original reference
- Reference code
- Solubility in moles per liter
- Solubility in grams per liter
- Temperature in degrees centigrade
- Evaluation (see below)
- Comments
- Recommended values (see below).

Software is used to convert the original solubility to moles per liter and grams per liter. Temperature is also converted to centigrade using software. All other entries are entered from the original article.

Evaluation

The evaluation is done nearly objectively according to the following point system:

PARAMETER	NUMBER OF POINTS		
	0	1	2
Temperature	Not given, ambient, or room temp.	Given with no range	Given with range
Purity of Solute	Not stated or as received	Stated with no range or as received with range	Stated with range or altered or calculated
Equilibration time/agitation	Not stated	Stated briefly	Described in detail
Analysis	Not stated	Stated briefly or stated in other paper	Described in detail
Accuracy and/or precision	1 significant figure or range > 20%	2 significant figure or range 5-20%	3 significant figure or range 1-5%

<u>Recommended Values</u>

Recommended values are provided when there is sufficient data available for a statistical evaluation. The values are based upon the mean of the individual values laying in the 20° to 40° centigrade range, each weighted according to the total evaluation score.

DATA VERIFICATION

<u>Automatic Checks</u>

Molecular weight is calculated from the empirical formula using the standard atomic weights of the elements. This computed molecular weight is then compared to the entered value. If the discrepancy is over 0.5, the corresponding record is marked by the computer. The error could come from incorrect entries of elements or entered molecular weight. After corrections are made the marked records are then rechecked.

In checking the reference code a program checks the code to make sure the first character is a capital letter and the rest numeric.

CAS registry number is manipulated according to the procedure in the Chemical Abstracts to check for errors.

<u>Manual Checks</u>

In order to perform the manual checks the data is printed out in reference and compound order. Using the original reference, the reference order printout is checked for completeness. And the other printout is checked for correctness.

DATA RETRIEVAL

Data Output

To prepare the corrected database output, the database is sorted by temperature, formula and compound name. After a few programs are run on the database, it is ready to be printed in a hardcopy format or onto tape. A sample of the hardcopy is shown in Table 4.

Screens

Software has been written to interactively retrieve wanted information via screens. The choices include the type of solubility units desired (Fig. 1) (original, moles per liter, or grams per liter), how the search is to be performed (Fig. 2) (via the compound name, reference code, formula, or CAS registry number), and how the data is to be sorted (Fig. 3).

TABLE 4 SAMPLE OUTPUT

FORM: C6 H6 CL6 RN: 319-85-7 MW: 290.83 MP: 314. BP:
NAME: beta-1,2,3,4,5,6-Hexachlorocyclohexane
SYN: beta-Benzene Hexachloride

REPORTED VALUES				REF	CALCULATED VALUES			EVAL	*
TEMP	U	SOLUBILITY	U		MOLES/L	GRAMS/L	T/C		
20.	C	5	ppm	C099	1.719E-05	5.000E-03	20.00	12000	
ns		0.5 E-4	gw	M061	1.719E-06	5.000E-04	ns	00000	
ns			pi	M161	pi	pi	ns	00000	
25.	C	240	ug	W025	8.252E-07	2.400E-04	25.00	10222	
RECOMMENDED VALUES					.	.			3

FORM: C6 H6 CL6 RN: 58-89-9 MW: 290.83 MP: 112.5 BP:
NAME: Lindane
SYN: gamma-1,2,3,4,5,6-Hexachlorocyclohexane
SYN: gamma-Benzene Hexachloride

REPORTED VALUES				REF	CALCULATED VALUES			EVAL	*
TEMP	U	SOLUBILITY	U		MOLES/L	GRAMS/L	T/C		
15.	C	2150	ppb	B083	7.393E-06	2.150E-03	15.00	22222	
25.	C	6800	ppb	B083	2.338E-05	6.800E-03	25.00	22222	
35.	C	11400	ppb	B083	3.920E-05	1.140E-02	35.00	22222	
45.	C	15200	ppb	B083	5.226E-05	1.520E-02	45.00	22222	
27.	C	12	ppm	B161	4.126E-05	1.200E-02	27.00	22220	1
35.	C	21	ppm	B161	7.221E-05	2.100E-02	35.00	22220	1
45.	C	27	ppm	B161	9.284E-05	2.700E-02	45.00	22220	1
50.	C	33	ppm	B161	1.135E-04	3.300E-02	50.00	22220	1
60.	C	45	ppm	B161	1.513E-04	4.400E-02	60.00	22220	1
15.	C	2.15	ppm	B162	7.393E-06	2.150E-03	15.00	10002	
25.	C	6.80	ppm	B162	2.338E-05	6.800E-03	25.00	10001	
35.	C	11.4	ppm	B162	3.920E-05	1.140E-02	35.00	10002	
20.	C	10	ppm	C099	3.438E-05	1.000E-02	20.00	12000	
24.	C	17000	ug	H116	5.845E-05	1.700E-02	24.00	22112	
19.	C	8.19	mg	I018	2.816E-05	8.190E-03	19.00	01112	2
25.0	C	7.52	ppm	M060	2.586E-05	7.520E-03	25.00	22222	
ns		1 E-4	gw	M061	3.438E-06	1.000E-03	ns	00000	
ns		7000	ppb	M110	2.407E-05	7.000E-03	ns	00000	1
25.	C	7.3	mg	M130	2.510E-05	7.300E-03	25.00	10001	
ns		10	mg	M161	3.438E-05	1.000E-02	ns	00000	
25.	C	7800	ug	W025	2.682E-05	7.800E-03	25.00	10212	
RECOMMENDED VALUES					**3.911E-05**	**1.139E-02**			3

* 1 Estimated from graph
2 Temperature range 18-20°C
3 Weighted mean values of 20-40°C

FIGURE 1   UNIT MENU

```
TYPE ONE OF THE FOLLOWING SOLUBILITY UNITS
(FOLLOWED BY A RETURN)

O     SOLUBILITY UNITS USED BY THE AUTHOR
G     SOLUBILITY UNITS IN GRAMS/LITER
M     SOLUBILITY UNITS IN MOLES/LITER
STOP   TO EXIT THIS PROGRAM.

OPTION?
```

FIGURE 2   SEARCH MENU

```
TYPE ONE OF THE FOLLOWING SEARCH OPTIONS
(FOLLOWED BY A RETURN)

N       NAME OF COMPOUND
R       REFERENCE NUMBER
F       FORMULA
RN      REGISTRY NUMBER
S       TO GO TO THE UNIT MENU
STOP    TO EXIT PROGRAM

OPTION?
```

FIGURE 3   SORT MENU

TYPE ONE OF THE FOLLOWING FOR THE TYPE OF SORT.

(FOLLOWED BY A RETURN)

R   SORTED BY REFERENCE

T   SORTED BY TEMPERATURE

S   TO RETURN TO THE LAST MENU

OPTION?

Correlation and Extrapolation in Chemical Engineering of Vapour Pressure Data Using Thermal Data

F. Mascarello

F. Hoffman-La Roche, ZFE 65/603, Grenzacherstrasse 124, 4002 Basel, Switzerland

Vapour pressures are among the most often measured physico-chemical properties. The measurements are most frequent and accurate in the pressure range 10-100 kPa. The determination of vapour pressures in the region below 1 kPa, on the other hand, are difficult, time-consuming and relatively inaccurate. Extrapolation based on vapour pressures measured in a limited range near the boiling point is usually unreliable.

However, accurate vapour pressures of relatively involatile substances at ambient temperature can be obtained by combining experimental data at higher pressures with other appropriate thermodynamic quantities in the temperature region of interest. Simultaneously, this approach can be used as a consistency test of experimental data. A demonstration of the procedure, adopted in our laboratory, will be given by using the Ambrose/Davies method of extrapolating vapour pressures.

This example demonstrates the possibility and necessity of combining physico-chemical property data by thermodynamic relations for correlating, extrapolating and/or consistency testing experimental data in a database.

C. Jochum M.G. Hicks J. Sunkel (Eds.)
Physical Property Prediction
in Organic Chemistry

Establishing Consistent Thermodynamic Data on Vaporization Equilibria for Organic Compounds

Vladimir Majer[1], Kvetoslav Ruzicka[2], Vlastimil Ruzicka, Jr.[2], and Milan Zabransky[2]

[1] Department of Chemistry, University of Delaware Newark, DE 19716, USA
[2] Department of Physical Chemistry, Institute of Chemical Technology, Prague, Czechoslovakia

Abstract

Vapor pressure data are abundant and accurate near the normal boiling point while they are scarce and unreliable in the low pressure range. Data on thermal properties (enthalpy of vaporization and difference in the heat capacities of an ideal gas and the liquid) are available, however at temperatures far below the normal boiling point. All three properties are related by exact thermodynamic relationships and can be correlated simultaneously. This procedure can be used in the evaluation of thermodynamic data on vaporization equilibria as a rigorous consistency test and for producing recommended data sets. When a suitable correlation equation is selected a single set of parameters permits generation of consistent data for several properties between the triple and normal boiling points. The procedure is especially useful for calculating vapor pressures and/or enthalpies of vaporization far below the normal boiling point. In combination with the group contribution methods for estimation of thermal properties, the principle of simultaneous correlation can serve as a base in formulation of a new approach to the prediction of vapor pressures and/or enthalpies of vaporization at low reduced temperatures. Such a procedure would not require use of any traditional parameters (critical constants, acentric factor, etc.) which are not available for most high boiling compounds.

C. Jochum M.G. Hicks J. Sunkel (Eds.)
Physical Property Prediction
in Organic Chemistry

INTRODUCTION

The aim of this paper is to show how the simultaneous approach to the properties describing vapor-liquid equilibria of pure compounds can lead to establishing consistent thermodynamic data and to developing new types of prediction schemes. Thermodynamic quantities characterizing the vaporization equilibria in one component system are the vapor pressure and the related thermal properties (enthalpy of vaporization and the difference in heat capacities of coexisting phases).

Vapor pressures (p_s) are among the most frequently measured physico-chemical quantities; the data are available in the open literature for more than 7000 compounds and new measurements are always in progress. Most observations have been carried out in the moderate pressure range (5 to 150 kPa) where the experimental procedure is not too complicated and data can be obtained with a good accuracy. Near the normal boiling point the good quality vapor pressures have errors below 0.05 per cent and some laboratories reduced the uncertainty in measurements to a few pascals. The situation is, however, quite unsatisfactory in the low pressure range (p_s below 1 kPa). The experimental procedure becomes much more complicated; the data are therefore scarce and usually subject to gross systematic errors. The values of individual authors differ often by 10 to 40 per cent and they are rarely consistent at their upper temperature limit with those in the moderate pressure range.

Thermal properties can be, on the other hand, obtained with a reasonable accuracy at temperatures deep below the normal boiling point (T_b) for a number of organic compounds.

Enthalpies of vaporization (ΔH_v) determined calorimetrically are available at present at one temperature or over a limited temperature range for about 650 organic compounds.The accuracy of most data varies between 0.2 and 2 per cent. Usually a value at 298.15 K is reported which is a temperature well below the normal boiling point for most organic substances.

The difference in the heat capacities of coexisting phases is closely related to the difference in the heat capacities of an ideal gas (C_p^{go}) and the liquid (C_p^{l}) which can be determined for a number of substances down to the triple point temperature. The latter difference will be in the text below denoted as heat capacity difference (ΔC_p). Values of C_p^{go} can be calculated from spectral data at any temperature with an error between 0.2 and 5 per cent. Experimental values of C_p^{l} are available from low temperature calorimetry for more than 1200 organic compounds with an error varying mostly between 0.2 and 1 per cent.

Vapor pressure, enthalpy of vaporization and heat capacity difference are related by exact thermodynamic relationships resulting from the Clapeyron and Kirchhoff equations. Thus, when a suitable correlation equation is selected, data on all three properties can be correlated together. Such a simultaneous correlation yields a smooth vapor pressure curve between the triple and normal boiling points with temperature derivatives consistent with the thermal data.

ADVANTAGES OF SIMULTANEOUS APPROACH TO DATA

1. The proposed procedure can serve in critical compilation and assessment of data characterizing vaporization equilibria. The simultaneous correlation of several thermodynamic properties makes it possible to test the data consistency and to detect possible systematic errors. This approach to evaluation is better than the previously used separate assessment of individual properties.

2. The simultaneous correlation of p_s, ΔH_v and ΔC_p data by a well selected relationship yields a single set of parameters from which consistent data on several properties between the triple and normal boiling points can be obtained.

3. Combination of interrelated properties available in different temperature ranges permits a reliable extrapolation controlled by the exact thermodynamic constraints. In this particular case the proposed method is extremely useful for two types of extrapolations:

3a. Vapor pressures can be calculated down to the triple point by combining accurate p_s data near the normal boiling point with the thermal data for the low pressure range. The values obtained are always consistent with those in the moderate pressure range. Considering the experimental difficulty and low credibility of vapor pressure measurements below 1 kPa,this approach seems to be, at present, the only effective way for obtaining reliable data in the low pressure range. This information is of both technological and scientific interest; the need has been accumulating over the last several years due to the increasing ecological problems all over the world. Vapor pressures of high boiling compounds at ambient temperature are required by environmental and other bodies dealing with pollution control.

3b. Similarly, enthalpies of vaporization at temperatures far below the normal boiling point can be determined by combination of vapor pressures near T_b with the heat capacity differences. Measurements of ΔH_v in this region are complicated and calorimetric data for only a limited number of compounds are available. Calculation via the Clapeyron equation using the extrapolation of vapor pressures alone fails completely due to high uncertainty in the determination of the temperature derivative of p_s. The simultaneous correlation of p_s and ΔC_p is especially useful for obtaining ΔH_v at 298.15 K for a variety of high boiling compounds. This value is required in many thermochemical and thermodynamic calculations for conversions between ideal gas and liquid states. Only very rough estimates are now available for many high boiling compounds.

Here we focus on the vapor-liquid equilibria in one component systems. It is, however, obvious that the same principle and methodology can be used for evaluation of properties characterizing vapor-solid equilibria. Quite analogous procedures using the identical programs can be utilized for obtaining vapor pressures above solids and enthalpies of sublimation over wide temperature range.

THERMODYNAMIC BACKGROUND

The Clapeyron equation relates the temperature derivative of vapor pressure to the enthalpy of vaporization and volumetric properties of coexisting phases

$$RT^2\left(\frac{d\ln p}{dT}\right)_s = \frac{\Delta H_v}{\Delta Z_v} = HZ \qquad (1)$$

where ΔZ_v stands for the difference in compressibility factors of coexisting phases and a newly defined quantity HZ approaches ΔH_v when ΔZ_v is close to unity. Another new quantity CZ can be defined as follows:

$$CZ = \left(\frac{dHZ}{dT}\right)_s = \frac{d}{dT}\left(RT^2\left(\frac{d\ln p}{dT}\right)\right)_s \qquad (2)$$

Using relationships for the temperature dependence of ΔH_v and ΔZ_v one can obtain an equation

$$CZ = \frac{C_p^g - C_p^l - 2(\partial \Delta Z/\partial T)_p HZ - (\partial \Delta Z/\partial p)_T HZ^2 p/(RT^2)}{\Delta Z} \qquad (3)$$

where heat capacity of vapor $C_p{}^g$ can be expressed from the relationships

$$C_p^g = C_p^{go} - RT \int_0^{p_s} [2(\frac{\partial Z^g}{\partial T})_p + T(\frac{\partial^2 Z^g}{\partial T^2})_p] \, d\ln p \qquad (4)$$

Combination of equations (3) and (4) provides an exact thermodynamic relation between vapor pressures and heat capacity difference ΔC_p. Below the normal boiling temperature T_b, pVT behavior of the saturated vapor can be approximated by the virial equation terminated after the second virial coefficient, and volume of the liquid phase can be neglected. The three previous equations then yield the relationship

$$CZ = \Delta C_p - T[p_s(d^2B/dT^2) + 2(dp/dT)_s(dB/dT) + B(d^2p/dT^2)_s] \qquad (5)$$

Since quantities HZ and CZ can be expressed either from the temperature derivatives of p_s or from ΔH_v and ΔC_v (corrected to the nonideality of the vapor phase) they serve in correlation as links between vapor pressure equation and thermal data.

PROCEDURE

The method of simultaneous correlation of vapor pressures and thermal data was first proposed by King and Al-Najjar (1974) and further developed by several authors (Ambrose and Davies, 1980; Mosselman et al., 1982, Rogalski, 1985; Ruzicka and Majer, 1986; King and Mahmud, 1986). The individual authors differed in types of input data considered, correlation equations adopted, and in the minimization procedure. There are several problems which have to be considered in the application of the procedure.

The selected correlation relationship must be flexible enough to adequately describe the vapor pressure and its first and second temperature derivatives over a temperature range of 200 K or more. Previous investigations (King and Al-Najjar, 1974; Ambrose and Davies, 1980; Ruzicka and Majer, 1986) indicated that four, or better, five parameter equations are necessary. Tests of both traditional (polynomial expansion in temperature with a logarithmic term) and modern (Wagner type) equations gave similar results (Ruzicka, 1986). The Cox equation can also be a suitable alternative as it proved to be useful in extrapolation of vapor pressures towards the triple point (Scott and Osborn, 1979).

Effect of the pVT behavior of the fluid has always to be taken into account during correlation. The equations relating vapor pressures and thermal data contain terms expressing pVT behavior of the fluid. The terms which reflect the vapor nonideality (compressibility factor and its derivatives) cannot be calculated accurately. The ideal gas approximation would bring about an error in determination of ΔH_v and ΔC_p from vapor pressures of approximately 5 and 20 (!) per cent, respectively, near the normal boiling point. The vapor nonideality, however, decreases progressively with decreasing vapor pressure and it can be completely neglected at sufficiently low values of p_s. Thus, according to the reliability of the volumetric data, the upper temperature limit for the use of ΔH_v and ΔC_p has to be established for each substance to avoid distortion of the results due to the inadequate description of the pVT behavior of the vapor. The investigation carried out previously (Ruzicka and Majer, 1986) indicates that for most substances the magnitude of the terms for the vapor nonideality is of the same order as the usual uncertainty in ΔH_v and ΔC_p at temperatures about T_b-40 and T_b-80, respectively. The vapor compressibility factor and its derivatives can be obtained at moderate pressures from the volume explicit virial expansion truncated after the second virial coefficient B. The analytical description for B and its derivatives has been obtained recently for about 100 compounds by simultaneous correlation of second virial coefficients and residual heat capacities of vapor (Malijevsky et al., 1986). For other compounds B values can be estimated by a universal Benson-type contribution method (McCann and Danner, 1984) requiring only one specific parameter (critical temperature). It was found (Majer et al., 1988) that this technique also yields realistic estimates of temperature derivatives of the second virial coefficient and is thus suitable for this application.

The exact approach to the determination of correlation parameters requires the use of the principle of maximum likelihood. This method is rather complex and convergence problems can sometimes arise in determining the parameters. A procedure using the minimization of a suitably selected function by method of weighted least squares can be used instead. This simpler approach can lead to practically identical results to the former one provided the weighting factors for all data points are properly adjusted. The simultaneous correlation must always be preceded by a preliminary assessment of the data uncertainty when expected errors in p_s, ΔH_v and ΔC_p are estimated.

The main practical interest in the simultaneous correlations is calculation of vapor pressures and/or enthalpies of vaporization well below the normal boiling point. It is desirable to determine what accuracy of the obtained values can be expected. The effect of errors in input quantities to the output of the procedure has to be analyzed. For example, some quantitative estimates should be obtained indicating how the uncertainty in thermal data is reflected in the extrapolated vapor pressures etc. Preliminary analysis for 1-alkanols (Ruzicka, 1986) showed that, even when the input data are of only medium quality, the vapor pressures can be calculated near the triple point with an error below 10 per cent which is better than can be expected for most direct measurements. Enthalpies of vaporization for high boiling compounds at 298.15 K can be obtained usually with an error around 1 per cent which is comparable with the expected accuracy of complicated calorimetric determinations in this region.

DATA REQUIREMENTS

Despite its advantages, the simultaneous correlation has, however, never been massively used in producing recommended values of vapor pressures and/or thermal data, and the individual properties were instead handled separately. The reason is obviously the great number of different input data which are needed. Collecting the values from the literature is a long and tedious procedure requiring experience in screening the data for several different properties and in assessing their reliability. In this connection all secondary sources containing recommended or compiled raw data and bibliographies of measurements are very helpful. A considerable effort has been made in data compilation over the last several years. Both in the West and in the Soviet block new publications and data files appeared recently making access to experimental data much easier and thus permitting application of this approach to a variety of compounds. Main recent compilations and data bases of relevant thermodynamic data are mentioned below.

Vapor pressures

Several compilations containing experimental values (or their bibliography) appeared over last ten years (Ohe S., 1976; Boublik et al., 1984; Dykyj and Repas, 1979; Dykyj et al., 1984, Lencka et al. 1984). From the open literature the most complete bibliography of measurements (more than 6600 compounds listed) can be found in the two volumes by Dykyj and coworkers. An extensive data base of evaluated experimental and recommended vapor pressures is available at the Thermodynamics Research Center of the Texas A&M University. This data file can be accessed on line through Technical Data Services, Inc., New York.

Enthalpies of vaporization

A project on critical compilation of enthalpies of vaporization was carried out at the Institute of Chemical Technology - Prague (1981-1985) under auspices of the IUPAC Sub-Committee on Thermodynamic Tables. The result of this project is an IUPAC publication (Majer and Svoboda, 1985) containing both raw experimental and recommended enthalpies of vaporization for over 600 organic compounds with comments on experimental methods and data quality. On the basis of this data base the retrieval and computation system ENTVAPOR has been produced for the IUPAC Committee on Chemical Data Bases (Majer and Davis, 1988). This system is available on a diskette and can be used on IBM compatible microcomputers.

Ideal gas heat capacities

The Thermodynamics Research Center of the Texas A&M University has been engaged over many years in compilation of spectral data and calculation of thermodynamic properties of ideal gas. Procedures are available for calculation of ideal gas heat capacity for a variety of compounds. Temperature correlation of a large number of C_p^{go} data available in the open literature was carried out recently at the Institute of Chemical Technology, Prague (Bures et al. 1987) using a nonpolynomial relationship permitting a reasonable temperature extrapolation.

Liquid heat capacities

The Chemical Thermodynamics Division of the National Bureau of Standards has been working on a compilation of thermodynamic properties of organic substances in the condensed phase. The book by Domalski et al. (1984) contains experimental C_p^l values at one temperature for a great number of compounds and can serve as a valuable bibliographic source (an update of the compilation is under preparation). The recommended heat capacities for selected oxygen containing compounds C1 to C4 were published by Wilhoit et al. (1985). An extensive project on critical compilation of liquid heat capacities as a function of temperature was started in 1984 at the Institute of Chemical Technology - Prague. This project was accepted as an official IUPAC project at the General Assembly in Boston 1987 (Zabransky et al., 1988). At present, data for more than 1300 substances are stored in a computer and process of assessment and correlation is under way. The output will become a book as well as the computation and retrieval system permitting direct accessing of recommended data on a microcomputer.

POSSIBLE APPLICATIONS OF THE PROCEDURE

The proposed procedure of simultaneous correlation of vapor pressures and thermal data can be used in three different ways.

1. Production of reference data

Recommended thermodynamic data on vaporization equilibria can be obtained for important organic compounds by combining selected experimental values of different properties. In the first step all available data on p_s, ΔH_v, $C_p^{\,l}$, $C_p^{\,go}$ below the normal boiling point have to be collected and their quality assessed. In the second step the consistency of data is examined by simultaneous correlation, which helps to reveal possible systematic errors. Finally the selected data sets are used in correlation for producing recommended values between the triple and normal boiling points (Ruzicka, 1986).

2. Correlation program as a part of data bank software

Industrial and scientific institutions have their own data bases for substances of their particular interest. They contain mostly recommended values on p_s, ΔH_v and heat capacities rather than raw data. These values are usually available over a limited temperature range in a form of smoothing equation whose parameters are stored in a computer. It is worthwhile to develop a modified correlation procedure which would use as input the data generated from smoothing equations for the individual properties, instead of direct experimental values (Ruzicka and Majer, 1986). This simplified version can be highly automated and designed in such a way that it can be used also by users who are not familiar with the method in detail. Such a program is aimed to be used for rapid calculation of p_s and/or ΔH_v at temperatures far below the normal boiling point and can be easily adapted in to a software of a data bank of thermodynamic properties.

3. Prediction methods for high boiling compounds

Enthalpy of vaporization at 298.15 K

This quantity is of great importance in conversion of thermodynamic properties between liquid and ideal gas states and usually cannot be obtained with the required accuracy. Credible values of enthalpies of vaporization from calorimetric measurements can be supplemented with the values determined for selected compounds from the simultaneous correlation in order to obtain a representative collection of reliable data for the largest variety of molecular structures. This data base can be used for formulation of Benson type contribution method for ΔH_v at 298.15.

Vapor pressures

For many high boiling compounds, the thermodynamic data on vaporization equilibria are scarce and of low reliability. The common corresponding states prediction methods are not suitable for high boiling compounds as their conventional parameters (critical data, acentric factor etc) are not available and in many cases cannot be reasonably estimated. The principle of simultaneous correlation can be used in formulation of a different type of estimation technique requiring practically no traditional parameters. Vapor pressure and enthalpy of vaporization are related to the heat capacity difference by their temperature derivatives. Therefore when ΔC_p can be estimated at one temperature (or better as a temperature function) p_s and also ΔH_v can be predicted after integration with reasonable accuracy between the triple and normal boiling points. In general one value of vapor pressure (usually normal boiling point) and one value of enthalpy of vaporization (estimated by group contribution method at 298.15 K) are only needed as additional input parameters. Nonideality corrections in the moderate pressure range can be also obtained from group contributions (McCann and Danner, 1984).

Estimation of heat capacity difference plays a crucial role in this approach. Temperature dependent values of C_p^{go} can be obtained from a group contribution method (Benson et al., 1969) and C_p^{l} can be estimated similarly for hydrocarbons (Luria and Benson, 1977). The same type of prediction scheme is being developed in the Chemical Thermodynamics Division of the National Bureau of Standards for condensed phases at 298.15 K. The contributions for hydrocarbons have been published (Domalski and Hearing, 1988) and procedure is being extended to other groups of compounds. When the IUPAC database on liquid heat capacities as a function of temperature will be completed the group contributions as a function of temperature will be determined.

A lot of effort has been made in compilation and evaluation of the thermal data related to vapor pressures. The simultaneous correlation opens new ways how to obtain data at temperatures where experimental values are lacking. When all this information is properly exploited there is a good chance that a universal contribution method can be developed in near future which will describe consistently thermodynamic properties characterizing vaporization equilibria between triple and normal boiling point.

REFERENCES

Ambrose D., Davies R.H. 1980; J Chem Thermodyn. 12, 871.

Benson S.W., Cruickshank F.R., Golden D.M., Haugen G.R., O'Neal H.E, Rodgers A.S., Shaw R., Walsh R. 1969; Chem. Rev. 69, 279.

Boublik T., Fried V., Hala E. 1984; The Vapour Pressures of Pure substances. Elsevier, Amsterdam.

Bures M., Holub, R., Leitner, J., Vonka P. 1987; Sb. Vys. Sk. Chem.-Technol. v Praze, N8, part I, II.

Domalski E.S., Evans W.H., Hearing E.D. 1984; Heat Capacities and Entropies of Organic Compounds in the Condensed Phase. J. Phys. Chem. Ref. Data Vol. 13, Suppl. 1.

Domalski E.S., Hearing E.D. 1988; J Phys. Chem. Ref. Data in press.

Dykyj J., Repas M. 1979; The Vapor Pressures of Organic Compounds. Veda, Bratislava.

Dykyj J., Repas M., Svoboda J. 1984; The Vapour Pressures of Organic Compounds. Veda, Bratislava.

King M.B., Al-Najjar H. 1974; Chem. Eng. Sci. 29, 1003.

King M.B., Mahmud R.S. 1986; Fluid Phase Equilibria 27, 309.

Lencka M., Szafranski A., Maczynski A. 1984; Verified Vapor Pressure Data, Organic Compounds Containing Nitrogen. PWN-Polish Scientific Publishers, Warsaw.

Luria M., Benson S.W. 1977; J. Chem. Eng. Data 22, 90.

Majer V., Davis W. 1988; Retrieval and Computation System ENTVAPOR, IUPAC Committee on Chemical Data Bases.

Majer V., Svoboda V. 1985; Enthalpies of Vaporization of Organic Compounds, A Critical Review and Data Compilation. IUPAC Chemical Series No. 32, Blackwell, Oxford.

Majer V., Svoboda V., Pick J. 1988; Heats of Vaporization of Fluids. Elsevier, Amsterdam.

Malijevsky A., Majer V., Vondrak P., Tekac V. 1986; Fluid Phase Equilibria 28, 283.

McCann D.W., Danner R.P. 1984; Ind. Eng. Chem., Process Des. Develop. 23, 529.

Mosselman C, van Vugt V.H., Vos H. 1982; J. Chem. Eng. Data 27, 246.

Ohe S. 1976; Computer Aided Data Book of Vapor Pressures, Chem. Soc. Jap., Tokyo 1976.

Rogalski M. 1985; Thermochem. Acta 90, 125.

Ruzicka K. 1986; Master thesis. Institute of Chemical Technology, Prague 1986.

Ruzicka K., Majer V. 1986; Fluid Phase Equilibria 28, 253.

Scott D.W., Osborn A.G. 1979; J. Phys. Chem. 83, 2714.

Wilhoit R.C., Chao J., Hall K.R. 1985; J. Phys. Chem. Ref. Data 14, 1.

Zabransky M., Ruzicka V. Jr., Majer V., Domalski E.S. 1988; Critical Compilation of Heat Capacities of Liquids, On Going IUPAC Project.

Critical Compilation of Heat Capacities of Liquids

Milan Zabransky[1], Vlastimil Ruzicka, Jr.[1], Vladimir Majer[2], and Eugene S. Domalski[3]

[1] Department of Physical Chemistry, Institute of Chemical Technology, 166 28 Prague 6, Czechoslovakia
[2] Department of Chemistry, University of Delaware Newark, DE 19716, USA
[3] Chemical Thermodynamics Division, National Bureau of Standards, Gaithersburg, MD 20899, USA

1. Introduction

Several compilations of heat capacities of liquids have been published over the last 20 years. However, some compilations are limited with regards to the number of compounds included, some by the degree of critical evaluation of the data, and some in the temperature range which the data cover. Besides these compilations, there are others which contain a number of tables having heat capacities for industrially important substances and other related properties. These multi-property tables mostly contain old data, with unclear identification of the sources and sometimes questionable levels of accuracy. Most of these tables lack a critical approach in the selection and evaluation of the data.

C. Jochum M. G. Hicks J. Sunkel (Eds.)
Physical Property Prediction
in Organic Chemistry

The project entitled "Critical compilation of heat capacities of liquids" deals with compilation and critical assessment of liquid heat capacities for a large variety of compounds. This data project is carried out under the auspices of the IUPAC Subcommittee on Thermodynamic Data.

2. The Purpose and Objectives

The purpose of the project is to compile and evaluate experimental heat capacities of pure substances in the liquid state. The project covers not only organic and inorganic compounds, but also organosilicon and organometallic compounds. There are several criteria for inclusion of a substance in the compilation. The first criterion requires the substance to have a melting point below 573 K. Then, only well-defined isotropic liquids are considered; undercooled liquids are excluded. The database contains isobaric heat capacities measured at pressure of 101.325 kPa and at saturation. In general, the database comes directly from experimental heat capacity measurements which have been reported in the literature. Except for a few compounds for which no other data are available, reference to comprehensive tables, calculated values, and other secondary data sources are not included. The bulk of the literature used has been published after 1920, which marks the beginning of the development of modern low-temperature calorimetry. Measurements published prior to this time are included if they contain reliable data or if no other data sources are available.

Objectives of the project:

(a) To compile all available data for the above specified compounds and establish the database of data taken from the literature (further referred to as the raw data).
(b) To evaluate data critically and generate sets of selected data.
(c) To correlate selected data and generate sets of recommended values.

3. Database of Raw Data

All available experimental heat capacities for compounds selected on the basis of the above criteria accompanied with some auxiliary information comprise the database of raw data. The database consists of 6 disk files; 2 files of experimental data, a file of compound names and codes, a file of literature references, a file of source documents and a file of calorimeters. The interelationships of files is given in Fig.1. On-line storage in the disk files is made by means of programs that allow some of the input data to be verified.

3.1. Files of experimental data (see Fig.2). The experimental data are stored either in the file of discrete data containing heat capacity values and corresponding temperatures or in the file of constants of a smoothing equation when direct data are not reported. The latter file includes parameters, the type of the smoothing equation and the temperature range over which the parameters are valid. The remaining parts of both files are identical and include the following items: Chemical Abstracts Service Registry Number of the compound; error of the measurement as indicated by the author(s) (if available); purity of the substance and method used for its determination; temperature unit; heat capacity unit; type of the heat capacity measured (C^l_p, C^l_{sat} or C^l_{avg} - the average value determined over the temperature range usually higher than 10 K); code for the literature reference; notes (if necessary).

3.2. File of compound names and codes (see Fig.3) contains the name of the compound (names conform to the Chemical Abstracts Service nomenclature) and possibly some synonyms, molecular formula, CAS Registry Number, group and member number. The CAS Registry Number is used as a unique identification code of the substance throughout the database. In order to present data in a clear form and to facilitate the retrieval of data for chemically similar substances, we introduced an additional identification code

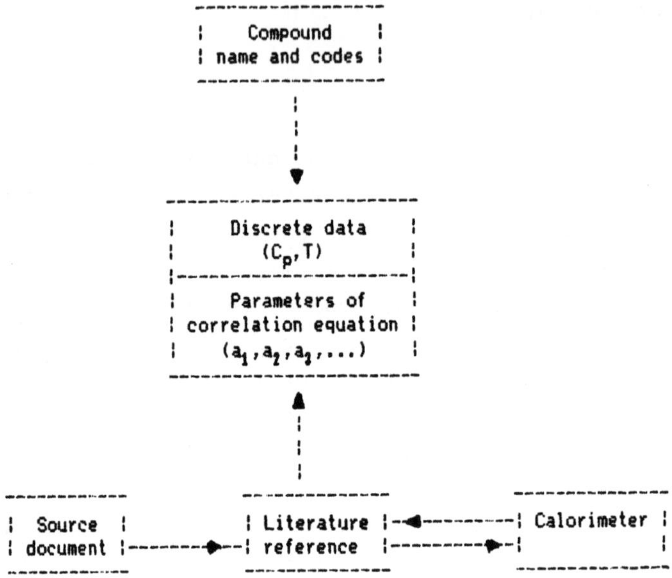

Fig.1 Interelationships of disk files Liquid Heat Capacity Database

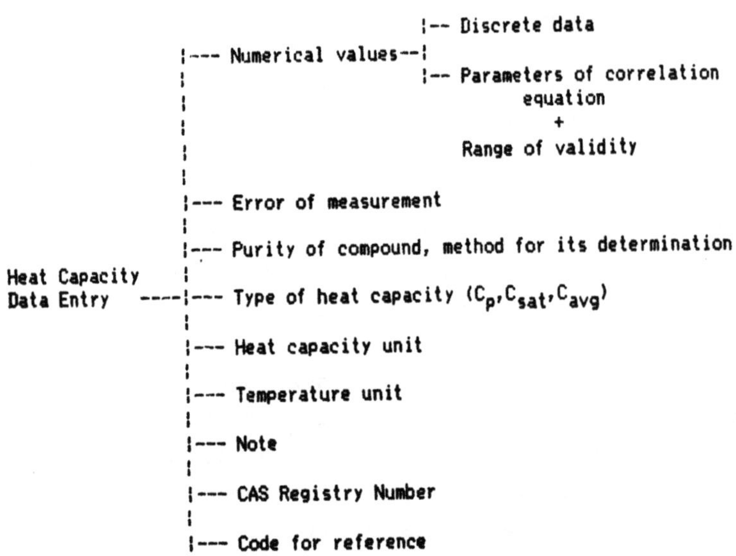

Fig.2 Items in the record of heat capacity data

for each substance consisting of a code number for a group and an order number within the group. The list of the groups of substances with the numerical codes used is shown in Tab.1. Within the groups, compounds are ordered according to Hill's system.

3.3. File of literature references (see Fig.4) contains name(s) of author(s), code for the source document, identification data (e.g. volume, page, year of publication), code for the calorimeter.

3.4. File of source documents includes abbreviations of names of journals, titles of books, reports, conference proceedings, etc. The abbreviations of the titles of documents conform to the rules of the Chemical Abstracts Service.

3.5. File of calorimeters (see Fig.5) contains a concise characteristics of the apparatus used in the heat capacity measurement. The following items are stored: type of the calorimeter; temperature range of operation (specified either numerically or by a code); measuring accuracy (specified either numerically or by a code); code for the literature reference; notes (if necessary).

4. Present State of the Compilation of Data

In May 1988, the compilation covered more than 1400 compounds. The number of data sets of discrete data was almost 2700. Each data set contains from 1 to 120 experimental points. In addition, there were about 150 data sets of parameters of a smoothing equation. The term data set denotes a set of experimental values (either discrete heat capacities and temperatures or parameters of a smoothing equation) reported by an author(s) for one substance in one original source. More than 1200 literature references and almost 300 calorimeters were

```
                  |---- Name, synonyms
                  |
                  |---- Formula
Compound Entry ---|
                  |---- CAS Registry Number
                  |
                  |---- Group and member number
```

Fig.3 Items in the record of compounds

```
                   |---- Name(s) of author(s)
                   |
Literature         |---- Identification (volume, year, page, number)
Reference Entry ---|
                   |---- Code for source document
                   |
                   |---- Code for calorimeter
```

Fig.4 Items in the record of literature references

```
                   |---- Type of calorimeter
                   |
                   |---- Measurement conditions (temperature range)
Calorimeter -------|
Entry              |---- Measurement accuracy
                   |
                   |---- Note
                   |
                   |---- Code for reference
```

Fig.5 Items in the record of calorimeters

stored in the corresponding files. At present, the database is almost complete and the compilation will be finished at the end of 1988.

5. Correlation Procedure

The data are correlated by a polynomial of the form:

$$C^l = \sum_{j=0}^{m} A_j (T/100)^j, \qquad (1)$$

where C^l is either C^l_p or C^l_{sat} in J/(K.g), T is in K and A_j are parameters of the correlation equation. Other types of correlation equations are also being tested.

Two correlation procedures are adopted:

(a) correlation of the data over the whole temperature range

(b) division of the overall temperature range into several parts and separate correlation of experimental values from each subinterval. The correlation program ensures that the derived parameters give identical C^l values as well as their first temperature derivatives at the boundaries of these subintervals (with the exception of compounds that exhibit a discontinuous function for $C^l = C^l(T)$).

The correlation is performed using the weighted least-squares method; the minimization function has the form:

$$S = \sum_{i=1}^{n} [(C^l_{exp} - C^l_{cal})^2/\delta^2 C^l]_i \quad (2)$$

where the summation is over all the values included in the

correlation. The variance of each value, $\delta^2(c^1)_i$, is estimated on the basis of the assumed experimental error of the data set used in the correlation. The input information is the percentage error, $\delta_r c^1$, as given by the author for the whole data set or estimated by the compiler. The variance of the i-th data point is expressed as:

$$\delta^2(c^1)_i = [(c^1)_i \cdot \delta_r c^1/100]^2 \qquad (3)$$

The correlation is carried out in three steps:

(a) All available c^1 values along with the supplementary input data for a substance are read in and a preliminary joint correlation is performed. Prior to the correlation, data sets considered a priori as unreliable are discarded if more accurate data are available for the same temperature range.

(b) Individual values or parts of data sets that showed little consistency with the other data are discarded, and the values of $\delta_r c^1$ can be altered in order to modify the weights of individual data sets. There are options to change both temperature limits of the correlation and/or divide the temperature range in two or more subintervals. A statistical procedure (the Fischer test) is used to establish the degree of the polynomial for the overall and partial correlations.

(c) The final correlation for the selected experimental values is performed and a class (I to VI) characterizing the accuracy of the correlated data is assigned. The criterion for judging the quality of the correlation is the weighted standard deviation, s_w:

$$s_w = [S_{min}/(n-m-1)]^{1/2} \ . \qquad (4)$$

When experimental data are consistent within the expected error limits, s_w should be smaller or near unity. In addition, use of the following statistical criteria is made:

the standard deviation:

$$s = \left(\left[\sum_{i=1}^{n}(C^{l}_{cal} - C^{l}_{exp})_i{}^{2}\right]/(n-m-1)\right)^{1/2} \qquad (5)$$

the percent standard deviation:

$$s_r = \left(\left[\sum_{i=1}^{n}((C^{l}_{cal} - C^{l}_{exp})C^{l}_{exp})_i{}^{2}\right]/(n-m-1)\right)^{1/2}.100 \qquad (6)$$

the bias

$$s_b = \left(\left[\sum_{i=1}^{n}(C^{l}_{cal} - C^{l}_{exp})_i\right]/n \qquad (7)$$

and the difference (+/-) between the experimental points with positive and negative deviation from the recommended values.

The experimental data that are included in determining the parameters of the correlation equation are selected on the basis of the following criteria:

1. accuracy of the measurement claimed by the author
2. reputation of the laboratory and reliability of the author's of the measurements
3. consistency of the data with the values from other sources (if available)
4. purity of the substance
5. type of calorimeter
6. time of origin of measurement (measurements published after the year 1940 are preferred)
7. scatter of the data.

The following levels of accuracy are assigned to the recommended data:

I highest quality data for compounds used in calibration experiments (uncertainty below 0.1 %)

II very reliable data (uncertainty below 0.25 %)

III reliable data (uncertainty below 0.5 %) IV medium quality data (uncertainty below 1 %)

V data of low reliability (uncertainty below 3 %)

VI unreliable data with possibility of gross systematic errors (uncertainty above 3 %)

6. Data Presentation

Several tables are used to present information on experimental and recommended data for each compound.

6.1. Table of experimental data (see Tab.2) contains basic information on the sources of the heat capacities. The table includes the following items: the abbreviated reference in the form YYAAA/BBB N (where YY are the last two digits of the year of publication, AAA and BBB are the first three letters of the last name of the first and the second author, respectively, N is a digit from 1 to 9 distinguishing papers published by the same author(s) within the same year. For a year before 1900 the reference code is preceded by an asterisk); temperature range; number of experimental points; error of measurement, $\delta_r C^l_{exp}$, as indicated by the author(s) (if available); purity and analytical method used for its determination (if available); type of the heat capacity presented; type of calorimeter used and relevant references; notes.

Tab.1 List of groups of substances

0.1.	Elements
0.2.	Inorganic Compounds
1.	Compounds of Carbon and Hydrogen
1.1.	Saturated hydrocarbons
1.2.	Saturated cyclic hydrocarbons
1.3.	Unsaturated hydrocarbons
1.4.	Aromatic and unsaturated cyclic hydrocarbons
2.	Compounds of Carbon, Hydrogen and Halogen
2.1.	Fluorine derivatives
2.2.	Chlorine derivatives
2.3.	Bromine derivatives
2.4.	Iodine derivatives
2.5.	Mixed halogen derivatives
3.	Compounds of Carbon, Hydrogen and Nitrogen
3.1.	Amines
3.2.	Nitriles
3.3.	Heterocyclic nitrogen compounds
3.4.	Miscellaneous nitrogen compounds
4.	Compounds of Carbon, Hydrogen and Oxygen
4.1.	Ethers
4.2.	Alcohols and phenols
4.3.	Carbonyl compounds
4.4.	Acids and anhydrides
4.5.	Esters
4.6.	Heterocyclic oxygen compounds
4.7.	Miscellaneous oxygen compounds
5.	Compounds of Carbon, Hydrogen and Sulphur
5.1.	Sulphides
5.2.	Thiols
5.3.	Heterocyclic sulphur compounds
6.	Other Compounds
6.1.	Compounds of carbon, hydrogen, halogen and oxygen
6.2.	Compounds of carbon, hydrogen, nitrogen and oxygen
6.3.	Compounds of carbon, hydrogen, oxygen and sulphur
6.4.	Miscellaneous compounds
7.	Organic Compounds Containing Other Elements than Halogens, Oxygen, Nitrogen and Sulphur
7.1.	Organosilicon compounds
7.2.	Organic compounds containing phosphorus and boron
7.3.	Organometallic compounds
7.4.	Salts of organic acids

Tab. 2

Name: Pentane
Formula: C5H12

Group no.: 11-010
CAS registry no.: 109-66-0
Molar mass: 72.15

Experimental liquid heat capacity data

citation	temp.range K	no. poin	error %	purity %	purity methd	type capac	calorimeter type-citation	note
				Temperature range: 148.6- 468.1 K				
30PAR/HUF 1	149.9- 290.0	14	1.00	not specified		C_p	IP- 25PAR	no
40MES/KEN	151.3- 286.4	19	nosp	99.961	melpt	C_p	AD- 39AST/EID 1	no
67MES/GUT	148.6- 302.9	25	0.10	99.86	melpt	C_{sat}	AD- 47HUF	no
71AMI/ALI	313.1- 468.1	19	2.00	not specified		C_p	not specified	yes
75GRI/RAS	299.9- 383.6	9	1.00	not specified		C_p	AD- 75RAS/GRI	yes
85CZA	299.0	1	nosp	not specified		C_p	IP- 79CZA	no

citation	note
71AMI/ALI	calculated from Cv measured on saturation line
75GRI/RAS	all values (except the first one) at pressure higher than saturation pressure

Tab. 3

Name: Pentane
Type: saturation

CAS registry no.: 109-66-0

Correlated liquid heat capacity data

citation	temp.range K	no.poin consid	$\delta_r C^l$ %	d_w	d	d_r %	bias	+/-
			Temperature range: 148.6- 383.6 K					
67MES/GUT	148.6- 302.9	25	0.10	0.259	0.001	0.026	0.000	3
75GRI/RAS	299.9- 383.6	9	1.00	0.572	0.014	0.572	-0.008	-5
30PAR/HUF 1	149.9- 290.0	0	1.00	0.601	0.013	0.601	-0.010	-12
40MES/KEN	151.3- 286.4	0	nosp	0.221	0.015	0.662	0.003	-5
71AMI/ALI	313.1- 383.1	0	2.00	0.913	0.046	1.827	0.040	8
85CZA	299.0	0	nosp	0.486	0.033	1.457	-0.033	-1

6.2. Table of correlated data (see Tab.3) provides information on the data sets both selected for inclusion in determining the parameters of the correlation equation and rejected from the correlation. The table contains the following items: the abbreviated reference; temperature range; number of values considered in the determination of the correlation parameters (zero signifies that the data set was not included among the selected data sets); $\delta_r C^l$ (this value is either equal to $\delta_r C^l{}_{exp}$ or is assigned by the compiler); weighted average deviation d_w; average deviation d; percent average deviation d_r; bias and the +/- points difference. This table is absent for those substances where only one data set was considered.

6.3. Table of correlation parameters (see Tab.4) gives results of the correlation of the heat capacity ($C^l{}_p$ and/or $C^l{}_{sat}$) as a function of temperature. It consists of: the total number of all experimental data points available; the total number of values used in the correlation; weighted standard deviation s_w; standard deviation s; percent standard deviation s_r; bias s_b; +/- difference; level of accuracy assigned by the compiler (I to VI); the temperature range of the parameters' validity; parameters of the correlation equation.

6.4. Table of recommended values (see Tab.5) contains heat capacity values calculated on the basis of the determined parameters, over the indicated range. Molar and specific heat capacities (J K-1 mol-1 and J K-1 g-1, respectively) at equally spaced temperatures (K) are presented.

6.5. Deviation plot (see Fig.6), provides the user with an overview of the deviations from the recommended values for all experimental data (both included in and rejected from the correlation) measured by various authors. The temperature is plotted along the x-axis and the relative percentage deviation along the y-axis. Points measured by different authors are distinguished by different symbols. Points that lie outside the range of the ordinate in the plot are accompanied by the numerical value of the deviation. Some points that overlap the accompanying figures are omitted.

Tab. 4

Name: Pentane

CAS registry no.: 109-66-0

Parameters of the correlation equation $C^l = f(T)$

type	no.data total	points consid	s_w	s	s_r %	bias	+/-	level of accuracy
C_{sat}	87	34	0.392	0.008	0.314	-0.002	-2	III

range,K	parameters A_0	A_1	A_2	A_3
148.6- 383.6	2.580230E+00	-8.558111E-01	3.353755E-01	-2.613452E-02

Tab. 5

Name: Pentane
type: saturation

CAS registry no.: 109-66-0

Table of recommended data

Temp.,K:	273.15	298.15	150.00	160.00	170.00	180.00	190.00
c,J/(K.g):	2.212	2.317	1.963	1.962	1.966	1.974	1.986
C,J/(K.mol):	159.6	167.2	141.6	141.6	141.9	142.4	143.3
Temp.,K:	200.00	210.00	220.00	230.00	240.00	250.00	260.00
c,J/(K.g):	2.001	2.020	2.042	2.068	2.097	2.128	2.163
C,J/(K.mol):	144.4	145.7	147.4	149.2	151.3	153.6	156.1
Temp.,K:	270.00	280.00	290.00	300.00	310.00	320.00	330.00
c,J/(K.g):	2.200	2.240	2.281	2.326	2.372	2.420	2.469
C,J/(K.mol):	158.7	161.6	164.6	167.8	171.1	174.6	178.2
Temp.,K:	340.00	350.00	360.00	370.00	380.00		
c,J/(K.g):	2.520	2.573	2.626	2.681	2.737		
C,J/(K.mol):	181.8	185.6	189.5	193.5	197.5		

Pentane [109-66-0]

type: saturation

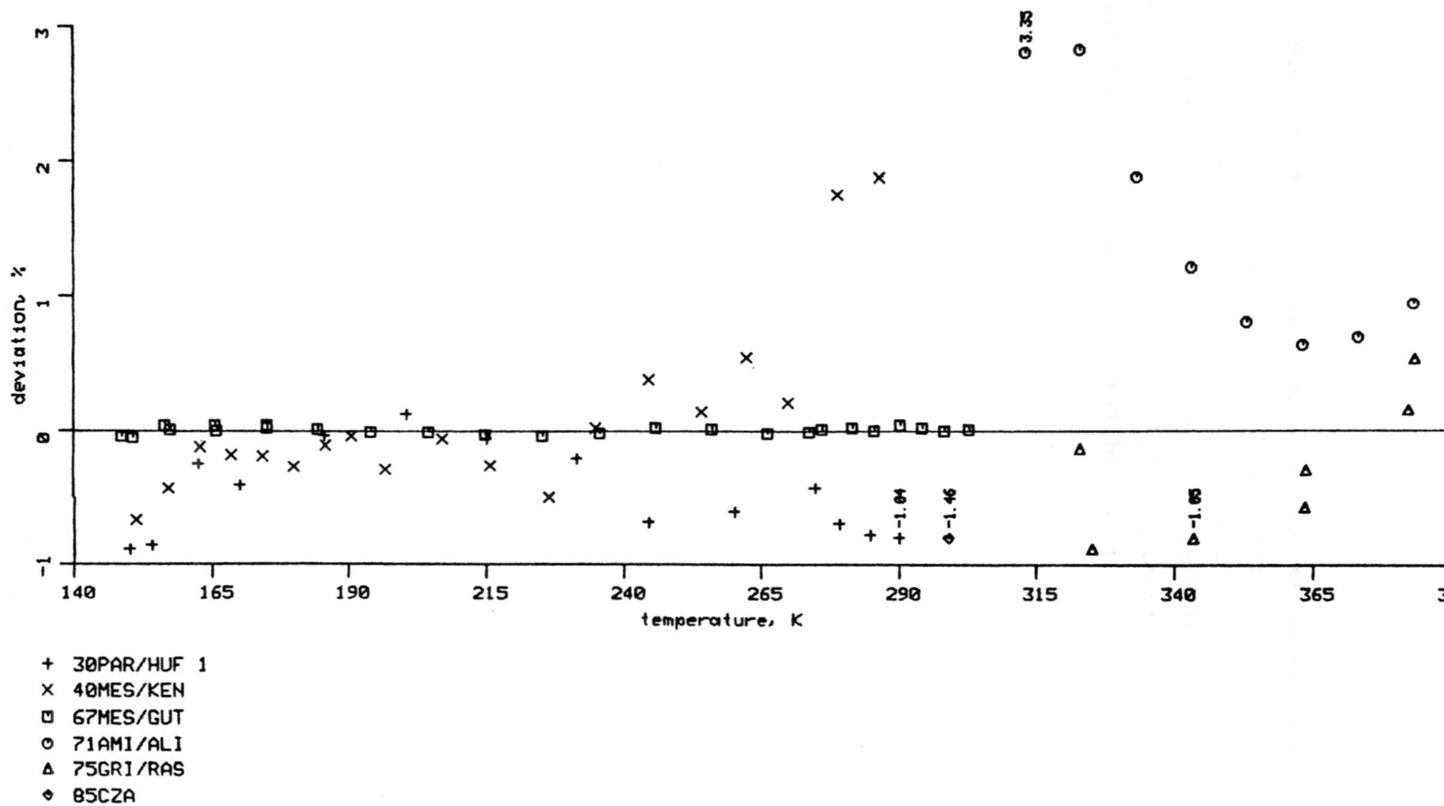

7. Future prospects

The project's output will become a textbook presenting a survey of published experimental data as well as recommended data. A computerized system permitting computation and retrieval of recommended data on a microcomputer will be prepared. When the project is completed, group contributions for liquid heat capacities as a function of temperature will be determined.

PETRA: Software Package for the Calculation of Electronic and Thermochemical Properties of Organic Molecules

Peter Löw and H. Saller

Chemodata Computer-Chemie, Dr. Troll-Strasse 14, 8030 Gröbenzell, Federal Republic of Germany

The progress in the field of chemistry as a scientific discipline, especially in this century, has consisted in setting up models, concepts and rules so as to create order and understanding for the enormous amount of chemical observations and factknowledge regarding connectivity and reactions.

Especially in the area of organic chemistry, models have been developing concerning the molecular procedure of chemical reactions, the vast subject of reaction mechanisms.

In order to be able to describe the chemical properties of molecules, concepts such as charge distribution, electronegativity, resonance effect, polarizability, have been defined.
These concepts are mostly empirical in nature and cannot be described via theoretical derivations; or are of a mathematical form which cannot be accurately solved.

During the last years methods have been developed to describe these concepts quantitatively and to design algorithms and procedures in order to be able to apply them in the form of computer programs (1).

PETRA (Parameter Evaluation for the Treatment of Reactivity Application) is a program package allowing the easy calculation of electronic and thermochemical properties in organic molecules.

For qualitative concepts, e.g. bond polarity, inductive effect, resonance stabilization or polarizability, the chemist obtains values which allow him a <u>quantitative</u> approach to the evaluation of problems, e.g. reactivity evaluations, structure-activity relationships, heats of reactions.

C. Jochum M. G. Hicks J. Sunkel (Eds.)
Physical Property Prediction
in Organic Chemistry

On developing the individual procedures, utmost care has been taken in order to demonstrate the usefulness of the calculated values via correlation with experimental electronic or thermochemical properties of organic molecules.

PETRA contains procedures for the calculation of the following atomic, bond and molecular properties:

- Charge Distribution in Sigma- and Pi-Bonds
- Bond Polarity
- Effective Polarizability
- Resonance Stabilization of Charges
- Standard Heats of Formation
- Ring Strain and Resonance Energy
- Bond Dissociation Energies
- Reactivity of Bonds

PETRA is easy to operate without specific previous knowlwedge. The program only requires the constitution of a molecule, i.e. the atoms contained in the molecule and the bonds between them. Molecules are fed in on a graphic screen by a mouse, or alphanumerically. Series of molecules may be evaluated in one program run.
The output of the results is provided either graphically or in the form of an easy to read data file.

Charge Calculation

The empiric concept for the evaluation of charges applied in PETRA is based on the partial equalization of orbital electronegativities which can be evaluated from ionization potentials and electron affinities (PEOE-procedure). On bond formation orbital electronegativities are partially equalized by the transfer of charges.

The evaluation of pi-charges in conjugated pi-systems represents an extension of this approach. In this process the various resonance structures of a pi-system are generated and the distribution of pi-charges is obtained by assigning "weights" to the individual structures.

A series of experimental data has been correlated with charge values in order to show the significance of the charges evaluated: C-1s ESCA shifts, dipole moments, 1H-NMR and 13C-NMR shifts, C-H and C-C coupling constants. The charge calculation supplies the sigma-, pi-, and total charge and a sigma- and pi-electronegativity value for each atom of the molecule.

Polarizability

Based on an additivity scheme and applying the polarizabilities of each atom, a procedure has been developed for the quantitative treatment of charge stabilization _within_ a molecule by means of polarizability effects. This procedure has been modified in order to make proper modelling of the distance dependence of the stabilization effect possible. For each atom of the molecule PETRA evaluates this effective polarizability and in addition the mean molecular polarizability.

The applicability of these effective polarizabilities evaluated has been proven via correlation with experimental values of proton affinities of amines, alcohols, ethers and thioesters as well as gas phase acidities of alcohols.

Thermochemistry

PETRA evaluates standard heats of formation of organic molecules on the basis of an additivity pattern with energy increments for substructures with 2, 3, and 4 atoms. For the purpose of predictions, the energy increments for the substructures contained in a molecule are added.

Up to now, about 90 different types of bonds (1,2 interactions) between the atoms H, C, N, O, F, Si, P, S, Cl, Br and I have been taken into consideration.
Over and above, ring strain- and aromatic delocalization energies can be obtained.

For those classes of compounds that could be fully parameterized the standard deviation between experimental and calculated values average between 0.5 and 1.0 kcal/mol. By means of calculating the standard heats of formation for the educt and product of a reaction, heats of reactions may be evaluated. Based on substructure parameters for radicals, PETRA evaluates bond dissociation energies (BDE).

Reactivity of Bonds

The physico-chemical parameters evaluated by PETRA, allow predictions on the reactivity of each bond by means of a reactivity function. These values make possible to recognize reactive centres and bonds in a molecule directly. As a result, the reactivity of the molecule can be estimated and the various reaction possibilities can be predicted.

Applications

PETRA may be applied for checking and predicting data. By establishing correlations with experimental physical properties the prediction of unknown values is made possible.

Meaningful variables for the investigation of Quantitative Structure Activity Relationships (QSAR) are made available by PETRA. In combination with statistics programs the results produced by PETRA permit the evaluation of models for the treatment of the biological activity of organic molecules. These models can be used to predict and estimate the activity of new molecules.

PETRA can be most usefully applied in organic chemistry. The variables calculated contain valuable information on the course of chemical reactions (reaction mechanisms). With the reactivity evaluations reaction possibilities as well as the stability of a molecule may be predicted.

References

1. J.Gasteiger, M.G.Hutchings, B.Christoph, L.Gann, C.Hiller, P.Löw, M.Marsili, H.Saller and K.Yuki, Topics Curr. Chem., 137, 19 (1987).

Workshop Review and Epilogue

Martin G. Hicks

Beilstein Institut, Varrentrappstrasse 40–42, 6000 Frankfurt/Main 90, Federal Republic of Germany

The view of the overwhelming majority of participants, including the Beilsteiners, was that the workshop was a resounding success. This success was not just a measure of the excellent food and accommodation provided by the Hotel Schloss Korb, but was also a reflection of the content of the scientific programme. The wish expressed by all for another workshop could not be correlated with the weather, since due to the incessant rain we were virtual (albeit willing) prisoners in a real castle.

When we organized the workshop we did not expect to be presented with a list of easy answers to solve all of our problems. What we hoped for was to gain an overview of just what is possible in all areas of data calculation and, with the aid of advice from acknowledged experts, devise a scheme for short and long term solutions at Beilstein.

For some of the participants the first problem to be solved was just how to go about meeting the Beilstein group at the non-existent platform 1 of München station. Scientific method finally prevailed and after several iterations and group contributions convergence was achieved at the local minimum of platform 11 !

The workshop itself got off to a good start with a lecture from Dr. Jan Czermak (Bundesministerium für Forschung und Technologie), who showed us what an enlightened attitude the BMFT

C. Jochum M. G. Hicks J. Sunkel (Eds.)
Physical Property Prediction
in Organic Chemistry

has, not just in funding this workshop, but in their wide ranging plans for the support of information science in Germany.

The next two lectures from Beilstein colleagues Dr. Clemens Jochum and Dr. László Domokos addressed the problem of data estimation as it relates to Beilstein. Clemens Jochum's lecture demonstrated the importance of data calculation for Beilstein, both as a means to assist error checking and correction and as a help in combined searches for data. László Domokos, who gave us a statistical breakdown, demonstrated very clearly the particular problems that face us. Although Beilstein has the biggest collection of data on organic compounds in the world, the amount of data per compound is very small. The data matrix is of such magnitude that if we were to start estimating now and fill up the data base at the rate of one datum per second, we would not be finished until 2039 !

Then Dr. Steve Heller (Agricultural Research Service) gave us a great deal to think about in general; how acceptable is calculated data to the user, and what will the effect on the interpreted quality of the data base be? He then gave actual examples of the use of estimation methods - in the prediction of toxicity data for pesticides.

The Beilstein staff particularly appreciated the talks by Dr. Ken Marsh (Thermodynamics Research Center) and Dr. Brian Pedley (University of Sussex), as they both illustrated the care which needs to be taken when abstracting and evaluating data. They are also both in the process of implementing their data bases in computer readable form, which will be of great benefit to all users, especially if they are really brought up to date and implemented with graphical structure input and display interfaces.

Various speakers illustrated the practical use, and that means money saving, of estimation methods. Dr. Bill Milne (National Institutes of Health) has developed a system for assisting in the pre-selection of compounds for screening for anti-cancer activity. With the large numbers of compounds which need testing and the high cost of testing he illustrated a practical use for estimation techniques (which is also welcomed by the mouse population of Bethesda).

By far the majority of the lectures were on different uses of group contribution methods. Dr. Eugene Domalski's (National Bureau of Standards) work using the Benson approach showed us just what careful work can achieve, very good agreement being obtained with experimental data for heats of formation, heat capacity and entropy. He also warned of the limitations of the method in that molecules which have other energy factors associated with them e.g. strain energy, are not at present accurately described.

We were very pleased to be able to welcome speakers from industry, who although they are very skeptical, nevertheless expressed the view that estimation methods have a very important role in the industrial environment.

Dr. Paul Mathias (Air Products) gave an excellent overview of the estimation methods for vapour pressure and phase equilibria. His final conclusion, that although group contribution methods can work surprisingly well they are not discriminating enough for general applicability and are thus often limited to certain series, was repeatedly endorsed throughout the workshop. In his opinion the likely way forward is to develop methods with a sound theoretical base using, for example, Molecular Mechanics and Quantum Mechanics.

Dr. Heiner Landeck (Linde AG) described the requirements of his firm for methods for estimating the physical properties of complex mixtures. He described the failings of present methods and stressed the real interest in industry for the development of sufficiently accurate methods.

The healthy skepticism was further endorsed by Prof. Hugo Kubinyi (BASF), who described the use of QSAR techniques in drug design and questioned their use except for gaining a greater understanding of the systems under examination. Multivariate statistics play a major role in evaluating in-vitro tests which are becoming more important as they begin to replace animal screening. The question of accurate measurement of K_{OW} was brought up later during Sam Yalkowsky's talk.

Dr. Andreas Barth (FIZ-Karlsruhe) reviewed the utilities offered by STN which, since it will host the Beilstein file, is an important example of a numerical data base which will have to decide its policy on estimated data. He suggested links with other data bases to provide the missing data and the use of external programs to calculate the missing parameters off-line. A valid point was raised in that for certain data such as vapour pressure it is possible to store a function and not just the data.

Dr. Tim Clark (University of Erlangen-Nürnberg) gave a very interesting and refreshingly honest report about the use of Molecular Orbital methods and Molecular Mechanics in calculating data for organic compounds. He showed that although there are some restrictions that need to be taken into account, in the best cases the quality of the calculated data can often be just as good as that obtained by experiment. Many of the present restrictions and limitations will eliminated by use of new hardware much of which is just becoming available.

Prof. Klaus Lucas (University of Duisburg) gave a very good overview of Statistical Thermodynamic methods. He discussed how the properties of real fluids composed of rigid molecules can be calculated when the intermolecular forces are modelled.

Tim Clark and Klaus Lucas together illustrated one of the many successes of the workshop in that Quadrupole Moments, required by Klaus Lucas, which he had previously found difficult to obtain, could be almost routinely provided by Tim Clark as practically a a side product from one of his ab-initio calculations.

Dr. Johnny Gasteiger (Technical University of München) gave us one of his usual good talks, one of the kind which are familiar to us at Beilstein. This time the slant of the lecture was not synthesis planning but an overview of what can be achieved with some of the data generated in programs such as EROS. Not only is calculation of heats of formation and strain energies possible but also the calculation of various electronic parameters (polarisabilities, partial charges etc.) allow the estimation of other properties such as chemical shifts and dipole moments.

These methods are now implemented in a package called PETRA which was demonstrated throughout the workshop by Dr. Heinz Saller and Dr. Peter Löw (Chemodata). Unfortunately time didn't permit a cross check between the data bases of say Marsh and Pedley and the calculated data from this program.

The irrepressible Prof. Mario Marsili (University of L'Aquila) gave a description of how he calculates molecular volumes and surface areas. The Boolean encoding of the volumes facilitates fast and easy manipulation which is of great importance in Molecular Modelling. His ability to map structures onto another provides a measure of structure similarity.

Further examples of the practical use of estimation methods were given by Dr. Reiner Brüggemann (Gesellschaft für Strahlen- und

Umweltforschung) and Dr. Aleksander Sabljic (Institut Rudjer Boskovic) who both talked about the increasingly important area of Environmental Effects. Their talks again illustrated the lack of necessary data in the literature and brought up the difficult question of defining the environment. It was good to hear from Dr. Brüggemann that, in his complex property-property system which uses chains of calculations to get the required data, errors are fully taken into account.

The industrially important area of mixture-properties was discussed by Prof. Jürgen Gmehling and Dr. Jürgen Rarey-Nies (University of Dortmund). They described the Dortmund data bank which contains evaluated data on mixtures and pure components, and the methods they use for calculating properties. Jürgen Gmehling described a modified version of UNIFAC, a group contribution method for describing liquids which, using the data in the data base can now predict solution behavior much more accurately. Jürgen Rarey-Nies described the program package RECVAL which is used to obtain molar excess Gibbs enthalpies by simultaneous fitting of different excess properties.

The methods and data which they use and have are in constant use providing continuous checking for their accuracy.

Dr. Howard Rytting (University of Kansas) and Dr. Sam Yalkowsky (University of Arizona) described their different approaches to phase equilibria and the problem of solubility. Dr. Rytting uses group contribution methods to calculate partitioning behaviour and solubility. Dr. Yalkowsky described the Arizona database of solubilities and also how to calculate aqueous solubilities from melting points and partition coefficients.

Whether the Gods of Chemistry were picking on the lone mathematician Dr. Peter Jochum (Softron) is questionable, but the incessant thunder and lighting during his talk made us wonder. We didn't have to wonder though about the quality of his excellent talk which described the use of various interpolation methods and their applicability to our problem.

Wednesday started well with Dr. Peter Willett (University of Sheffield) describing the problem of defining structure similarity. After the similarity measure has been calculated, fragment code or Maximal Common Substructure (MCS), the compound can be added to the data base employing one of two techniques either ranking or clustering. Ranking involves identifying the nearest neighbours of a compound and clustering consists of defining the group of compounds to which it belongs. His general conclusion was that ranking was the better method to use. The use of similarity is something that we are familiar with at Beilstein (Beilstein system number, Lawson number) and the increase in efficiency brought about by MCS could be useful.

In a similar vein Dr. Mattias Otto (University of Freiberg) described his "fuzzy" set theory. Instead of calculating and storing data, why not use the information already available to make estimates of the compounds which are likely to have the same sort of properties. The question is, are there enough data for this method and how well can it cope with unrelated properties?

Prof. Peter Jurs (Penn State University) described two very nice pieces of work illustrating the calculation of structure properties, firstly the boiling points of olefins and secondly the GC retention times for PCB's. One feature of his PCB work is that he used 3D structures, calculated by molecular mechanics, which take into consideration many more effects, steric etc., than some of the more simple group contribution methods. This endorsed the advice of Mathias and Clark. This work highlighted once more what is possible within a group of compounds, but the problem of general applicability of such methods has yet to be solved.

Although a little way from the general theme of the workshop Prof. Shin-ichi Sasaki (Toyohashi University of Technology) showed us the tremendous work he has done in building up his "total system for molecular design". Consisting of three parts, Bioactive Structure Prediction (Tutors), Synthetic Route Design (Aiphos) and Structure Elucidation (Chemics). Even the occasional slide in Japanese didn't hinder us from appreciating the work.

A strong Czechoslovakian representation helped to make the workshop truly international.

Prof. Pavel Chuchvalec (Institute of Chemical Technology, Prague) gave an interesting talk describing the data base of pure component properties which is used for calculating mixture data that he has set up. He also illustrated one of the uses of data estimation in which we are interested, namely data checking, in his use of the Antoine equation to check vapor pressures.

His colleague, Dr. Vlastimil Ruzicka gave us analogous examples with a data base of heat capacities.

Dr. Vladimir Majer (University of Delaware) discussed a data base of thermal data related to vapour pressure and the checking of the data by estimation methods. He described how simultaneous correlation could be used to calculate data at other temperatures.

Similarly Dr. F. Mascarello (Hoffmann-La Roche) discussed the need to combine physical data by thermodynamic relationships to enable checking and extrapolation of vapour pressure data.

Apart from the stimulating lectures, which were without exception well attended, demonstrating how valuable people found them, the evenings provided us with the opportunity for some closer questioning. One of the most satisfying things to come out of the workshop was the contacts that were made between scientists from different areas of work who would otherwise never even have heard about each other. These contacts were also fruitful, providing a basis for future collaborations. Never before has it been possible in such a short space of time to get such a wide ranging overview of the area of data estimation.

For us at Beilstein the workshop was invaluable, we all found the excellent lectures highly enlightening. Taken as a whole, they have provided us with the comprehensive overview that we wanted. The workshop has not miraculously solved all Beilstein's problems - we didn't really expect it to - but it has at least provided us with signposts to the directions in which solutions lie. Of course the problems with which we are grappling are common to all branches of science and the solutions will be a synthesis of work from a number of different disciplines, so that we feel sure that this workshop has already paid dividends to all those who attended. We hope that this years workshop will be the first of many.

As far as we are concerned there are a few basic conclusions to be drawn:

We need to assess which methods are most applicable for which purposes.

We need to know which estimated data we should offer. What do the end-users want ?

If estimated data is to be incorporated into the data bank. How apparent should it be to the end-user ?

The advantages of speed and simplicity of methods such as those using Group Contributions, which indeed work very well in many cases, are often offset by their limitation to discrete series.

The general problem of defining a structure by a two dimensional connection table or fragment code which doesn't take into consideration 3D or electronic effects remains a weakness of some estimation methods. With the advent of new technology, more theoretically based methods will become implementable, by providing a better description of the molecule they could allow estimation methods to become more general: this workshop could herald the start of a new era in data estimation.

Once more we would like to thank the BMFT for their sponsorship and also all those who attended the workshop, especially the lecturers who played a major role in making it a success.

J. Gasteiger, Technische Universität München, Garching (Hrsg.)

Software-Entwicklung in der Chemie 1

Proceedings des Workshops „Computer in der Chemie"
Hochfilzen/Tirol, 19.-21. November 1986

1987. XII, 257 Seiten. Broschiert DM 54,-.
ISBN 3-540-18465-1

Inhaltsübersicht: Kommunikation über Computer. - Die Datenstruktur des Beilstein für organische Verbindungen. - Graphische Eingabe von chemischen Reaktionen unter Verwendung von GKS auf Mikrocomputern. - Eine Verarbeitung der R, S- und E, Z-Nomenklatur zur Spezifikation der Stereochemie. - PIMM - Ein Kraftfeldprogramm zur Berechnung von Molekülen und Molekülkomplexen. - RESY - Struktur-Retrieval-System, Baustein eines integrierten Chemie-Systems. - Ein automatisierter Molekülbaukasten. - Molekulardynamische Simulation von Flüssigkeiten. - Lokale Wechselwirkungspotentiale zur Simulation von Wasser und wässrigen Lösungen. - Ablauf der dezentralen Datenerfassung für das Gmelin-Online-System: Von der Diskette zur Datenbank. - Programme zur Auswertung kristallographischer Daten in Datenbanken. - Computerunterstützte Strukturaufklärung organischer Verbindungen: Automatische Interpretation von 13C-NMR-Spektren. - Multidimensionale Spektroskopie. - UNIFAC - ein wichtiges Werkzeug für die chemische Industrie. - Dortmunder Datenbank - Organisation, Stand und Anwendungsmöglichkeiten. - Mustererkennung bei dynamischen Prozessen durch Simultanspektroskopie. - Rechnernetze in der instrumentellen Analytik. - Sprachverwirrungen bei integrierten Mikro-Mainframe Entwicklungen. - INFUCHS: Konzept und Entwicklungsstand eines Fakteninformationssystems für umweltrelevante Chemikalien. - SPEKTREN II - ein strukturorientiertes spektroskopisches Informationssystem auf der Basis einer relationalen Datenbank.

Springer-Verlag Berlin
Heidelberg New York London
Paris Tokyo Hong Kong

Springer

MIX
Papier aus verantwortungsvollen Quellen
Paper from responsible sources
FSC® C105338

If you have any concerns about our products,
you can contact us on
ProductSafety@springernature.com

In case Publisher is established outside the EU,
the EU authorized representative is:
Springer Nature Customer Service Center GmbH
Europaplatz 3, 69115 Heidelberg, Germany

Printed by Libri Plureos GmbH
in Hamburg, Germany